Fluid Mechanics: Selected Concepts

Fluid Mechanics: Selected Concepts

Editor: Dayana Foster

New York

Published by NY Research Press
118-35 Queens Blvd., Suite 400,
Forest Hills, NY 11375, USA
www.nyresearchpress.com

Fluid Mechanics: Selected Concepts
Edited by Dayana Foster

© 2023 NY Research Press

International Standard Book Number: 978-1-64725-467-4 (Hardback)

Cataloging-in-Publication Data

Fluid mechanics : selected concepts / edited by Dayana Foster.
   p. cm.
Includes bibliographical references and index.
ISBN 978-1-64725-467-4
1. Fluid mechanics. 2. Fluid dynamics. I. Foster, Dayana.
TA357 .F58 2023
620.106--dc23

Contents

Preface

The world is advancing at a fast pace like never before. Therefore, the need is to keep up with the latest developments. This book was an idea that came to fruition when the specialists in the area realized the need to coordinate together and document essential themes in the subject. That's when I was requested to be the editor. Editing this book has been an honour as it brings together diverse authors researching on different streams of the field. The book collates essential materials contributed by veterans in the area which can be utilized by students and researchers alike.

Fluid refers to a state of matter wherein the matter yields to sideways or shearing forces. Gases, plasmas and liquids are considered to be fluids. They can be found in stationary as well as moving forms. The physics of stationary fluids is studied under fluid statics, whereas fluid mechanics is a branch of physics involved in the study of mechanics related with fluids and the forces they generate. It has applications in many fields such as mechanical engineering, biomedical engineering, chemical engineering, civil engineering, oceanography, biology, geophysics, astrophysics and meteorology. Some of the prominent relationships which are studied under this discipline are kinematic, regulating, stress, constitutive, and conservation relationships. This book explores some important selected concepts related to fluid mechanics. The various studies that are constantly contributing towards advancing technologies and evolution of fluid mechanics are examined in detail. The book aims to serve as a resource guide for students and experts alike and contribute to the growth of the discipline.

Each chapter is a sole-standing publication that reflects each author's interpretation. Thus, the book displays a multi-facetted picture of our current understanding of application, resources and aspects of the field. I would like to thank the contributors of this book and my family for their endless support.

Editor

A New Wall Model for Large Eddy Simulation of Separated Flows

Ahmad Fakhari (ID)

Institute for Polymers and Composites, Department of Polymer Engineering, Campus of Azurém, University of Minho, 4800-058 Guimarães, Portugal; ahmadfakhari@gmail.com

Abstract: The aim of this work is to propose a new wall model for separated flows which is combined with large eddy simulation (LES) of the flow field in the whole domain. The model is designed to give reasonably good results for engineering applications where the grid resolution is generally coarse. Since in practical applications a geometry can share body fitted and immersed boundaries, two different methodologies are introduced, one for body fitted grids, and one designed for immersed boundaries. The starting point of the models is the well known equilibrium stress model. The model for body fitted grid uses the dynamic evaluation of the von Kármán constant κ of Cabot and Moin (Flow, Turbulence and Combustion, 2000, 63, pp. 269–291) in a new fashion to modify the computation of shear velocity which is needed for evaluation of the wall shear stress and the near-wall velocity gradients based on the law of the wall to obtain strain rate tensors. The wall layer model for immersed boundaries is an extension of the work of Roman et al. (Physics of Fluids, 2009, 21, p. 101701) and uses a criteria based on the sign of the pressure gradient, instead of one based on the friction velocity at the projection point, to construct the velocity under an adverse pressure gradient and where the near-wall computational node is in the log region, in order to capture flow separation. The performance of the models is tested over two well-studied geometries, the isolated two-dimensional hill and the periodic two-dimensional hill, respectively. Sensitivity analysis of the models is also performed. Overall, the models are able to predict the first and second order statistics in a reasonable way, including the position and extension of the downward separation region.

Keywords: wall layer model; LES; separated flow; body fitted; immersed boundary

1. Introduction

Wall bounded turbulent flows, especially at high Reynolds numbers, require a high resolution near the wall, because of the need to solve the thin viscous sub-layer [1–3]. In LES, this resolution is comparable with that required by direct numerical simulation (DNS). As far as Reynolds number increases, the cost of a wall-resolving LES augments with $Re^{2.5}$ (for a discussion, see Piomelli and Balaras [1]). Moreover, for applications where wall roughness is the rule rather than the exception, it is not feasible to describe the wall boundary in a deterministic sense.

The most practical way of simulating a wall-bounded, high Reynolds number flow, is to consider coarse grid, and model the wall shear stress with a wall function. This function is designed to mimic the stress induced by the wall, produced by the adherence condition and ruled by the turbulent field. Several wall layer models have been developed in the past (for a discussion, see Piomelli and Balaras [1] and Piomelli [2]). They are designed to work under certain conditions, and in particular they may suffer in the presence of massive separation. This subject is still challenging for improvement.

In equilibrium stress models, the grid is generated in such a way that the first near-wall computational node is placed in the log region of the boundary layer. So, based on the law of the wall, the instantaneous or mean horizontal velocity can be fitted to determine the wall stress. These models are valid only under the equilibrium assumption and work for both smooth and rough walls.

Deardorff [4] and Schumann [5] first used the concept of wall layer modeling in conjunction with LES. Applying LES on plane channel flow and annuli at large Reynolds numbers, they assumed the existence of an equilibrium-stress layer near the wall and used the outer flow velocity to calculate the wall stress based on the logarithmic law.

Deardorff [4] considered a second order velocity derivative in vertical direction and forced plane averaged stream-wise and spanwise velocity gradients to follow logarithmic law. Schumann [5] obtained the plane averaged wall shear stress from plane averaged velocity at the first grid point, using the iterative method so that the shear velocity is computed from the law of the wall from the averaged velocity. These methods are the least expensive wall layer models. They also provide roughness corrections easily from the logarithmic law modification, which is an important feature in environmental and oceanographic flows. The models suffer in cases of marginal separation, or strong pressure gradient [2].

In zonal approaches (two-layer models, TLMs), stress derives from the solution of a separate set of equations on a finer mesh close to the wall. This model was proposed by Balaras and Benocci [6] who applied turbulent boundary layer equations inside the boundary layer (inner layer) and LES on the outer region. In this model, a finer grid is considered between the wall and the first computational node of the coarser grid. A uniform pressure field is applied from LES for the inner layer, and velocity field from LES is considered at the edge of the Reynolds Averaged Navier-Stokes (RANS) region as its boundary condition. The coupling between the inner and outer regions of the boundary layer is weak, and these models have problems when a perturbation extends from the wall to the outer layer. Balaras and Benocci [6] used an algebraic eddy viscosity to parametrize the near-wall region. In addition to plane channel flow, this method was tested on square ducts and also rotating channel, the cases in which equilibrium stress model was not valid or even failed; this method showed a good accuracy in simulating these flows. These methods, able to simulate equilibrium as well as non-equilibrium flows were not designed to deal with separated flows.

Cabot [7] applied this approach to simulate flow over a backward-facing step with only 10% grid points fewer than those used in former wall-resolved LES. He found that stream-wise pressure gradient has an important effect in boundary layers of TLMs. In this case, mean velocity and skin friction coefficient were predicted well; however, the backward-facing step is just a simple case where separation is induced by the sharp discontinuity in the streamlines.

Cabot and Moin [8] used the wall layer model in conjunction with LES to simulate the flow over a backward-facing step. They considered a zonal approach using thin boundary layer equations (TBLEs) in the inner layer and LES (dynamic SGS model) in the outer region. This way, they were able to use eight uniform cells from the bottom of the wall up to half the height of the step, greatly reducing the total number of cells with respect to a stretched-grid case previously used for the wall-resolved LES.

Finally, in hybrid methods, a single grid is used and the turbulence models are different in inner and outer layers. Detached eddy simulation (DES) is a hybrid approach proposed by Spalart et al. [9] as a method to compute massive separation. According to DES, a stretched grid is used close to the wall to resolve the boundary layer. RANS is applied in the inner and LES in the outer region.

Since there is no zonal interface between the inner and the outer layer, velocity profile is smooth everywhere. Some weaknesses of this method are logarithmic layer mismatch between the two regions and high grid resolution dependency near the wall. In addition, time and length scales of the eddies in the inner layer (modelled by unsteady RANS equations with the Spalart-Allmaras 1-equation turbulent model) are much larger than those computed in the outer region by LES; this leads to much lower resolved stress by LES than the modelled stress (by RANS) at the RANS/LES interface and even farther Piomelli [2]. A review of DES and the two modifications of it, which are delayed detached eddy simulation (DDES) and improved delayed detached eddy simulation (IDDES), can be seen in Mockett et al. [10].

The total shear stress in plane channel flow only depends on the distance from the wall, therefore a reduction of eddy viscosity farther than the RANS/LES interface affects the velocity profile and generates a high gradient at the transition into LES region. This phenomenon is called the DES buffer layer. To eliminate this, Keating and Piomelli [11] added stochastic force in the interface region to accelerate resolved eddy generation, thus obtaining better results. Temmerman et al. [12] also computed the constant required to calculate eddy viscosity in the RANS region in such a way to equate the eddy viscosity of RANS and LES at the interface of their hybrid RANS/LES application.

While the hybrid RANS/LES methods work satisfactorily in flows with instabilities such as those with adverse pressure gradient and concave curvature, they are weak in attached flows, and, in general, in flows with a low level of instability; in these cases a false merging region may appear at the RANS/LES interface. Since the grid near the wall must be resolved, they are the most expensive ones in the wide family of wall layer models. In addition to the three main wall layer frameworks herein discussed, there are other models which are not mentioned here.

The near-wall modelisation must be implemented in the algorithms that solve the governing equations using different strategies. A body fitted (also called boundary fitted) case is the case in which a grid is generated to follow the geometry. Since the grid boundaries coincide with the solid surface of a solid body, boundary conditions are imposed on certain grid nodes or cell faces. This facilitates the computation of all the vectorial quantities as well as of their gradients.

In complex geometry, where a structured grid can hardly follow the boundary complexities, alternative approaches must be employed. Among others, the immersed boundary method (IBM) has proved to be simple, effective and accurate (Mittal and Iaccarino [13]). The domain grid can be either Cartesian or curvilinear. In this approach the cells are cut by the solid surface and the cell boundaries in the general cases do not coincide with the solid boundaries. Hence the fluxes at the solid surfaces are difficult to define. Finally, the velocities in near-wall computational nodes are reconstructed and thus a local frame of reference must be introduced at each near-wall node to identify directions normal and tangential to the immersed boundary.

A number of papers have been published discussing different implementations of immersed boundary methods (see Mittal and Iaccarino [13]). Among others, Roman et al. [14] extended IBM to the general case of curvilinear coordinates, and simulated both steady and unsteady flows in complex geometric configurations showing the flexibility of this methodology when compared to the Cartesian counterpart. However, few papers have dealt with the development of wall layer models in conjunction with immersed boundaries. This is a matter of great importance when flows at a high Reynolds number over complex geometry must be simulated.

Tessicini et al. [15] applied wall layer model LES in the presence of immersed boundaries to simulate the flow past a 25°, asymmetric trailing edge of a model hydrofoil. The Reynolds number, based on free stream velocity and the hydrofoil chord (C), was chosen equal to $Re_C = 2.15 \times 10^6$. They used turbulent boundary layer equations for modelling the wall layer. In general, their results were good, although they had a deviation at the second off-the-wall node, which was considered as the outer boundary for the wall model. This discrepancy was sensitive to the distance of the second node from the wall, and was more evident where the distance was larger.

Posa and Balaras [16] proposed a new near-wall reconstruction model to account for the lack of resolution and provided correct wall shear stress and hydrodynamic forces. They used a zonal approach (TLM), boundary layer equations with a finer grid in the near-wall region (called in this case the full boundary layer FBL) and LES in the outer region. They validated their model to simulate flow around a cylinder and a sphere.

Then, they considered a coarser grid, and set one node inside the boundary layer and neglected the advective term in boundary layer equations. This was justified by the fact that, if the first point off the wall is located inside the boundary layer, neglecting this term does not provide significant errors. They assumed a constant pressure gradient between the first and second nodes, and obtained tangential velocity with two approaches: The reduced diffusion model (RDM) and the hybrid reconstruction method (HRM). The Reynolds number in the cylinder case was $Re = 300$ based on free stream velocity and the cylinder diameter, and in the sphere was $Re = 1000$. The pressure coefficient for all cases was predicted well, while the skin friction was underestimated in all cases, but an improvement was observed using RDM and HRM in comparison with linear reconstruction. The prediction of separation was also improved a lot using these two approaches.

Chen et al. [17] also proposed a wall layer model based on turbulent boundary layer equations at high Reynolds numbers for implicit LES in the presence of immersed boundaries. First, they tested it on a turbulent channel flow in the range of Reynolds numbers (based on shear velocity) from $Re_\tau = 395$ to $Re_\tau = 100,000$. They used a minimum of 20 cells inside the boundary layer for lower Reynolds numbers and 40 cells for higher values. Inside the boundary layer (inner region), the pressure gradient and friction are dominant, thus advection is neglected in cases where there are slow changes in wall parallel direction and sharp changes in wall normal direction. The fact that the changes in wall normal direction are sharper at higher Reynolds numbers requires more resolution for capturing the flow in this direction.

They also simulated flow over a backward-facing step at a Reynolds number based on the step height $Re_h = 5000$ in a Cartesian grid. Finally they simulated the flow over periodic hills at $Re_H = 10,595$ based on hill height, using different resolutions, and they obtained comparatively good results. To summarize, different models have an acceptable accuracy under some special conditions; therefore, more general models are needed to work well in different situations characterized by attached as well as separated flow conditions.

In the present study, attempts are made to overcome some of the difficulties mentioned above, proposing wall layer model to be used in conjunction with LES, suited for both attached and detached flows. Since in engineering applications a combination of body fitted solid walls and immersed boundaries is often employed, a comprehensive model must include the two aspects. Here the models are designed to work in body fitted cases and in the presence of immersed boundaries, respectively. The solver for the simulations is the LES model (LES–COAST) developed over the years at the Laboratory of Environmental and Industrial Fluid Mechanics of the University of Trieste, Italy, and which has successfully been used in a number of environmental real-scale applications (see among others Roman et al. [18] and Galea et al. [19]). The paper is organized as follows: In Section 2 the governing equations are described and the numerical method employed; in Section 3, the wall layer model for body fitted grids, its improvement and application to a case characterized by the presence of a boundary layer together with massive separation, namely the case of the flow over an isolated hill, are shown; in Section 4 the model in the presence of the immersed boundary is presented, together with its application to the case of single hill and to the case of periodic hill. Concluding remarks are given in Section 5.

2. Governing Equations

The incompressible form of the filtered Navier-Stokes equations read as:

$$\frac{\partial \overline{u}_j}{\partial x_j} = 0, \tag{1}$$

$$\frac{\partial \overline{u}_i}{\partial t} + \frac{\partial (\overline{u}_i \overline{u}_j)}{\partial x_j} = -\frac{1}{\rho_0}\frac{\partial \overline{p}}{\partial x_i} - \frac{\partial \tau_{ij}}{\partial x_j} + \nu \frac{\partial^2 \overline{u}_i}{\partial x_j \partial x_j}. \tag{2}$$

The filter operation (denoted by overbar) allows reproduction of the space–time evolution of the large scales of motion, which are anisotropic and energetic, while the effect of the small sub-grid scales (SGS) is contained in the SGS stress terms (τ_{ij}) in the momentum equation. It is here modeled through an SGS eddy viscosity model:

$$\tau_{ij} - \frac{\delta_{ij}}{3}\tau_{kk} = -2\nu_T \overline{S}_{ij}, \tag{3}$$

where $\overline{S}_{ij}$ is the resolved-scale tensor, defined as:

$$\overline{S}_{ij} = \frac{1}{2}\left(\frac{\partial \overline{u}_i}{\partial x_j} + \frac{\partial \overline{u}_j}{\partial x_i}\right), \tag{4}$$

ν_T is the SGS turbulent viscosity, also known as eddy viscosity. The standard Smagorinsky model [20] is based on the equilibrium assumption. Considering the length scale $l \sim \overline{\Delta}$, the eddy viscosity can be written as:

$$\nu_T = (C_s \overline{\Delta})^2 |\overline{S}|, \tag{5}$$

where C_s is the constant of the model (the Smagorinsky constant), and $|\overline{S}| = \sqrt{2\overline{S}_{ij}\overline{S}_{ij}}$ is the contraction of the strain rate tensor of the large scales, $\overline{S}_{ij}$. Finally, the SGS stresses are calculated as:

$$\tau_{ij} = -2\nu_T \overline{S}_{ij}. \tag{6}$$

The Smagorinsky constant is commonly considered to range between 0.065 and 0.2; Lilly [21] found a theoretical value of 0.18, but the Smagorinsky constant depends on the type of the flow, e.g., in shear flows it must be declined to 0.1 [22]. A comprehensive study of the Smagorinsky constant in plane channel flows can be found in Stocca [23]. The filter width is proportional to the grid size in all directions, and is equal to $\overline{\Delta} = (\Delta x \Delta y \Delta z)^{1/3}$. Near the walls, where the eddy viscosity is expected to vanish, the van Driest damping function is adopted.

In the dynamic Smagorinsky model proposed by Germano et al. [24], the coefficient is calculated dynamically by defining an additional test filter (denoted by a caret) with a width $\widehat{\Delta}$ larger than the grid filter width $\overline{\Delta}$; here $\widehat{\Delta} = 2\overline{\Delta}$. The dynamic Smagorinsky coefficient is computed in a least-square sense over tensor components:

$$C = -\frac{1}{2}\frac{\langle L_{ij}M_{ij}\rangle}{\langle M_{ij}M_{ij}\rangle}, \tag{7}$$

where $<>$ denotes spatial average and L_{ij} resolved turbulent stresses:

$$L_{ij} \equiv \widehat{\overline{u}_i\overline{u}_j} - \widehat{\overline{u}}_i\widehat{\overline{u}}_j, \quad M_{ij} \equiv 2(\Delta^2 \widehat{|\overline{S}|\overline{S}_{ij}} - \widehat{\Delta}^2|\widehat{\overline{S}}|\widehat{\overline{S}}_{ij}) \tag{8}$$

and then eddy viscosity can be calculated as below.

$$\nu_T = C\widehat{\Delta}^2 |\overline{S}|. \tag{9}$$

The advantage of the dynamic model over the standard one is that the eddy viscosity vanishes near the walls and in laminar flows [24]; so, the model allows to treat transitional and wall bounded flow without the need to use special treatments of the constant. In order to follow the solid boundaries, the numerical model uses the curvilinear-coordinate form of the Navier-Stokes equations frame. Spatial discretization in the computational space is carried out using second order central finite differences. Temporal integration is carried out by using the second order accurate Adams-Bashforth scheme for the convective term and the implicit Crank-Nicolson scheme for the diagonal viscous terms. A collocated grid is considered where pressure and Cartesian velocity components are defined at the cell centers, and the volume fluxes are defined at the midpoints of the cell faces. For the numerical model see Zang et al. [25], whereas for the implementation of the SGS models see Armenio and Piomelli [26]. A new parallel version of the model has been developed over the years and a version suited for environmental applications (LES-COAST) is available [27].

3. Wall Layer Model for Body Fitted Geometry

A basic equilibrium wall stress model is present in the solver. The wall stress is obtained from instantaneous horizontal velocity at the first off-wall centroid based on the law of the wall:

$$u_p^+ = \begin{cases} \frac{1}{\kappa} ln(y_c^+(1)) + B & if \ y_c^+(1) > 11 \\ y_c^+(1) & if \ y_c^+(1) \leq 11 \end{cases} \tag{10}$$

where

$$u_p^+ = \frac{u_p}{u_\tau} = \frac{\sqrt{u^2 + w^2}}{u_\tau}$$

is the modulus of instantaneous non-dimensional velocity at the first off-wall computational node P, which has a distance $y_c(1)$ from the wall; $\kappa = 0.41$ is the von Kármán constant, and B varies between 5 and 5.5 [2]. Here $B = 5.1$ was used for all the simulations. The friction velocity u_τ is calculated from the velocity u_p at each near-wall computational node, and depends on the distance from the wall $y_c^+(1)$, either from the linear or logarithmic law. Then, wall shear stress τ_w is calculated from friction velocity; $\tau_w = \rho u_\tau^2$. More details of the procedure can be found in Fakhari [28].

In addition to this, the knowledge of the contraction of the resolved strain rate tensor is required in order to use an SGS eddy viscosity at the wall in the Smagorinsky model. Since the tangential velocity at the wall is not determined, the evaluation of the leading terms of $\overline{S}_{ij}$ becomes increasingly wrong with decreasing grid resolution.

To overcome this problem, the leading terms of the strain rate tensor $\overline{S}_{ij}$ are set analytically based on the location of $y_c(1)$. If the first point P is in the logarithmic region, the leading terms of the strain rate tensor are:

$$\overline{S}_{12} = \frac{u_\tau}{\kappa y_c(1)} \frac{\overline{u}}{u_p} + \frac{\partial \overline{v}}{\partial x} , \quad \overline{S}_{32} = \frac{u_\tau}{\kappa y_c(1)} \frac{\overline{w}}{u_p} + \frac{\partial \overline{v}}{\partial z} . \tag{11}$$

Consequently the value of the eddy viscosity near the wall adjusts consistently with the imposed stress. The wall layer model mentioned above is for smooth surfaces. If we deal with a rough surface with the roughness height y_0, the velocity profile is:

$$u_p^+ = \frac{1}{\kappa} ln(\frac{y_c(1)}{y_0}) \quad y_c(1) > y_0. \tag{12}$$

The same procedure based on Equation (11) can be used to modify the leading terms of $\overline{S}_{ij}$.

This wall layer model is very economic and reasonably accurate in attached flows. Its main drawback is the simulation of separated flows [1,2] as will be discussed with details in Section 3.2. The logarithmic law does not capture separation; therefore, in detached flows a stretched grid near-wall is required to have more resolution there to set the near-wall computational nodes inside the viscous layer.

Cabot and Moin [8] used wall layer models with LES to simulate flow over a backward-facing step. They considered a zonal approach using the thin boundary layer equation (TBLE) in the inner layer and LES with dynamic SGS model in the outer region. They used 8 uniform cells from the bottom of the wall up to half the height of the domain, much less than in a former stretched-grid which was used for a wall-resolving LES. The authors introduced a dynamic treatment of the von Kármán constant to equate the stress predicted by TBLE in the inner layer ($\kappa y \widehat{\overline{u}}_\tau \widehat{\overline{S}}_{ij}$) in a least-square sense to the total one (resolved + SGS) computed by LES in the outer region ($-\widehat{\overline{u}_i \overline{u}_j} - \widehat{\tau}_{ij}$). The dynamic κ was used to calculate the eddy viscosity in the inner region, $\nu_t = \kappa y u_\tau D$, $D = [1 - exp(-y^+/A^+)]^2$ in which $A^+ = 17$, but in that case since $y^+ \gg A^+$ they considered D as unity. In Cabot and Moin [8] the use of dynamic κ was justified by the fact that their wall model (TBLE) would carry shear stress both in the RANS eddy viscosity and the RANS convective terms, thus they needed a reduced RANS eddy viscosity, and by using the dynamic κ which had the values less than the von Kármán constant, and reduced the shear stress.

3.1. Model Optimization for Body Fitted Geometry

In this work the context is different. It is known that the equilibrium stress wall function works well in equilibrium flows, and has a drawback in the presence of separation [1,2]. In addition, in separated regions the velocity profile does not follow the standard log law; therefore, a modification of the standard log law is required. Here, the aim of using dynamic κ is not to reduce stress, but to have it deviate from the standard log law. The dynamic κ is calculated here to match analytical stress from the boundary layer and the stress computed with dynamic Smagorinsky using a coefficient:

$$\kappa = -coeff \cdot \frac{< y_c(1) \widehat{\overline{u}}_\tau \widehat{\overline{S}}_{ij} (\widehat{\overline{u}_i \overline{u}_j} + \widehat{\tau}_{ij}) >}{< y_c(1)^2 \widehat{\overline{u}}_\tau^2 \widehat{\overline{S}}_{ij} \widehat{\overline{S}}_{ij} >}, \tag{13}$$

in which $<>$ denotes averaging over the directions of homogeneity. First, a simulation of a periodic open channel flow was carried out in order to obtain a cross-sectional plane of data instantaneously to be used as the inflow over the hill, as will be discussed in Section 3.2. For the simulation of periodic open channel flow, a plane averaged approach was performed to compute $coeff$ in such a way that the dynamic κ becomes equal to the von Kármán constant. In LES–COAST, $coeff = 0.4$ was obtained, and this will be explained in Section 3.2.

3.2. Application of the Wall Layer Model for Body Fitted Geometry

The wall layer model is here tested on a detached flow case. The accuracy of the wall layer model is checked and then the model is improved to increase the accuracy of the wall function. Most of the wall layer models have weakness in capturing the points of flow separation and re-attachment, separately, especially when the slope of a solid wall changes gradually. Thus, this work focuses on flow simulation over a single two-dimensional hill, which, from one side, is a very challenging problem for wall layer models and, from the other side, literature data are available.

The geometry and boundary conditions are collected from ERCOFTAC classic database, environmental flows area, case 69 (http://ERCOFTAC/case69) [29]. The hill height is $H = 0.117$ m and the amplitude of the hill is $a = 3H$ (Figure 1a). The wind tunnel experiments were carried out by Khurshudyan et al. [30]. A wall-resolved LES of turbulent boundary layer over a 2D hill was also performed by Chaudhari et al. [31] and Chaudhari [32] for two different aspect ratios $a = 3H$ and $5H$. Since they faced difficulty in validating the reattachment point for the aspect ratio of $a = 3H$ with the experiment, this case was specifically chosen for this study as a challenging problem.

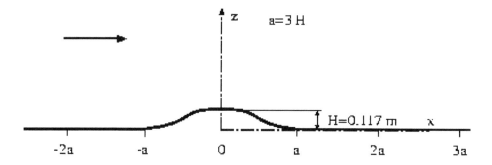

(a) Hill geometry in which H is the hill height, a is the aspect ratio, with courtesy of ERCOFTAC, Environmental Flows [29].

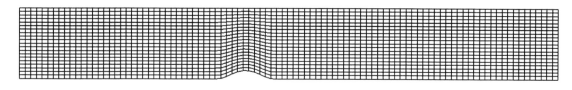

(b) Computational domain.

Figure 1. The hill geometry and the domain grid with resolution of $96 \times 20 \times 32$ cells in streamwise, wall normal and spanwise directions, respectively, for simulation of flow over hill using dynamic κ (body fitted approach).

A domain dimension with streamwise length $L_x = 7.12\,\delta$, height δ and width (in spanwise direction) $L_z = \delta$ is considered where $\delta = 1$ m.

The boundary conditions are given as follows: At the inlet, inflow is obtained from simulation of a periodic open channel flow with the same cross-sectional grid resolution; a cross-sectional plane was considered to collect the instantaneous data at each time step after the channel flow reached a steady state and the collected data was used as inflow for the flow simulation over the hill. The roughness characteristic height of the wall for this case is $y_0 = 1.57$ mm, shear velocity $u_\tau = 0.178$ m/s and the Reynolds number based on shear velocity is $Re_\tau = u_\tau \delta / \nu = 1187$; at the outlet, radiative boundary conditions were given; at the upper boundary the free slip condition is imposed; periodic conditions are applied along the spanwise direction. The dynamic SGS model wis employed for simulations with body fitted grid, with the constant obtained from spanwise averaging; simulations were run with a fixed Courant number equal to 0.2.

The results of this simulation, which was initially performed using $96 \times 32 \times 32$ cells in streamwise, wall normal and spanwise directions, respectively, show that the wall layer model described in Section 3, with a fixed value of the von Kármán constant, is not able to capture flow separation when a uniform grid is employed (where $y_c^+(1) = 18.5$). Conversely, stretching the grid to set the first computational node off the wall inside the viscous layer allows obtaining flow separation. Figure 2 shows a comparison between the experiment by Khurshudyan et al. [30] and three simulations: The case with uniform grid, and two cases with stretched grids in which the first centroids are at $y_c^+(1) = 4$ and $y_c^+(1) = 7$, respectively. All the grids share the same number of grid cells and the resolution in streamwise and spanwise directions, while the two grids are stretched vertically using the Vinokur approach [33] to have the largest cells at the top of the domain with $\Delta y^+ \approx 64$. The free surface velocity is used as U_{ref} to normalize the velocity plots.

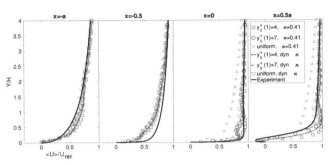

(a) Mean stream-wise velocity upstream, over and slightly downstream from the hill.

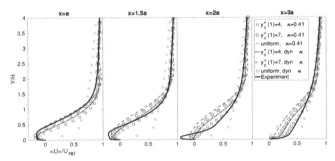

(b) Mean stream-wise velocity downstream from the hill.

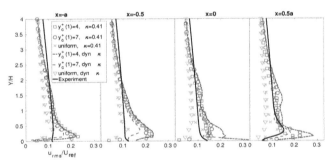

(c) RMS of streamwise velocity fluctuations upstream, over and slightly downstream from the hill.

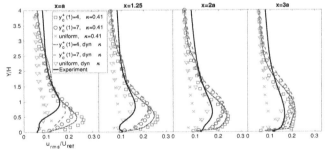

(d) RMS of streamwise velocity fluctuations downstream from the hill.

Figure 2. Mean streamwise velocity and root mean square (RMS) of its fluctuations obtained from simulation using 32 cells in the wall normal direction for the cases with $y_c^+(1) = 4$ and $y_c^+(1) = 7$ compared to the experiment by Khurshudyan et al. [30]; the hill starts from $x = -a$ and ends at $x = a$.

For the cases with fixed value of κ, the starting point of flow separation is at $x = 0.5a$ using the stretched grids; the reattachment point is predicted well in these cases. This test reveals that the resolutions in streamwise and spanwise directions are sufficient to capture boundary layer separation. Overall, separation is predicted with a grid that for the vertical resolution near the wall resembles that of a wall-resolving LES. This is opposite the expectations from a wall layer model, which is supposed to work well with a poor resolution in the wall region.

The improved model which makes use of the dynamic evaluation of the von Kármán constant κ is tested on the uniform mesh. In this case separation is captured; at $x = 0.5a$ the first centroid is too far from the wall to capture the starting point of separation, while at the other locations, separation is predicted correctly as well as boundary layer reattachment.

Applying the dynamic evaluation of κ in the case of the stretched grids reveals that in the case with $y_c^+(1) = 4$ the mean velocity profile at $x = 0$ demonstrates a better acceleration at the first centroid compared to the case with a fixed value of κ; moreover, at $x = 0.5a$, which is the starting point of separation, the negative velocity displays a larger magnitude at the few cells above the wall, which is closer to the experimental values, while a bit farther from the wall, the case with constant κ is in a better agreement with the experiment. In the separation region and even after reattachment, the mean velocity at the near-wall centroid is smaller than using the fixed value of κ, and the reattachment point is predicted correctly for $y_c^+(1) = 4$. The case with $y_c^+(1) = 7$ demonstrates higher velocity at the near-wall computational node using dynamic κ in the beginning of separation ($x = 0.5a$), while at $x = a$ and beyond, the velocity gets smaller than the fixed κ, and the reattachment point is delayed.

The root mean square (RMS) of streamwise velocity fluctuations exhibits an overestimation close to the wall for stretched grids using both fixed and dynamic κ, while the uniform grid with dynamic κ slightly underestimates this quantity in most of the regions. This overestimation is due to the fact that the equilibrium stress model for the rough surface computes the shear velocity and wall shear stress from the logarithmic law (Equation (12)), while the near-wall computational nodes in the stretched grids are located inside or near the viscous sub-layer.

Figure 3 contains mean velocity profiles related to the present simulations, the wall-resolved LES by Chaudhari et al. [31] and the experiment by [30] at three critical locations: The hill crest, the starting point of separation and just beyond the reattachment point. For the wall-resolved LES case, the data were obtained by extraction of significant points from their plots. The mean velocity profile shows that at the top of the hill, the flow acceleration for wall-resolved LES agrees well with the experiment. Moreover, the point of separation is captured well, but the flow experiences an early reattachment at $x = 2a$.

The wall-resolved LES underestimates root mean square of streamwise velocity fluctuations except at the start of separation $x = 0.5a$ close to the wall, as depicted in Figure 3b.

Figure 3a shows that, overall, the present results exhibit a reasonable accuracy, also in view of the grid herein employed. It is to be noted that the reference wall-resolved LES by Chaudhari et al. [31] with $495 \times 70 \times 136$ numerical grids (48 times more expensive than the current wall layer model simulation) had an early prediction of reattachment point, while in this simulation there is a better agreement of the reattachment point with the experiment.

Successively, a different grid resolution was applied in the vertical direction ($96 \times 20 \times 32$ cells in streamwise, wall normal and spanwise directions, respectively) on the same geometry with the same boundary conditions. The domain grid is shown in Figure 1b, with the first computational node situated in the logarithmic region at $y_c^+(1) \approx 30$.

The simulations were carried out as follows. First, an open channel flow at $Re_\tau = 1187$ based on the friction velocity u_τ and the channel height was simulated, with a grid resolution of $40 \times 20 \times 32$ cells in streamwise, wall normal and spanwise directions, respectively. Before the simulation, a test was carried out to compute the dynamic (Smagorinsky) model constant and also $coeff$ of Equation (13) applying a plane average approach, and using $coeff = 0.4$ as a target value of the von Kármán constant κ.

Thereafter, in another attempt, spanwise averaging of the numerator and denominator of Equation (13) with $coeff = 0.4$ and also the dynamic model constant was performed. Note that κ must be positive since it is used in the denominator to calculate friction velocity and strain rate tensor (see Section 3). Setting the minimum value of 0.06 avoided unrealistic values of velocity due to the near-zero or negative value of κ. Moreover, larger threshold values were tested to check the sensitivity to this lower bound; the analysis showed that it does not affect the results, hence the method is robust from this aspect.

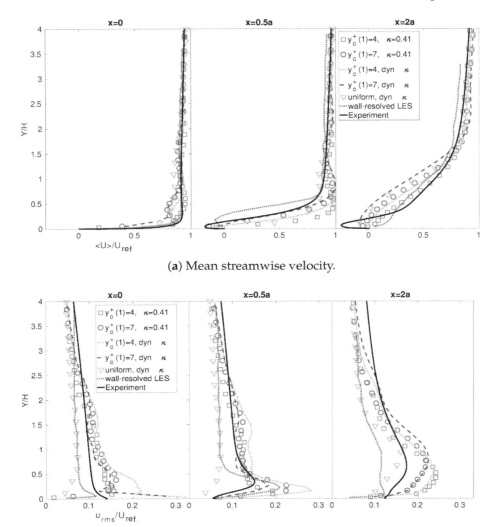

(a) Mean streamwise velocity.

(b) Root mean square of streamwise velocity.

Figure 3. Mean streamwise velocity and RMS of its fluctuations at three critical positions from simulation using 32 cells in the wall normal direction for the cases with $y_c^+(1) = 4$ and $y_c^+(1) = 7$ using constant and dynamic k, and uniform grid with dynamic κ compared to resolved LES by Chaudhari et al. [31] and the experiment by Khurshudyan et al. [30] at the hill crest, the beginning of separation and just before the reattachment points.

Figure 4a shows time- and spanwise-averaged velocity profiles at different positions along the streamwise direction. As expected (Figure 4a), the current resolution is too coarse, since the first computational node is out of the separation region; this explains the absence of separation at $x = 0.5a$. However, separation occurs at $x = 2a$ which has a delay with respect to both the wall-resolved LES and the experiment, and reattachment is obtained at $x = 3a$ which is closer to the experiment, while the wall-resolved LES exhibits an earlier reattachment.

Figure 4b shows the root mean square of streamwise velocity fluctuations. Generally there is an underestimation of the velocity fluctuations especially right before and at the starting point of the separation region. Chaudhari et al. [31] also underestimated the RMS of the streamwise velocity fluctuations uphill, and overestimated it in the separation region.

As was discussed in the previous section, the use of dynamic κ in this method is different than in Cabot and Moin [8]. To understand this better, instantaneous values of the dynamic κ along x for the flow over the single hill of this simulation in Figure 5. Since κ is in the numerator to calculate the shear velocity and wall shear stress ($u_\tau = u * \kappa / ln(y/y_0)$), on top of the hill, there is an increase of κ, which gives a better acceleration rather than the standard log law ($\kappa = 0.41$). Then, at the starting point of separation, κ reduces and therefore the wall shear stress goes close to zero. Thereafter, inside the separation region, at most of the spots κ is large, leading to a negative value of the wall shear stress with a larger magnitude, and finally it goes close to 0.41 after reattachment. In this way, the equilibrium stress model is applicable on separated flows.

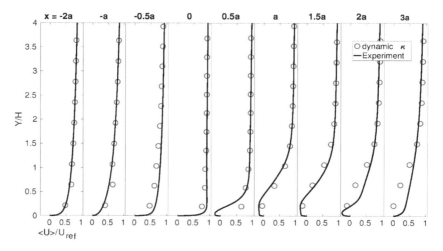

(a) Mean streamwise velocity.

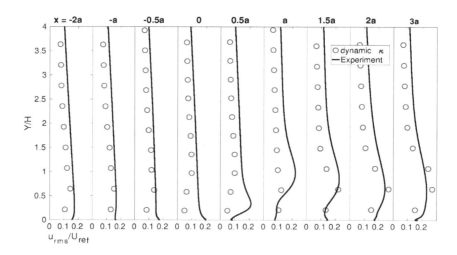

(b) Root mean square of streamwise velocity fluctuations.

Figure 4. Mean streamwise velocity and RMS of its fluctuations obtained from simulation using 20 cells in the wall normal direction, applying dynamic κ compared to the experiment by Khurshudyan et al. [30]; the hill starts from $x = -a$ and ends at $x = a$.

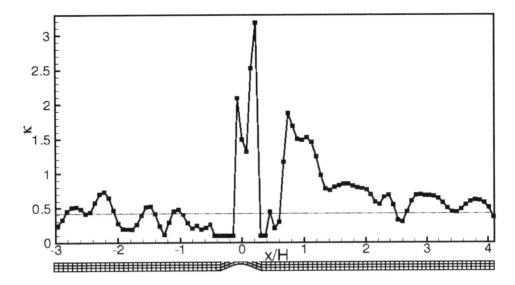

Figure 5. Instantaneous values of dynamic κ on a stream line in the current simulation of flow over the single hill.

4. Wall Layer Model for Immersed Boundary Methodology (IBM)

This part of the work takes advantage of the model developed in Roman et al. [34]. The velocity reconstruction method is briefly recalled here. With reference to Figure 6, the velocity at the projection point (PP) is interpolated from velocities at its surrounding points (empty square points in Figure 6). Then, shear velocity at the PP point is computed. Finally, based on the distance of the first centroid off-the-wall, i.e., the IB node, from wall in wall units, the velocity at the IB node is calculated using the law of the wall:

$$\overline{u}_{IB} = \begin{cases} \overline{u}_{PP} - \dfrac{1}{\kappa}\sqrt{\dfrac{\tau_w}{\rho}} ln(\dfrac{d_{PP}}{d_{IB}}) & if \ d_{IB}^+ > 11 \\[2ex] \dfrac{d_{IB}u\tau^2}{\nu} & if \ d_{IB}^+ \le 11 \end{cases} \qquad (14)$$

Considering that the velocity is known at the PP point, a parabolic interpolation is carried out to compute the wall normal velocity at the IB node:

$$\overline{u}_{n,IB} = \overline{u}_{n,PP} \dfrac{d_{IB}^2}{d_{PP}^2}. \qquad (15)$$

A reconstruction of shear stress at the cell face is also done using a RANS-like eddy viscosity. The eddy viscosity at the IB nodes is calculated analytically from mixing the length theory from the equation

$$\nu_t = C_w \kappa u_\tau d_{IB}, \qquad (16)$$

where κ is the von Kármán constant, and C_w is an intensification coefficient to be determined such that the Reynolds shear stress is a fraction of the wall shear stress near the wall (where ν_T is set). Consequently, this coefficient can be written as:

$$C_w = \dfrac{\tau_F d_F}{\tau_w d_{IB}}, \qquad (17)$$

where the index F denotes a quantity calculated at the cell face, d_F is the distance between the cell face and the immersed surface. A more detailed description can be found in Roman et al. [34] and Roman [35]. This wall layer model is very economic since the velocity is just interpolated from the projection point PP, and works well in attached flows using a very coarse grid [34]. Since the interpolation is based on the logarithmic law, and the velocity direction at the IB node always follows the velocity at the PP, this method has a drawback in separated flows even setting the IB node inside the viscous layer.

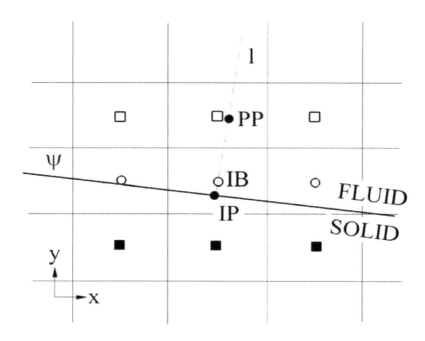

Figure 6. Discretization of a fluid–solid interface with the immersed boundary method by Roman et al. [34]. Solid squares, empty squares and empty circles represent solid, fluid and IB nodes respectively. Reproduced with permission from AIP Publishing [4714120594491].

Simulation of the flow over the hill, applying Cartesian and curvilinear grids, using the Roman et al. [34] model in its original formulation, does not allow to obtain acceptable results, since the flow does not separate even with increasing resolution close to the immersed boundary and with setting the of the IB node inside the viscous layer [28]. Here, first the Roman et al. [34] model on a standard open channel flow is re-analysed and successively the flow over the hill is investigated.

4.1. Calibration of the Wall Layer Model with IBM

In order to check the accuracy of the current IBM, an open channel is considered with a resolution of $48 \times 20 \times 32$ cells in the streamwise, wall normal and spanwise directions, respectively, and a dimension of $7.12\delta \times 1\delta \times 1\delta$ in which δ is the height of the open channels. The lower wall is reproduced with immersed boundaries and the simulation is carried out imposing a non-dimensional driving pressure gradient $(dp/dx = 1)$. The friction Reynolds number is $Re_\tau = 2000$, giving the IB node as located at $y^+ = 50$.

The velocity profile scaled with the friction velocity collapses over the theoretical line as has already been shown in Roman et al. [34]. However, considering the non-scaled averaged velocity, the method underestimates the velocity compared to the logarithmic law at the same rate as the shear velocity, thus demonstrating a loss of momentum. In particular, the wall shear stress is underestimated by approximately 16% in the simulation. This has to be attributed to the current implementation of IBs, which does not conserve momentum perfectly. This is a common issue in a number of IB methodologies, although more recent implementations appear more conservative (Kim et al. [36], Meyer et al. [37] and Rapaka and Sarkar [38]).

After this observation, several cases of open channel geometries are created sharing the same dimension and resolution; the only difference stands in the fact that the immersed boundary surface was moved slightly to check whether the position of the IB node affects the results. Five cases are considered:

Cases 1 to 4 in which the IB node was located in the logarithmic region, and case 5 in which it was in the viscous layer. The positions of the immersed boundary for the five cases are displayed in Figure 7 and details are given in Table 1. Simulations are hence carried out for all cases at $Re_\tau = 2000$, with the same boundary conditions as the original case. The results show an underestimation of friction velocity for cases 1 to 4 in which the IB node is set in the logarithmic region, and an overestimation of shear velocity for case 1 which has the IB node in the viscous layer (see Table 1), indicating that the IBM of Roman et al. [34] is sensitive to the position of the immersed surface with respect to the grid line, and it is geometry dependent. Figure 8a displays a non-scaled mean velocity profile for all cases compared to the theoretical velocity based on the law of the wall.

Apart from the shear velocity deviation, the non-dimensional velocity profile compared well with the wall law for the first four cases (Figure 8b). A calibration of the IBM is carried out to avoid momentum loss. For this purpose, the coefficient which is used to calculate eddy viscosity at the IB nodes (C_w) in Equation (16) is varied to reach the target value of the wall stress.

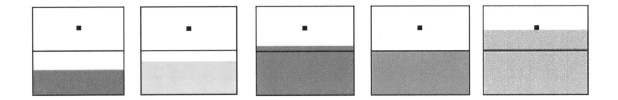

Figure 7. Five different channels using the immersed boundary as the lower wall in different positions, cases 1 to 5 in order from left to right. Filled squares display IB node.

Table 1. Different geometries description of IB cases and shear velocity obtained from simulation.

Case	IB Surface Position	d_{IB}^+	u_τ
1	just above centroid	101	0.8883
2	between centroid and grid line	82	0.8816
3	just above grid line	43.5	0.8541
4	overlapping grid line	50	0.8393
5	just below centroid	5	1.2740

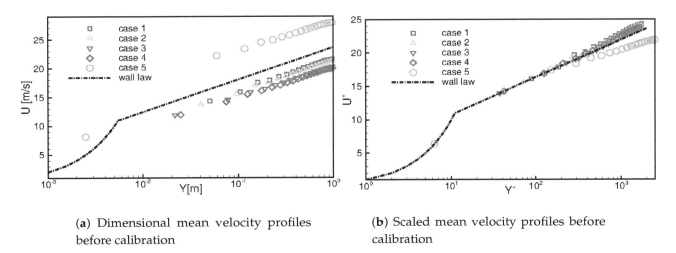

(a) Dimensional mean velocity profiles before calibration

(b) Scaled mean velocity profiles before calibration

Figure 8. *Cont.*

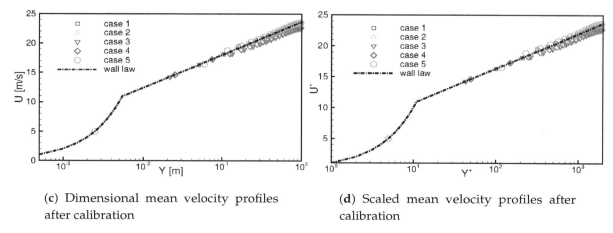

(c) Dimensional mean velocity profiles after calibration

(d) Scaled mean velocity profiles after calibration

Figure 8. Mean velocity for IBM cases before and after calibration versus law of the wall.

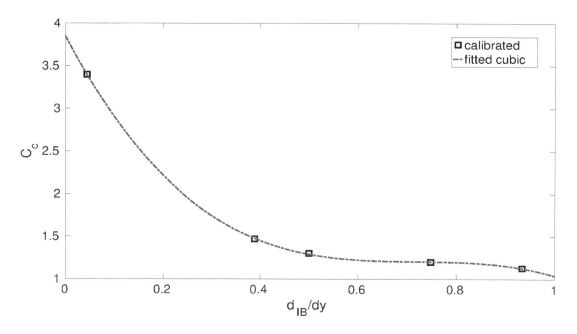

Figure 9. Cubic line fitted to the calibrated values computed for the 5 cases.

Table 2 shows the optimal coefficients and friction velocities obtained after running the simulations for all five cases. Further, it has been verified that changes in the coefficients by 10% produced very small variations in the shear velocity, thus showing the robustness of the method.

Table 2. Optimal coefficient for different positions of the IB.

Case	d_{IB}/dy	Coefficient ($C_{optimal}$)	u_τ
1	0.93318	1.13	0.9987
2	0.74818	1.20	1.0043
3	0.38909	1.47	0.9966
4	0.5	1.3	0.9974
5	0.04386	3.4	0.9942

The simulations are then repeated at $Re_\tau = 4000$ to check the accuracy of the calibration; it appeared insensitive to variations in the Reynolds number [28]. The coefficient can be written as a function of the non-dimensional distance from the wall (d_{IB}/dy), which is within the interval of 0–1. As it can be seen in Figure 9, a cubic law can be fitted to the numerical points is written below.

$$C_c = -6.9576 \left(\frac{d_{IB}}{dy} \right)^3 + 15.051 \left(\frac{d_{IB}}{dy} \right)^2$$
$$- 10.902 \left(\frac{d_{IB}}{dy} \right) + 3.8493 \tag{18}$$

The dimensional and non-dimensional mean velocity profiles after calibration for all the cases are shown in Figure 8c,d, respectively. All cases are non-dimensionalized with shear velocity obtained from the mean velocity at the IB nodes based on the law of the wall. An improvement can be observed after calibration, especially for case 5 in which the IB node is located close to the immersed boundary. Finally the RANS-like eddy viscosity at the IB nodes can be calculated from the calibrated coefficient C_c, which is obtained from the cubic fitting:

$$\nu_T = C_c k u_\tau d_{IB}. \tag{19}$$

Calibration of the coefficient allows to improve the prediction of the wall shear stress, hence, the velocity profile. The results of the simulations for all cases show that the mean velocity and root mean square of velocity fluctuations are in good agreement with the law of the wall and also the DNS of plane channel flow of Hoyas and Jimenez [39] for a channel at $Re_\tau = 2000$ [28]. Only for cases 3 and 4 in which IB nodes are at the beginning of the log-region ($d_{IB}^+ \leq 50$) is there a deviation of the mean velocity profile. However, this is a case that can be hardly encountered in real-scale flows, characterized by very high values of Re. It is important to point out that calibration may not be necessary in the presence of IB methodologies strictly conservative for momentum.

While the wall layer model predicts the velocity profile well, it underestimates the wall normal velocity fluctuations. Thus, random fluctuations near the wall are added to improve the model. In hybrid LES/RANS and DES, stochastic forcing is needed to reduce the effect of excessive damping coming from the RANS eddy viscosity [11,12]. In the present model, stochastic forcing is more likely required because of the grid coarseness, which is, however, a standard in simulations for engineering purposes.

Taylor and Sarkar [40] used random stochastic forcing in the wall normal direction after finding that the near-wall model (NWM) LES with the dynamic eddy viscosity model (DEVM) underrated vertical velocity fluctuations in their body fitted plane-channel-flow simulation.

This forcing term was applied to the right-hand side of the momentum equation in the wall normal direction to increase velocity fluctuations in this direction and control the Reynolds stress in order to allow it to fit the logarithmic velocity profile. This term is written as:

$$f_y(x, y, z) = \pm R * A(y), \tag{20}$$

where R is a random number between 0 and 1, and $A(y)$ is an amplitude function. At each time step, the latter can be obtained from summation of the amplitude at the previous iteration, and another term including the error function:

$$A(y)^{n+1} = A(y)^n + \frac{u_\tau \epsilon(y)}{\tau}, \tag{21}$$

where the error function is the difference between the resolved shear stress and the theoretical value based on the logarithmic law:

$$\epsilon(y) = \frac{ky}{u_\tau}\left(\frac{d\langle u\rangle}{dy}^2 + \frac{d\langle w\rangle}{dy}^2\right)^{\frac{1}{2}} - 1, \tag{22}$$

where τ is a relaxation time. This approach is correct if the first computational node is in the logarithmic region since in Equation (22) the error is established upon ky/u_τ, which is the theoretical resolved shear stress based on the logarithmic law. However, this is not a serious issue, since the model is designed to work for high Re number flows. Finally, the sign of the function is computed to reduce the correlation between streamwise and vertical velocity fluctuations (u' and v') if the error is positive, and to enhance this correlation if the error is negative:

$$sgn\left(f_y(x,y,z)\right) = -sgn\left(\frac{d\langle u_s\rangle}{dy}\right) * sgn\left(\epsilon(y)\right)$$
$$* sgn\left(u'_s(x,y,z)\right), \tag{23}$$

where u_s is the velocity in the direction of the mean wall shear stress.

The stochastic random force in case 4 is imposed instantaneously at every iteration of the simulation. Since the IB node is located in the logarithmic region, the horizontal velocity is interpolated based on the logarithmic law from the projection point PP; adding this force to the first two nodes does not improve the velocity profile. Considering one more point to impose the forcing term on, it makes the deviation from the log law disappear (Figure 10).

This happens because the stochastic forcing allows to obtain a more accurate prediction of the resolved Reynolds shear stress (associated to the resolved velocity fluctuations) as can be seen in Figure 11.

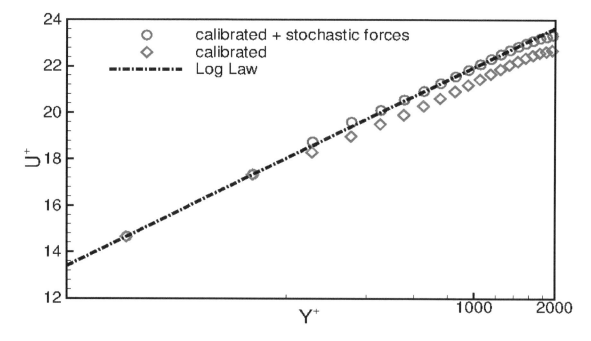

Figure 10. Mean velocity profile before and after adding stochastic forces for case 4 compared to the log law.

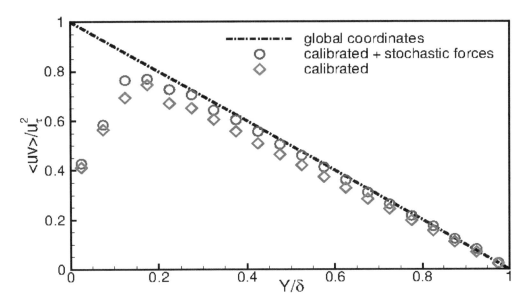

Figure 11. Reynolds shear stresses in global coordinates compared to the calibrated IBM for case 4.

The relaxation time plays an important role in the simulation. It has to be large enough to let the flow adapt itself to the force. On the other hand, if the relaxation time is too large the effect of force is diluted. In the present simulations, $0.0006\, u_\tau/\delta \leq 1/\tau \leq 0.002\, u_\tau/\delta$ gave the best results. Values larger than $0.002\, u_\tau/\delta$ increased the velocity continuously since the flow did not have time to adjust itself to the force; values smaller than $0.0006\, u_\tau/\delta$ displayed a very small improvement in velocity profile.

This method works very well when the IB node is located in the logarithmic region and in attached flows since a logarithmic law for the velocity profile is expected. However, in the present form, it is not able to predict massive separation.

Steady two-dimensional separation is known to be related to the sign of the streamwise pressure gradient. After neglecting advective terms, the boundary layer equation in the streamwise direction simplifies as:

$$\nu \left(\frac{\partial^2 u}{\partial y^2} \right)_{wall} = \frac{1}{\rho} \frac{\partial p}{\partial x}. \tag{24}$$

Considering the flow over a curved surface, a negative streamwise pressure gradient develops upward, associated to an acceleration of the flow. Thereafter, the change of curvature makes the pressure gradient become unfavourable, and from the continuity equation it contributes to an enhancement of the boundary layer thickness. Hence the second derivative of velocity is positive and therefore there is a change of sign of the velocity curvature inside the boundary layer. At a point called inflection, the second derivative of streamwise velocity becomes zero; $\partial^2 u/\partial y^2 = 0$. If the unfavourable pressure gradient is strong enough, the flow next to the wall changes direction giving rise to a recirculation region. As already noted, in this method the velocity at the IB node is calculated from the friction velocity at the projection point. This means that u_{IB} is always in the same direction of U_{PP}, and the flow is not allowed to separate, since the above-described physical mechanism is not accounted for. For this reason, the method of calculation of the velocity at the IB point is modified as follows. The boundary layer equation close to the wall, after neglecting advective terms is:

$$\nu_T \left(\frac{\partial^2 u}{\partial y^2} \right)_{IB} \sim \left(\frac{1}{\rho} \frac{\partial p}{\partial x} \right)_{IB}. \tag{25}$$

Using the central difference scheme (CDS) in order to discretize the second derivative of velocity for a non-uniform grid (Ferziger and Peric [41]) considering PP, IB and IP nodes, it can be written as:

$$\left(\frac{\partial^2 u}{\partial y^2}\right)_{IB} \approx \frac{u_{PP}(d_{IB}) + u_{IP}(d_{PP-IB}) - u_{IB}(d_{PP})}{\frac{1}{2}(d_{PP})(d_{PP-IB})(d_{IB})}, \tag{26}$$

in which d_{PP-IB} is the distance between IB and PP points. Finally the streamwise tangential velocity at the IB node can be written as a function of the tangential pressure gradient instead of the friction velocity;

$$u_{IB} = u_{PP}\left(\frac{d_{IB}}{d_{PP}}\right) - C\left(\frac{\partial p}{\partial x}\right)_{IB} \frac{(d_{PP-IB})(d_{IB})}{\nu_T}, \tag{27}$$

where the eddy viscosity is also the calibrated value obtained from Equation (19).

Using a coefficient (C) of the order of 10^{-3} inside the second term containing the pressure gradient at the IB avoids problems in regions characterized by the presence of large values of the pressure gradient with small values of turbulent eddy viscosity. Many simulations have been carried out to find out the best criterion for using this new scheme. Finally, applying Equation (27) under the following conditions:

$$\left(\frac{\partial p}{\partial x}\right)_{IB} > 0, \tag{28}$$

$$d_{IB}^+ < 40, \tag{29}$$

gave the best results for simulating separated flows. The two criteria physically mean that the new scheme to calculate tangential velocity at the IB is used if the tangential pressure gradient at the IB is unfavourable, and the distance of the IB node from the immersed boundary is such that the velocity point does not belong to the log region of the velocity profile.

4.2. Flow Simulation over a Single Hill Using IBM

This new model is applied for simulation of the flow over a two-dimensional hill. First an open channel is constructed in which the immersed boundary is set as the lower wall. After the periodic open channel flow reached a steady state, instantaneous data are collected from a cross sectional plane at different time steps. Then the bottom of the domain including the hill shape is created as the immersed boundary, and the data that are obtained from the plane channel flow are used as inflow for the flow simulation over the hill.

The simulation is performed using two different grids (Cartesian and curvilinear) at $Re_\tau = 1187$. Applying the new scheme, it is feasible to capture separation in the simulation with a grid resolution of $128 \times 40 \times 32$ cells in the streamwise, wall normal and spanwise directions, respectively. With this resolution, the IB node upward and downward from the hill is located at $d_{IB}^+ \approx 15$. The mean velocity profiles for both grids are compared with the body fitted approach and the experiment by Khurshudyan et al. [30] as shown in Figure 12a. In the simulation for both grids, the starting point of separation is predicted well at $x = 0.5a$. The reattachment points for both cases are anticipated before $x = 2a$; therefore, at $x = 2a$ the mean velocities are positive, although very small close to the immersed boundary.

In addition, for the Cartesian grid a simulation without calibration is performed to check the effect of eddy viscosity calibration on flow separation; as it can be observed in Figure 12a, the case without calibration is not able to capture the starting point of separation properly, and in general it underestimates separation.

The root mean square of streamwise velocity fluctuations of the present simulations (Figure 12b) are in a good agreement with the experimental data. While there is an overestimation of fluctuations beyond the starting point of separation, the behavior of this quantity is similar to that of the experiment. Comparing the results of the present model with the one developed for the body fitted grid, it appears that the present IBM anticipates the starting point of separation in better agreement with the wall-resolved LES and the experiment, while the prediction of the reattachment point is similar to the wall-resolved LES, earlier than in the body fitted approach and the experiment. The present IBM gives more energetic velocity fluctuations than the body fitted approach. Moreover, comparing this result with the wall-resolved LES by Chaudhari et al. [31], it can be stated that the mean velocity profile has similar behavior: Predicting the starting point of separation well, and undergoing an early reattachment. The RMS of streamwise velocity fluctuations is also similar to the wall-resolved LES; depicting marginal underestimation before and after the separation region especially far from the wall, and an overestimation inside the separation region near the wall.

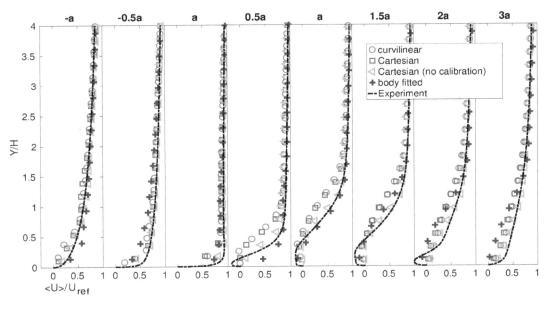

(a) Mean velocity.

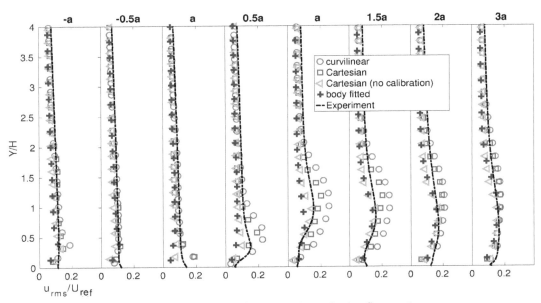

(b) Root mean square of streamwise velocity fluctuations.

Figure 12. The results for the simulation of flow over a single hill using new the IBM scheme in Cartesian and curvilinear grids compared to the body fitted approach and experiment by Khurshudyan et al. [30]; the hill starts at $x = -a$ and ends at $x = a$.

4.3. Flow Simulation over 2D Periodic Hills

Finally, the new scheme is tested on flow over two-dimensional periodic hills with polynomial shape (Almeida et al. [42]). It is Case 81 of the ERCOFTAC classic database (http://ERCOFTAC/cases/case81)[43]. The shape of the domain is shown in Figure 13.

The edges of the domain are at the top of the hill. The hill height is $h = 28$ mm, and the crests of the two successive hills are separated by $L_x = 9h$. The height of the channel is equal to $L_y = 3.035h$, and the channel width is $L_z = 4.5h$. The reference for comparing the results are the mean velocity and RMS of the velocity fluctuations obtained from resolved LES by Temmerman and Leschziner [44] available in ERCOFTAC, and for separation and reattachment points, obtained from resolved LES by Fröhlich et al. [45] from the NASA Langley Research Center database (http://nasa/LES/2dhillperiodic) [46].

The computational domain is created in two different ways, using Cartesian and curvilinear grids, respectively (see Figure 14). The latter is shaped in such a way as to follow the hill geometry. In both cases, the bottom surface is created as an immersed boundary. The Reynolds number based on the hill height and bulk velocity on top of the first hill is equal to $Re = U_b h/\nu = 10595$, where ν is the molecular viscosity. The flow is periodic in the x and z directions, using immersed boundary at the bottom and free-slip surface at top.

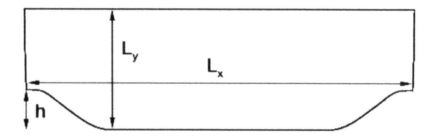

Figure 13. Domain characteristics for the simulation of flow over 2D periodic hills, with courtesy of ERCOFTAC [43].

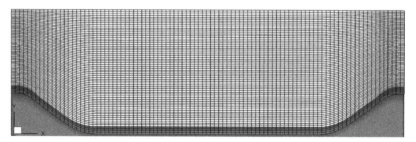

(**a**) Curvilinearn grid.

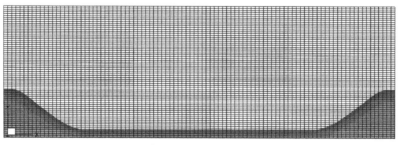

(**b**) Cartesian grid.

Figure 14. Two different grids for simulation of periodic hills using immersed boundaries, $96 \times 64 \times 32$ cells in x, y and z directions, respectively.

A domain with grid resolution of $96 \times 64 \times 32$ cells out of immersed boundary in streamwise, wall normal and spanwise directions, respectively, is constructed. This is equivalent to the coarsest grid of Chen et al. [17], who simulated flow over periodic hills using immersed boundary and applied a zonal approach (TLM); in particular, the authors used the turbulent boundary layer equations close to the immersed boundary and LES in the interior region.

The robustness of the method to reproduce separation is tested under a large variation of the threshold values employed in Equation (29). The results of this test for the threshold in the range of 30–50 are shown in Figure 15. The mean velocity profile in Figure 15a shows that in all cases separation is captured at $x = 1a$, but using $d_{IB}^+ < 50$ makes the reattachment point shift back slightly with respect to the other cases and the wall-resolved LES. The mean velocity profile for $d_{IB}^+ < 40$ is marginally closer to the resolved LES compared to $d_{IB}^+ < 30$ (which overestimates it narrowly). In addition to the mean velocity profile, Reynolds shear stresses for $d_{IB}^+ < 40$ demonstrates better agreement to the wall-resolved LES in the separation region and after reattachment (Figure 15b). Overall, it can be stated that the method is robust for large variations of the threshold value chosen to switch from one to the other law for the velocity at the first grid point. Large variations of this value produce marginal variations in the results.

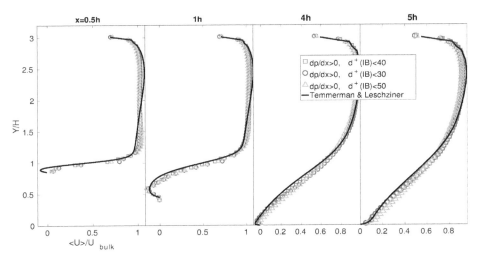

(a) Mean streamwise velocity.

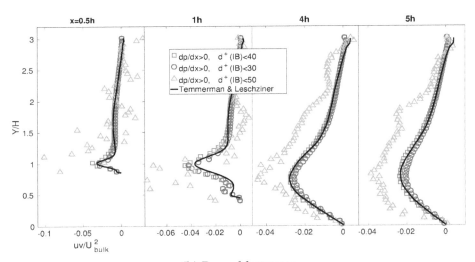

(b) Reynold stresses.

Figure 15. Mean streamwise velocity and Reynolds shear stresses for a Cartesian grid compared to resolved LES by Temmerman and Leschziner [44], applying three different criteria to use the new scheme; $(\partial p / \partial x)_{IB} > 0$ and: $d_{IB}^+ < 40$ (square), $d_{IB}^+ < 30$ (circle), $d_{IB}^+ < 50$ (delta).

Figure 16 depicts statistics of the simulations of flow over periodic hills compared to the wall-resolved LES by Temmerman and Leschziner [44] at ten different streamwise locations. Velocities are made non-dimensional with U_b which is the bulk velocity on the first hill crest; moreover, Reynolds stress and turbulent kinetic energy are non-dimensionalized with U_b^2.

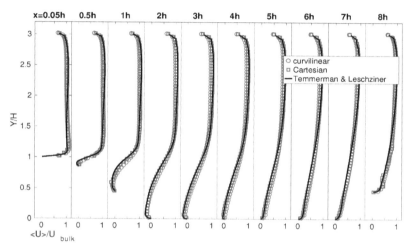

(a) Mean streamwise velocity profiles.

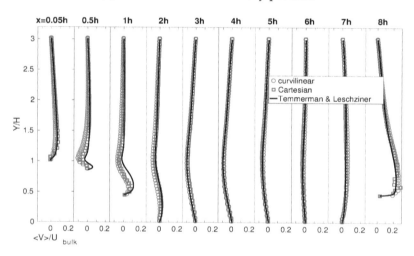

(b) Mean vertical velocity profiles.

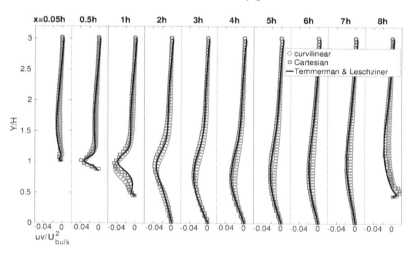

(c) Reynolds shear stress profiles.

Figure 16. *Cont.*

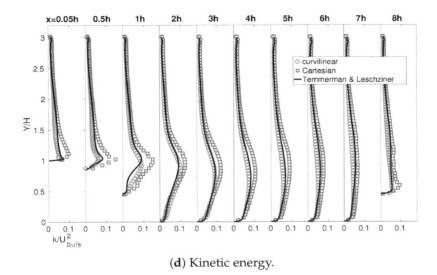

(**d**) Kinetic energy.

Figure 16. Results for simulation of flow over periodic hills at different positions, using Cartesian and curvilinear grids compared to the resolved LES by Temmerman and Leschziner [44].

Figure 16a shows mean streamwise velocity profiles (averaged in time and the spanwise direction) at different streamwise locations. They are overall in good agreement with the wall-resolved LES by Temmerman and Leschziner [44]. While the mean velocity for the Cartesian grid at $x = 0.5h$ matches the profile obtained from the wall-resolved LES, the velocity is not negative at the IB node, while the curvilinear grid displays a negative velocity at the IB node in this location. Separation is captured in both Cartesian and curvilinear grid cases, and the reattachment point for the Cartesian grid is a bit earlier than the resolved LES, while the flow in the curvilinear grid reattaches earliest before $x = 4h$.

The mean vertical velocity profiles in Figure 16b show a slight underprediction for both grids downhill, but in the other locations they matched the wall-resolved LES data.

Figure 16c displays the Reynolds shear stress for both cases (based on the resolved fluctuations). It behaves similarly to that of the resolved LES, while a marginal underestimation is observed before the end of downhill at $x = 1h$ near the immersed boundary, especially in the case with curvilinear grid.

As illustrated in Figure 16d, turbulent kinetic energy for the Cartesian grid is slightly overpredicted with respect to the wall-resolved LES, while the curvilinear grid demonstrates a closer value to the reference data. Overall, the behavior of this quantity is similar to that of the wall-resolved LES for both cases.

Table 3 displays the separation and reattachment points in the current simulation, the wall-resolved LES of Fröhlich et al. [45] and WMLES (using two-layer model) of Chen et al. [17]. In the present simulation, the boundary layer separation occurs at $x \approx 0.54h$ with the Cartesian grid and at $x \approx 0.47h$ with the curvilinear grid; in the wall-resolved LES it happens at $x \approx 0.22h$. Therefore a delay is observed to predict the separation point. Moreover, the intensity of the recirculation bubble predicted by the optimized IBM simulations is smaller than that of the wall-resolved LES (comparing the present simulation values with the large negative values of the reference simulation). The wall-resolved LES shows the reattachment point at $x \approx 4.69h$, while in the current simulation the boundary layer reattaches a bit earlier in both cases; the Cartesian grid predicts the reattachment point at $x \approx 4.20h$, while the curvilinear grid anticipates it at $x \approx 3.81h$.

Overall, the agreement of the simulation result is acceptable, considering the simplicity of the model, which does not require the solution of an additional equation in the wall layer and the number of grid points employed in the tests. Moreover, in comparison to the simulations of [17], the new wall layer model on the Cartesian grid predicts the separation point better than their case with the same resolution, and on the curvilinear grid it is even closer to the reference data than the WMLES of [17], but the reattachment point in their case is predicted closer to the reference. The other statistics obtained from the current simulation, especially the mean vertical velocity profile, are in better agreement with the wall-resolved LES than the simulations of [17].

Table 3. Resolution, separation point (x_s) and reattachment point (x_r) in resolved LES Fröhlich et al. [45], the present optimized IBM simulations and the wall model LES by Chen et al. [17].

Case	Resolution	x_s	x_r
Resolved LES (Fröhlich et al.)	$196 \times 128 \times 186$	$0.224h$	$4.69h$
IBM Cartesian ($n_2 = 64$)	$96 \times 64 \times 32$	$0.54h$	$4.20h$
IBM curvilinear ($n_2 = 64$)	$96 \times 64 \times 32$	$0.47h$	$3.81h$
WMLES (TBLE/LES at $y^+ = 30$)	$96 \times 64 \times 32$	$0.65h$	$4.00h$
WMLES (TBLE/LES at $y^+ = 15$)	$192 \times 72 \times 48$	$0.50h$	$4.42h$

5. Summary and Conclusions

In the present paper, a wall layer modeling for massive separation has been developed within the LES context. Two main frameworks have been considered; namely, the case where the solid wall is reproduced using a body fitted curvilinear grid and the case where the solid surface is simulated with immersed boundaries. The research aimed at developing a simple wall layer model to be used in conjunction with LES for engineering applications where the near-wall resolution is generally coarse. The starting point was the equilibrium wall layer model, which gives good results in attached flows [47]. In the case of body fitted curvilinear grids, the equilibrium wall layer model was improved using the dynamic κ procedure of Cabot and Moin [8] in a new fashion. The dynamic κ was applied on the log law without using TBLE in order to deviate the wall shear stress and the two components of strain rate tensor from the standard log law.

When the solid surface was simulated with immersed boundaries, a different strategy was accomplished. The starting point was the equilibrium wall layer model of Roman et al. [34], in which the velocity at the near-wall nodes was reconstructed from a farther point (the projection point) based on the law of the wall, and a RANS-like eddy viscosity was given at the IB node. To reconstruct the fluctuations in the wall normal direction well, especially near wall, random stochastic forcing was applied to the first three near-wall cells. Moreover, the model was complemented with a procedure designed at reproducing massive separation without the need to use zonal approaches. Specifically, after removing the advective terms from the boundary layer equation, the tangential streamwise velocity at the IB node (u_{IB}) was obtained from the pressure gradient instead of calculating it from shear velocity at the projection point based on the law of the wall: When the pressure gradient was unfavourable and the IB node was not in the log region, the tangential velocity was obtained through a balance between the wall normal diffusion term and the streamwise pressure gradient. In some sense, this represents a simplified procedure with respect to the zonal approach where the boundary layer equation is solved within the near-wall layer. In order to check what grid topology is best suited for reproducing this kind of flow, both Cartesian and curvilinear grids were considered in conjunction with immersed boundaries.

The models were tested and validated over two two-dimensional geometries, both being well studied in the literature through physical and high-resolution LESs: First the isolated hill (case 69 of the ERCOFTAC classic database) was considered; then, the periodic hill which is case 81 of ERCOFTAC classic datasbe was considered.

The model for body fitted geometry was generally in good agreement with the reference data; the starting point of separation was delayed with respect to the wall-resolved LES and the experiment, but the reattachment point was closer to the experiment, while the wall-resolved LES displayed an early reattachment. Applying 32 and 20 cells in the vertical direction correspondent to $\Delta y^+ \approx 37$ and 60, respectively, the results were comparatively acceptable. However, decreasing the vertical resolution further would delay the starting point of separation more. The model developed for immersed boundaries gave a mean velocity profile and other statistics in acceptable agreement with the reference data on both isolated and periodic hills, also in consideration of the coarseness of the grids employed; for the isolated hill, 40 cells in the vertical direction were required to cover the hill with four cells applying the Cartesian grid, which was correspondent to $\Delta y^+ \approx 30$; higher resolution was not used to avoid the near-wall computational node being located in the viscous sub-layer, and lower resolution did not capture the flow separation.

For the periodic hill, 64 cells in the vertical direction gave $\Delta y^+ \approx 105$ applying the Cartesian grid. Lower resolution did not display flow separation because of an excessive under-resolution of the hill geometry.

Acknowledgments: The author wholeheartedly thanks Vincenzo Armenio from Università degli Studi di Trieste, Italia, and Federico Roman from IEFLUIDS S.r.l., Università degli Studi di Trieste, Italia, for their effective collaboration to accomplish this work.

References

1. Piomelli, U.; Balaras, E. Wall-Layer Models For Large-Eddy Simulations. *Annu. Rev. Fluid Mech.* **2002**, *34*, 349–374. [CrossRef]
2. Piomelli, U. Wall-Layer Models For Large-Eddy Simulations. *Prog. Aerosp. Sci.* **2008**, *44*, 437–446. [CrossRef]
3. Bae, H.J.; Lozano-Durán, A.; Bose, S.T.; Moin, P. Dynamic slip wall model for large-eddy simulation. *J. Fluid Mech.* **2019**, *859*, 40-432. [CrossRef] [PubMed]
4. Deardorff, J.W. A Numerical Study of Three-Dimensional Turbulent Channel Flow at Large Reynolds Numbers. *J. Fluid Mech.* **1970**, *41*, 453–480. [CrossRef]
5. Schumann, U. Subgrid-Scale Model For Finite Difference Simulation of Turbulent Flows in Plane Channels And Annuli. *J. Comput. Phys.* **1975**, *18*, 376–404. [CrossRef]
6. Balaras, E.; Benocci, C. *Sub-Grid Scale Models in Finite-Difference Simulations of Complex Wall Bounded Flows*; AGARD CP: Neuilly-Sur-Seine, France, 1994; Volume 551, pp. 2.1–2.5.
7. Cabot, W.H. Near-Wall Models in Large-Eddy Simulations of Flow Behind a Backward-Facing Step. In *Annual Research Briefs—+*; Center for Turbulence Research, Stanford University: Stanford, CA, USA, 1996; pp. 199–210.
8. Cabot, W.; Moin, P. Approximate Wall Boundary Conditions in The Large-Eddy Simulation of High Reynolds Number Flows. *Flow. Turbul. Combust.* **2000**, *63*, 269–291. [CrossRef]
9. Spalart, P.R.; Jou, W.H.; Strelets, M.; Allmaras, S.R. *Comments On The Feasibility of LES For Wings and on a Hybrid RANS/LES Approach*; Advances in DNS/LES; Liu, C., Liu, Z., Eds.; Greyden Press: Columbus, OH, USA, 1997; pp. 137–148.
10. Mockett, C.; Fuchs, M.; Thiele, F. Progress in DES For Wall-Modelled LES of Complex Internal Flows. *Comput. Fluids* **2012**, *65*, 44–55. [CrossRef]
11. Keating, A.; Piomelli, U. A Dynamic Stochastic Forcing Method As A Wall-Layer Model For Large-Eddy Simulation. *J. Turbul.* **2006**, *7* 1–24. [CrossRef]
12. Temmerman, L.; Hadziabdic, M.; Leschziner, M.A.; Hanjalic, K. A Hybrid Two-Layer URANS-LES Approach For Large Eddy Simulation at High Reynolds Numbers. *Int. J. Heat Fluid Flow* **2005**, *26*, 173–190. [CrossRef]
13. Mittal, R.; Iaccarino, G. Immersed Boundary Methods. *Annu. Rev. Fluid Mech.* **2005**, *37*, 239–261. [CrossRef]
14. Roman, F.; Napoli, E.; Milici, B.; Armenio, V. An Improved Immersed Boundary Method For Curvilinear Grids. *J. Comput. Fluids* **2009**, *38*, 1510–1527. [CrossRef]
15. Tessicini, F.; Iaccarino, G.; Fatica, M.; Wang, M.; Verzicco, R. Wall Modelling For Large-Eddy Simulation Using an Immersed Boundary Method. In *Annual Research Briefs*; NASA Ames Research Center/Stanford University Center of Turbulence Research: Stanford, CA, USA, 2005; pp. 181–187.
16. Posa, A.; Balaras, E. Model-Based Near-Wall Reconstructions For Immersed-Boundary Methods. *J. Theor. Comput. Fluid Dyn.* **2014**, *28*, 473–483. [CrossRef]
17. Chen, Z.L.; Hickel, S.; Devesa, A.; Berl, J.; Adams, N.A. Wall Modelling For Implicit Large-Eddy Simulation and Immersed-Interface Methods. *J. Theor. Comput. Fluid Dyn.* **1994**, *28*, 1–21.
18. Roman, F.; Stipcich, G.; Armenio, V.; Inghilesi, R.; Corsini, S. Large Eddy Simulation of Mixing in Coastal Areas. *J. Heat Fluid Flow* **2010**, *31*, 327–341. [CrossRef]

19. Galea, A.; Grifoll, M.; Roman, F.; Mestres, M.; Armenio, V.; Sanchez-Arcilla, A.; Mangion, L.Z. Numerical Simulation of Water Mixing And Renewals In The Barcelona Harbour Area. *Environ. Fluid Mech.* **2014**, *14*, 1405–1425. [CrossRef]

20. Smagorinsky, J. General Circulation Experiments with the Primitive Equations. *Mon. Weather Rev.* **1963**, *91*, 99–164. [CrossRef]

21. Lilly, D.K. The representation of small-scale turbulence in numerical simulation experiments. In Proceedings of the IBM Scientific Computing Symposium on Environmental Sciences, New York, NY, USA, 14–16 November 1967.

22. Blazek, J. *Computational Fluid Dynamics: Principles and Applications*, 3rd ed.; Butterworth-Heinemann: Oxford, UK, 2015.

23. Stocca, V. Development of a Large Eddy Simulation Model for the Study of pOllutant Dispersion in Urban Areas. Ph.D. Thesis, University of Trieste, Trieste, Italy, 2010.

24. Germano, M.; Piomelli, U.; Moin, P.; Cabot, W.H. A dynamic subgrid-scale eddy viscosity model. *Phys. Fluids* **1991**, *3*, 1760–1765. [CrossRef]

25. Zang, J.; Street, R.L.; Koseff, J.R. A Non-Staggered Grid, Fractional Step Method For Time-Dependent Incompressible Navier-Stokes Equations in Curvilinear Coordinates. *J. Comput. Phys.* **1994**, *14*, 459–486. [CrossRef]

26. Armenio, V.; Piomelli, U. A Lagrangian Mixed Subgrid-Scale Model in Generalized Coordinates. *J. Flow Turbul. Combust.* **2000**, *65* , 51–81. [CrossRef]

27. Petronio, A.; Roman, F.; Nasello, C.; Armenio, V. Large Eddy Simulation Model For Wind-Driven Sea Circulation In Coastal Areas. *Nonlin. Process. Geophys.* **2013**, *20*, 1095–1112. [CrossRef]

28. Fakhari, A. Wall-Layer Modelling of Massive Separation in Large Eddy Simulation of Coastal Flows. Ph.D. Thesis, University of Trieste, Trieste, Italy, 2015.

29. Flow over Isolated 2D Hill. Available online: http://cfd.mace.manchester.ac.uk/cgi-bin/cfddb/prpage.cgi?69&EXP&database/cases/case69/Case_data&database/cases/case69&cas69_head.html&cas69_desc.html&cas69_meth.html&cas69_data.html&cas69_refs.html&cas69_rsol.html&1&0&0&0&0/ (accessed on 22 November 2019).

30. Khurshudyan, L.H.; Snyder, W.H.; Nekrasov, I.V. *Flow and Dispersion Of Pollutants over Two-Dimensional Hills*; Environment Protection Agency Report No EPA-600/4-81-067; Environment Protection Agency: Research Triangle Park, NC, USA, 1981.

31. Chaudhari, A.; Vuorinen, V.; Hämäläinen, J.; Hellsten, A. Large-eddy simulations for hill terrains: Validation with wind-tunnel and field measurements, *J. Comput. Appl. Math.* **2018**, *37*, 2017–2038. [CrossRef]

32. Chaudhari, A. Large-Eddy Simulation of Wind Flows over Complex Terrains for Wind Energy Applications. Ph.D. Thesis, Lappeenranta University of Technology, Lappeenranta, Finland, 2014.

33. Vinokur, M. On One-Dimensional Stretching Functions For Finite-Difference Calculations. *J. Comput. Phys.* **1983**, *50*, 215–234. [CrossRef]

34. Roman, F.; Armenio, V.; Fröhlich, J. A Simple Wall-Layer Model For Large Eddy Simulation With Immersed Boundary Method. *J. Phys. Fluids* **2009**, *21*, 101701. [CrossRef]

35. Roman, F. A Numerical tool for Large Eddy Simulations in Environmental and Industrial Processes. Ph.D. Thesis, University of Trieste, Trieste, Italy, 2009.

36. Kim, J.; Kim, D.; Choi, H. An Immersed-Boundary Finite-Volume Method for Simulations of Flow In Complex Geometries. *J. Comput. Phys.* **2001**, *171*, 132–150. [CrossRef]

37. Meyer, M.; Devesa, A.; Hickel, S.; Hu, X.Y.; Adams, N.A. A Conservative Immersed Interface Method For Large-Eddy Simulation of Incompressible Flows. *J. Comput. Phys.* **2010**, *229*, 6300–6317. [CrossRef]

38. Rapaka, N.R.; Sarkar, S. An Immersed Boundary Method For Direct And Large Eddy Simulation of Stratified Flows in Complex Geometry. *J. Comput. Phys.* **2016**, *322*, 511–534. [CrossRef]

39. Hoyas, S.; Jiménez, J. Reynolds Number Effects On The Reynolds-Stress Budgets In Turbulent Channels. *J. Phys. Fluids* **2008**, *20*, 101511. [CrossRef]

40. Taylor, J.; Sarkar, S. Near-Wall Modelling For LES of an Oceanic Bottom Boundary Layer. In Proceedings of the Fifth International Symposium on Environmental Hydraulics, Tempe, AZ, USA, 4–7 December 2007.

41. Ferziger, J.H.; Peric, M. *Computational Methods For Fluid Dynamics*, 3rd ed.; Springer: Berlin, Germany, 2002.

42. Almeida, G.P.; Durao, D.F.G.; Heitor, M.V. Wake Flows Behind Two Dimensional Model Hills. *Exp. Therm. Fluid Sci.* **1992**, *7*, 87–101. [CrossRef]

43. Flow over Periodic Hills. Available online: http://cfd.mace.manchester.ac.uk/cgi-bin/cfddb/prpage.cgi?81&LES&database/cases/case81/Case_data&database/cases/case81&cas81_head.html&cas81_desc.html&cas81_meth.html&cas81_data.html&cas81_refs.html&cas81_rsol.html&1&1&1&1&1&unknown (accessed on 22 November 2019).
44. Temmerman, L.; Leschziner, A. Large Eddy Simulation of Separated Flow in a Streamwise Periodic Channel Construction. In Proceedings of the International Symposium on Turbulence and Shear Flow Phenomena, Stockholm, Sweden, 27–29 June 2001.
45. Fröhlich, J.; Mellen, C.P.; Rodi, W.; Temmerman, L.; Leschziner, M.A. Highly Resolved Large-Eddy Simulation of Separated Flow In a Channel In a Channel With Streamwise Periodic Constructions. *J. Fluid Mech.* **2005**, *526*, 19–66. [CrossRef]
46. LES: 2-D Periodic Hill. Available online: https://turbmodels.larc.nasa.gov/Other_LES_Data/2dhill_periodic.html (accessed on 22 November 2019).
47. Fakhari, A.; Cintolesi, C.; Petronio, A.; Roman, F.; Armenio, V. Numerical simulation of hot smoke plumes from funnels. In *Technology and Science for the Ships of the Future, Proceedings of NAV 2018: 19th International Conference on Ship & Maritime Research, Trieste, Italy, 20–22 June 2018*; IOS Press: Amsterdam, The Netherlands, 2018; pp. 238–245.

2

Optimization of the Orifice Shape of Cooling Fan Units for High Flow Rate and Low-Level Noise in Outdoor Air Conditioning Units

Se Min Park [1], Seo-Yoon Ryu [2], Cheolung Cheong [2,*], Jong Wook Kim [1], Byung Il Park [1], Young-Chull Ahn [2] and Sai Kee Oh [1]

[1] LG Electronics, Changwon 2nd Factory, Seonsan-dong, Changwon-city, Keongsang Namdo 51554, Korea; mach4780.park@lge.com (S.M.P.); jw01.kim@lge.com (J.W.K.); byungil.park@lge.com (B.I.K.); Saikee.oh@lge.com (S.K.O.)

[2] School of Mechanical Engineering, Pusan National University, Busan 46241, Korea; rjsoo0302@pusan.ac.kr (S.-Y.R.); ycahn@pusan.ac.kr (Y.-C.A.)

* Correspondence: ccheong@pusan.ac.kr

Abstract: Demand for air conditioners is steadily increasing due to global warming and improved living standards. The noise, as well as the performance of air conditioners, are recognized as one of the crucial factors that determine the air conditioners' values. The performance and noise of the air conditioner are mostly determined by those of its outdoor unit, which in turn depend on those of the cooling fan unit. Therefore, a cooling fan unit of high-performance and low noise is essential for air-conditioner manufacturers and developers. In this paper, the flow performance and flow noise of the entire outdoor unit with an axial cooling fan in a split-type air-conditioner were investigated. First, a virtual fan tester constructed by using about 18 million grids is developed for highly resolved flow simulation. The unsteady Reynolds-Averaged Navier–Stokes equations are numerically solved by using finite-volume computational fluid dynamics techniques. To verify the validity of the numerical analysis, the predicted P–Q curve of the cooling fan in a full outdoor unit is compared with the measured one. There was an excellent agreement between the two curves. The further detailed analysis identifies the coherent vortex structures between the fan blade tip and fan orifice, which adversely affect the flow performance and causes flow noise. Based on this analysis, the optimization of fan orifice was carried out using the response surface method with three geometric parameters: inlet radius, neck length, and outlet angel of the orifice. The optimum layout for the high flow rate is proposed under the understanding that the increased flow rate can be converted to noise reduction. The additional computation using the proposed optimum orifice shows that the flow rate is increased by 4.6% at the operating point. Finally, the engineering sample was manufactured by using the optimum design, and the measured data confirmed that the flow rate were increased by 2.1%, the noise reduction was made by 2.8 dBA, and the power consumption is reduced by 4.0% at the operating rotational speed.

Keywords: Aerodynamic noise; Fan noise; Outdoor unit of air-conditioner; Fan orifice; Virtual Fan-performance tester

1. Introduction

As environmental pollution becomes serious, there is a growing interest in a pleasant living environment. Accordingly, there is a steady increase in demand for home appliances that control indoor air. A typical example is an air conditioner. The air conditioner is used not only to cool the air by removing heat in the room but also to control the air quality by lowering the humidity. The air conditioner is used continuously regardless of the season so that the consumers are exposed directly to the air conditioner and are sensitive to its performance and noise. Therefore, related manufacturers and developers are continually investing in research to improve air conditioner performance and to reduce noise.

The noise of a split-type air conditioner is classified into the noise due to the outdoor unit and the indoor unit. Since the indoor unit is installed in the room, the consumer is directly subject to the noise of the indoor unit. However, the device generating noise is only the fan in the indoor unit, and thus the noise level of the indoor unit is relatively low. On the other hand, the outdoor unit has more equipment than the indoor unit, such as compressors, evaporators, and fans, which in turn cause more noise problems. The noise of outdoors often leads to a dispute between the neighbors so that noise-related regulations vary depending on the region. The overall performance of the split-type air conditioner is highly dependent on the performance of the outdoor unit, which in turn is closely related to the performance of the cooling fan in the outdoor unit. Therefore, to improve the performance of the air conditioner, a fan with a high flow rate is required. However, in general, the increase in fan airflow is directly related to the increase in noise. Therefore, it is essential for securing the market competitiveness of the manufacturer to develop high flow-performance and low-noise fans. In this study, the flow and noise performance in the outdoor unit of air-conditioner is investigated and improved by developing the high performance and low noise fan units.

Lee et al. [1] presented a hybrid computational aeroacoustics (H-CAA) method for efficiently predicting an internal aerodynamic noise of a centrifugal fan unit in a household refrigerator. Heo et al. [2] used the H-CAA method and developed low-noise centrifugal fans by introducing the inclined S-shaped trailing edge. These studies confirmed that the H-CAA method could be reliably used to predict the BPF noise of the fan unit. By extending these works, Heo et al. developed an efficient numerical method for predicting broadband noises as well as tonal noise of a centrifugal fan [3]. The proposed method is based on the Unsteady Fast Random Particle Mesh (U-FRPM) method. Utilizing these numerical methods, they [4] improved aerodynamic and aeroacoustics performance of an axial-flow fan unit by modifying its housing structure without changing the fan blade. Zhao et al. [5] investigated the shapes of the tip and trailing edge of axial fan blades to reduce fan noise in an outdoor unit. Jiang et al. [6] investigated the aeroacoustics of an axial fan in an outdoor unit. Fukano's model, combined with typical computational fluid dynamics (CFD) simulation, is used to predict the broadband noise. After the predicted results were compared with the measured data, the distance to blade trailing edge is proposed as an essential parameter to improve the accuracy of noise predictions in the Fukano's model. Wright et al. [7] presented the aerodynamic and acoustic properties of axial-flow fans with swept blades based on the theory of Kerschen and Envia for swept cascades. The experimental results confirmed that the swept-bladed fans show noise savings compared to the zero-sweep baseline model. Ye et al. [8] investigated the effects of various tip structures on the flow field, losses distribution, and noise characteristics and improved fan performance by changing the blade tip structure. It was shown that the grooved blade tips improved the efficiency by reducing mixing losses between the leakage flow and mainstream but increased the fan noise.

These studies [4–8] revealed that the characteristics of gap flow between the tip of the fan blade and the fan orifice (housing or shroud) are critical to the aerodynamic and aeroacoustic performance of fans. However, most of these studies concerned a fan unit alone or a fan in a simplified outdoor unit. The axial fan used in an outdoor unit has a complicated inflow structure, and even the same fan has different performance depending on the inflow structure affected by various devices. Notably, in the case of the outdoor unit, there are many components, not only an axial fan but an orifice, a heat exchanger, a heat sink, a motor, and a motor mount. These components can affect the flow performance as well as the noise of the outdoor unit. In this study, the whole outdoor unit, including these components, is investigated. As described above, among the components, the orifice is known to be one of the main parts affecting the flow and noise performances of an axial fan. As the cooling fan rotates in the outdoor unit, complicated tip vortex structures are created, which can adversely affect the flow performance and noise of fan [9–11]. The orifice organizes the flow by guiding the flow path between the orifice and the blade tip. In this study, the shape of an orifice is optimized to maximize the flow rate and thus to reduce the noise of an axial fan in the outdoor unit of an air conditioner.

2. Targeted Axial-Flow Fan in Outdoor Units

Figure 1a shows the geometries of the internal structure of a targeted outdoor unit. The outdoor unit consists of an axial-flow fan, an orifice, a grille, a heat exchanger, a motor, and a motor-mount, as shown in Figure 1b. The axial-flow fan unit is used to drive airflow through the outdoor unit and to cool the heat exchanger. Figure 1c shows the axial-flow fan. The axial-flow fan consists of three blades. The diameter of the fan is 400 mm and has a rake shape similar to the winglet of airplane to reduce the tip-wake and flow noise. Note that all of devices except the heat exchanger are considered in the current study. The effects of heat exchanger are accounted for by considering additional pressure loss due to it.

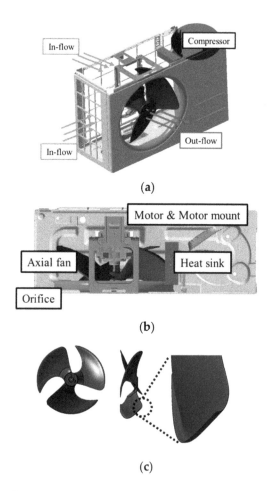

Figure 1. Internal structure of targeted outdoor unit and main parts: (**a**) internal structure of outdoor unit, (**b**) main parts, and (**c**) axial-flow fan.

3. Experimental Analysis of Targeted Outdoor Units

3.1. Experimental Set-Up for Characterizing Flow Performance

To characterize the flow performance of the targeted outdoor unit of air-conditioner, the fan performance experiment was conducted. The fan performance tester satisfying the regulations of AMCA (Air Movement and Control Association) 210-07 which provides methodology for testing the air moving devices is used. To improve the measurement accuracy, a large chamber of 1500(L) × 600(W)

× 600(H) mm, five different-sized nozzles and two screens were installed in the fan performance tester. The pressure difference between the inlet and outlet of nozzle is measured, and the flow rate is calculated from the measured differential static pressure. This system is schematically shown in Figure 2. Consequently, the combination of measured static pressure and flow rate is arranged to determine the P–Q curve. The more detailed information about the fan performance tester was provided in the reference [4]. The measured P–Q curve is presented in the next section where the measured P–Q curve is compared with the predicted one to confirm the validity of numerical results.

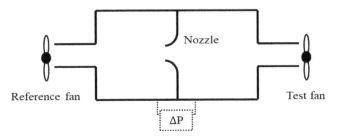

Figure 2. Schematic diagram of the actual fan-performance tester.

3.2. Experimental Set-Up for Characterizing Noise Performance

In this study, to characterize the acoustic performance of the targeted outdoor unit, the spectral sound pressure levels were measured in the anechoic chamber. The experiment was conducted in accordance with the national test standard, KS (Korean Industrial Standards) C 9306 for air-conditioners [12]. The tested fan system is fixed at the center of the chamber and the microphone was installed downstream at a distance of 1 m from the center of an axial fan. The acoustic pressure was measured with a sampling frequency of 6400 Hz and a frequency resolution of 1 Hz. The measured overall sound pressure level with increasing rotating speed of fan is represented in Figure 3.

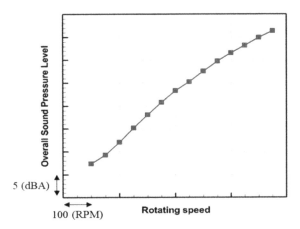

Figure 3. Measured overall sound pressure level versus rotating speed.

4. Numerical Methods and Validation

4.1. Governing Equations and Numerical Methods

To analyze the airflow driven by the axial-flow fan in the outdoor unit numerically, the following three-dimensional incompressible unsteady Reynolds-Averaged Navier–Stokes (RANS) equations are used as the governing equations:

$$\frac{\partial}{\partial x_j}(\rho u_j) = 0, \tag{1}$$

$$\frac{\partial}{\partial t}(\rho u_i) + \frac{\partial}{\partial x_j}(\rho u_j u_i) = -\frac{\partial p}{\partial x_i} + \frac{\partial}{\partial x_j}\left(\mu\left(\frac{\partial u_i}{\partial x_j} + \frac{\partial u_j}{\partial x_i}\right) - \rho\overline{u'_j u'_i}\right) \tag{2}$$

Equations (1) and (2) are numerically solved using a finite volume method based on the unstructured grids. The Re-Normalization Group (RNG) k-ε model is used as the turbulence model [13–16], and the standard wall function is used to model near-wall flows. For the numerical solutions of turbulent kinetic energy and turbulent dissipation rate, the first-order upwind discretization scheme is selected. Second-order upwind discretization scheme is used to solve the momentum term, the Semi-implicit Method for Pressure-Linked Equations (SIMPLE) scheme is used to solve the pressure-velocity coupling equations. The first order implicit method is selected for the transient discretization. The time step is 0.01/angular velocity. The under-relaxation factors are set to be 0.3, 0.7, 0.8, and 0.8 for pressure, momentum, turbulent kinetic energy, and turbulent viscosity, respectively. The volume flow rate at the discharge zone is used as the convergence criteria. The sliding mesh technique is used to simulate the rotating of the axial fan. These numerical methods are realized using the commercial CFD solver of ANSYS fluent.

Ffocws-Williams and Hawkings (FW–H) equation [17] in the following form is used for the prediction of radiated acoustic pressure:

$$4\pi c^2 H \rho'(\vec{x},t) = \frac{\partial}{\partial t} \int_S \frac{\rho_0 \vec{v} \cdot \vec{n}}{r|1 - M_r|} dS(\vec{\eta}) - \frac{\partial}{\partial x_i} \int_S \frac{p_{ij} n_j}{r|1 - M_r|} dS(\vec{\eta}) + \frac{\partial^2}{\partial x_i \partial x_j} \int_V \frac{T_{ij}}{r|1 - M_r|} d^3 \vec{\eta}. \quad (3)$$

Note that the volume source, the last term in the right-hand side of Equation (3), which is known to be negligible at low Mach number flow, is neglected.

4.2. Validation of Numerical Model

To investigate the detailed characteristics of the flow field driven by the axial fan in the outdoor unit more accurately, the virtual fan-performance tester (VFT) is designed. To determine the dimensions of the virtual fan tester and the number of grids for the given outdoor unit, a grid-refinement study was carried out. Figure 4 shows the three types of VFTs made on a basis of D, which is the diameter of the fan. The static pressure on the inflow and outflow planes which are located upstream and downstream, respectively, of the axial-flow fan in the outdoor unit is specified as the inflow and outflow boundary conditions. To reproduce the experimental conditions of the actual fan-performance tester and to predict the P–Q curve of the outdoor unit, the pressure difference between inlet and outlet boundaries is varied. The wall boundary conditions are applied on the four side-planes parallel to the fan rotation axis and on the middle plane separating the upstream and downstream flows. The rotation speed is set to be 1000 RPM.

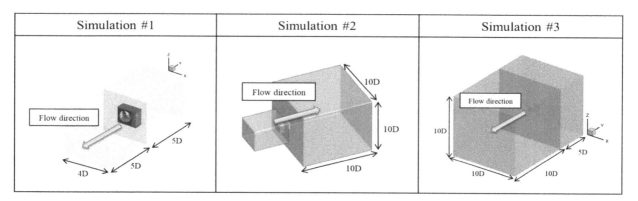

Figure 4. Various virtual fan testers used for grid-refinement study.

The numerical results predicted using the VFTs are compared with the measured one using the actual fan performance tester of which the specifications are described in Section 3.1. The validity of VFT is confirmed by comparing the numerical results with the experimental results. Figure 5 compares the P–Q curves predicted by using the VFTs with the measured data. The agreement is generally good between the predicted and measured results in the lower pressure region, but the discrepancy increases as the pressure increases. The difference seems to depend on the size of the virtual fan tester: as the size of the virtual fan tester increases, the agreement between the numerical and experimental results becomes closer. The final dimensions of the VFT and the number of grids are determined using these results, which show that Simulation #3 presents the reasonable agreement up to the operation point. The maximum relative error in the volume flow rate at the fixed static pressure is 4.0%. These results confirm the validity of the current numerical method. Note that the subsequent detailed flow analysis and the optimization were carried out for the operating point where the P–Q curve and the resistance curve, which is mainly due to the heat changer, meet together.

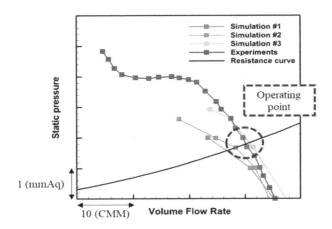

Figure 5. Comparison of predicted P–Q curves with experimental ones.

Aerodynamic noise due to fan-driven flow in the out-door unit is also predicted using the FW–H integral equation in Equation (3). The half-spherical shape of the FW–H integral surface was formed in the downstream direction. Figure 6 compares the predicted sound pressure spectrum with the measured one. There are good agreements at the blade-passing-frequencies between two results, though some differences are noticed in the broadband noise. The difference in the broadband high-frequency range is due to the intrinsic limitation of RANS equations in resolving the random components because the RANS equations are derived by averaging the Navier–Stokes equations. However, the difference between the numerical and experimental results is 2.1 dBA in terms of the overall sound pressure level because the tonal components dominantly contribute to the overall sound pressure level.

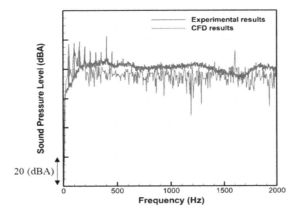

Figure 6. Comparison of predicted noise spectrum with experimental ones.

5. Numerical Results of Axial Fan System of Outdoor Unit

5.1. Analysis of Flow Field Driven by Axial Fan Systems of Outdoor Units

The flow field predicted for the targeted cooling fan system by using the VFT is investigated in detail. Since the computation domain includes the entire fluid region of the outdoor unit, the predicted flow field includes most of the essential characteristics of flow which involve the interaction of the driven flow with internal structures such as the motor-mount, the inlet-grille, the heat sink, and the orifice. Figure 7 shows the overall fluid velocity field on the vertical cross-sectional plane passing through the rotating center of the fan with the zoomed plot around the motor mount. It can be seen that the vortex flow is formed around the motor mount. However, this vortex flow is relatively weak in comparison to the vortex flow around the orifice.

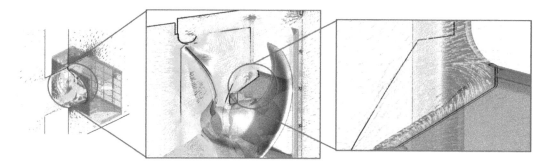

Figure 7. Vector fields of fluid velocities on vertical cross-sectional plane passing rotating center of fan blades with zoomed plot around motor mount.

Figure 8 shows the vector fields of fluid velocities on the vertical and horizontal cross-sectional planes passing through the heat sink with the zoomed plot around the heat sink. It can be seen that the fluid flow passes smoothly through the heat sink, which means the effective cooling of the heat sink through the airflow driven by the fan.

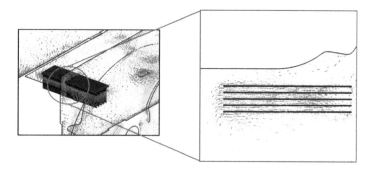

Figure 8. Vector fields of fluid velocities on vertical and horizontal cross-sectional planes passing heat sink.

Figure 9 shows the vector field of flow velocities on the horizontal cross-sectional plane passing the rotating center of the fan with zoomed plots around the orifice. It can be seen that the strongest vortex structures are generated near the orifice than other components. It is well known that the tip vortex of the fan blade affects the flow performance and flow noise of the fan most significantly. The orifice is generally used to suppress this tip vortex and thus to increase the flow performance and to decrease the flow noise. These identified vortex flows can reduce the flow performance and increase flow noise of the axial flow fan, which implies that the orifice shape needs to be re-designed in combination with the tip shape of the axial-flow fans.

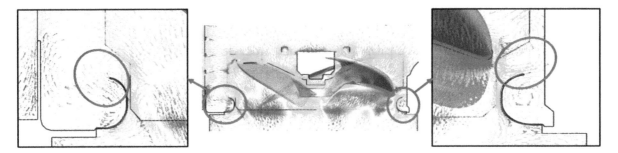

Figure 9. Vector fields of fluid velocities on the vertical cross-sectional plane passing rotating center of fan blades around the orifice.

Figure 10 shows the iso-contours of the static pressure on the same horizontal cross-sectional plane. The circular contours with lower pressure value at its center are identified around the orifice, which indicates the core of vortex generated by the interaction of the blade tip with the orifice.

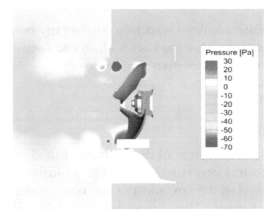

Figure 10. Iso-contours of static pressure on horizontal cross-sectional plane passing rotating center of fan blades.

Figure 11 shows the iso-contours of the magnitudes of the flow velocity vectors on the same horizontal cross-sectional plane. The lower values are identified around the orifice, which means the vortex adversely affects the flow performance of the fan.

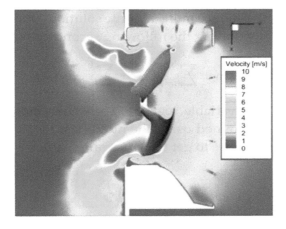

Figure 11. Iso-contours of velocity magnitudes on horizontal cross-sectional plane passing rotating center of fan blades.

5.2. Analysis of Aeroacoustic Performance of the Fan System of Outdoor Units

Figure 12 shows the directivity of the overall sound pressure level of outdoor unit. It can be seen that the highest sound pressure wave propagates in the direct downstream direction normal to the fan rotational plane. This is well-known dipolar characteristics of fan noise.

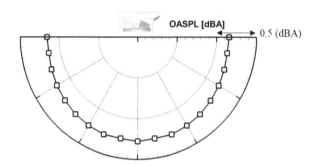

Figure 12. Directivity of overall sound pressure level.

Based on these flow and noise analysis results, a further investigation is carried out to find the optimum orifice shape, which can minimize the vortex observed between the orifice and the fan blades and thus improve the flow and noise performances.

6. Optimization of Orifice Shape

6.1. Optimization Method

The orifice shape in the axial fan system of the outdoor unit is optimized for the maximum flow coefficient and the minimum sound pressure levels. The volume flow rate of the fan is generally proportional to the rotational speed of the fan, and the acoustic power of the fan is proportional to the 5th to 6th power of fan rotational speed. This dimensional analysis suggests that the optimization for the higher flow rate is more practical. The increase of volume flow rate through the optimization can be converted to the reduction of noise. To maintain the cooling capability of the fan, the volume flow rate must be equal to or higher than that of the original model. Therefore, if the volume flow rate is increased, the rotational speed of the fan can be reduced to adjust the volume flow rate to the original one. The noise of the outdoor system can be reduced due to the reduction of rotational speed.

The optimization is performed using the response surface method (RSM), which is a collection of mathematical and statistical techniques for empirical model building, can solve multiple responses over the entire area of interest, and produce more accurate solutions than other methods [18–20]. In this study, the optimization is performed using the following second-order regression equation:

$$\eta = \beta_0 + \sum_{i=1}^{k} \beta_i x_i + \sum_{i \leq j}^{k} \beta_{ij} x_i x_j, \tag{4}$$

where η, β, and x means a dependent variable, regression constants, independent variables, respectively. If the regression constants are determined using the numerical results, optimization can be derived by determining the dependent variables to get the maximum value of the independent variable from Equation (4).

6.2. Geometric Parameters for Optimization of Orifice Shape

The orifice shape of this study can be divided into three sections depending on its role, as shown in Figure 13: inlet, neck, and outlet. First, in the inlet section, it is necessary to design a shape that can reduce the flow separation of the suction flow and prevent the reverse flow generated by the blade tip of the axial fan. The second is the neck section, which is the shape of a straight line. In this section, if the flow doesn't develop sufficiently, it becomes the main cause of the flow noise and the loss of flow rates. Third, it is an outlet section that determines the directions of discharge flow and recovers the static pressure loss at the inlet section by serving as a diffuser. The straight shape of the outlet is advantageous for discharging performance. Figure 14 depicts the definitions of three design factors selected for the optimization of the orifice. The selected parameters (R, H, θ) represents the shape of the inlet, neck, and outlet, respectively. Note that, although the original orifice shape includes the curved shape in the downstream field, a linear shape is chosen. The volume flow rate is chosen as an objective function. The ranges of three factors are determined to satisfy its geometric installation constraint. The levels of the factors are summarized in Table 1.

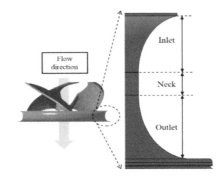

Figure 13. Shape of the orifice.

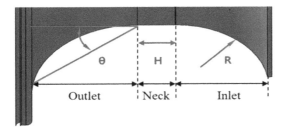

Figure 14. Design parameters for optimization of orifice shape.

Table 1. Factors and level for optimization of orifice shape

Factors	Level		
	−1	0	1
R (mm)	12.3	18	23.7
H (mm)	8	27.5	47
θ (deg)	13.1	28.8	44.5

6.3. Optimization Results

The numerical optimization based on RSM is carried out for the maximum volume flow rate using 15 cases. The same numerical methods as used for the flow and noise predictions of the existing fan system described in the previous sections are applied to each of these cases. Based on these results, the quadratic function of Equation (4) is obtained from the RSM using the Minitab software (version 17.0, Minitab Inc., United States of America,). From the quadratic function, the influence of the three parameters on the volume flow rate is obtained, as shown in Figure 15. Among the three parameters, R [mm] representing the inlet shape showed a tendency to include the maximum value of the volume flow rate within the selected design range. However, the distribution of H [mm] representing the neck shape and θ [deg] representing the outlet shape has the maximum volume flow rate at the boundary of the set region. These results show that, as the length of the neck section is longer, and the discharge area of the outlet section is smaller, the flow rate is more increased. Note that the lower and upper limit of H and θ, respectively, is determined considering the allowable limit of orifice installation in the outdoor unit.

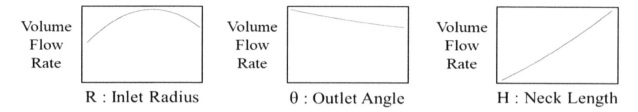

Figure 15. The relation between flow rate and each design factor.

Figure 16 shows the optimum shape of orifice, and the optimal design values for each section are (R, H, θ) = (19.2, 55, 10). The numerical simulation of flow field driven by the axial-flow fan combined with the optimized orifice in the outdoor unit is carried out, and the flow rate of the optimized orifice is found to be higher than the original orifice by about 4.6%. The detailed flow field obtained from the numerical simulation using the optimized orifice is shown in Figures 17 and 18. Figure 17 shows the iso-contours of flow velocity magnitude on the horizontal cross-sectional plane. It can be found that the outlet flow direction of the outdoor unit using the optimized orifice is directed more downstream than that of the original one shown in Figure 8. It is emphasized that this dramatic change of flow pattern is achieved only by changing the orifice shape without changing the fan blade.

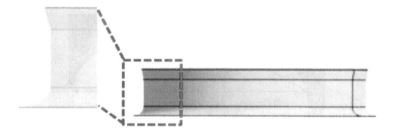

Figure 16. Optimized shape of orifice.

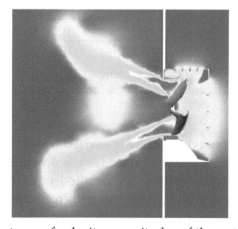

Figure 17. Contours of velocity magnitudes of the optimized orifice.

Figure 18 compares the flow velocity vectors around the orifice between the original and the optimized ones. It can be seen that the vortex structure generated at the inlet section of the original orifice almost disappears in the optimized orifice and the discharging flow is developed more sufficiently at the neck section of the optimized orifice than the original one. In addition, in case of original orifice shape, the tip vortex formed and separated from the blade tip interacts with the inlet of orifice, whereas the tip vortex in the optimized orifice system was developed in the neck section.

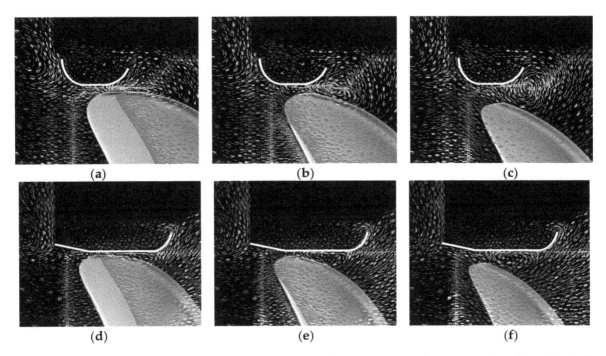

Figure 18. Snapshots of vector fields of fluid velocities: (**a–c**) for original orifice; (**d–f**) for optimized orifice.

As shown in Figure 17, there is a significant change in the outlet flow pattern between the optimized and original models. To find the reason for the difference, Figure 19 compares the flow velocity vectors between different outlet angles (θ) at the fixed other design factors. Although the other two parameters are the same for each other, the tip vortex pattern changes significantly according to the variation of the outlet angle. In the case of the orifice with larger outlet angle, the tip vortex formed from the blade tip is convected downstream and interacts with the orifice vortex. The tip vortex in the case of the orifice with smaller outlet angle remains in the vicinity of the inlet of orifice and decay gradually. This result implies that the change in the outlet flow pattern is induced by the combination of the optimum design parameters but not just by the linear shape of the outlet part.

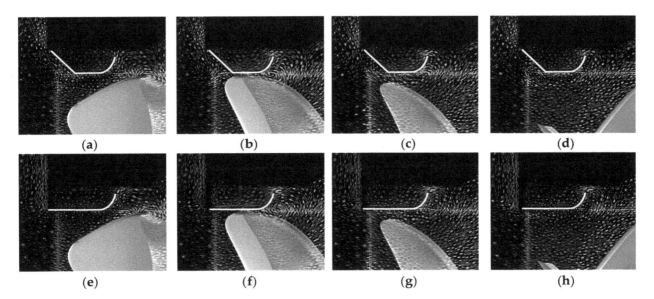

Figure 19. Snapshots of vector fields of fluid velocities at fixed inlet & neck design factors: (**a–d**) for the case of (R, H, θ) = (18, 28.8, 55); (**e–h**) for the case of (R, H, θ) = (18, 28.8, 0).

For more quantitative analysis of orifice roles, the radial and axial loadings of orifice surface were calculated, and its instantaneous distributions are shown in Figures 20 and 21, respectively. It can be seen in Figure 20 that the three distinct lobs that are related to the three fan blades' location are formed in the distribution of radial-direction loading of the optimized orifice, and its magnitude is higher than that of the original orifice. This implies that the optimized orifice guides the flow more effectively by pushing the airflow more in the centripetal direction. Contrary to the radial loading, as shown in Figure 21, the axial-direction loading of the optimized orifice shows circular shape, whereas that of the original orifice shows three lobs. The reason for this difference can be explained by using the difference in the tip vortex distributions between two orifices shown in Figure 18. The tip vortex in the optimized orifice interacts mainly with the neck part of the orifice, whereas the tip vortex in the original orifice interacts with the inlet part. Therefore, the tip vortex interacting with the neck part more affects the radial loading of the optimized orifice while the vortex interacting with the inlet part more affects the axial loading of the original orifice. However, the magnitude of axial loading is higher in the optimized orifice, which means that the optimized orifice makes more contribution to the airflow rate by pushing the air more in the axial direction.

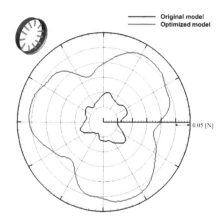

Figure 20. Comparison of radial loading on orifice between original and optimized ones.

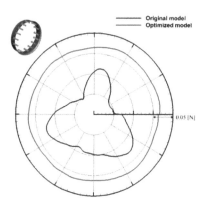

Figure 21. Comparison of axial loading on the orifice between original and optimized ones.

Finally, the sound pressure radiation from the fan-driven flow in the outdoor unit is computed and compared between the original and optimized orifices. Figure 22 shows the predicted sound pressure spectrum. Although the optimization is carried out for the maximum volume flow rate, it can be seen that the spectral sound pressure levels from the outdoor unit with the optimized orifice was less than those of the original orifice. The reason for this seems to be due to the change of vortex strength of which the reduction can increase the flow rate and at the same time reduce aerodynamic noise. Figure 23 shows the directivity of SPLs predicted for the original and optimized orifices. It can be seen that the overall SPLs of the outdoor unit with the optimized orifice are reduced in all of the downstream directions.

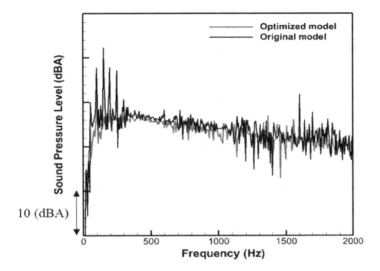

Figure 22. Comparison of predicted sound pressure level spectrum between original and optimized models.

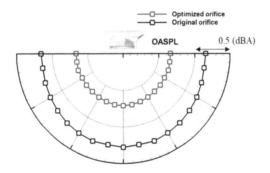

Figure 23. Comparing of predicted directivity between original model and optimized models.

7. Experimental Validation of Optimized Design Using Engineering Samples

Flow and noise experiments are performed to validate the actual performance of optimized orifice for the axial-flow fan in the outdoor unit. The experimental set-ups for the evaluation of flow and noise performances are the same as those described in Section 3. The experimental results are summarized in Table 2.

Table 2. Summary of experimental results at operational speed.

Performances	Original Orifice	Optimized Orifice
Volume flow rate	A	1.02A
Overall SPL (dBA)	B	B-2.8
Input power (%)	C	0.96C

As a result, the flow performance is increased by 2.1%, the sound pressure level is reduced by 2.8 dBA, and the input power is reduced by 4.0% at the targeted operating condition. Figure 24 compares the spectral sound pressure levels between the two models. It can be seen that the sound pressure levels of the optimized system are lower than those of the original one in the high-frequency broadband components.

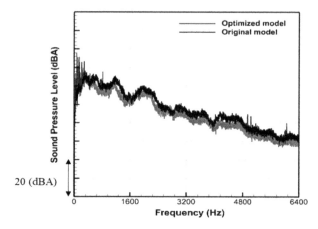

Figure 24. Comparison of sound pressure spectral levels predicted for optimized and original models.

To confirm the performances at the various operating conditions, the flow and noise experiments are carried out by varying the rotational speed of the fan. The volume flow rate and the sound pressure level measured are shown in Figure 25. It can be seen that the volume flow rate of the fan system using the optimized orifice is higher than the original one in the entire range of rotation speeds of the fan. In addition, the sound pressure levels of the optimized system are reduced in comparison with the original one in the entire range. Based on these measured data, it can be found that the aerodynamic noise of the outdoor unit with the optimized orifice is reduced by 2.9 dBA at the same volume flow rate. These measured data confirm the validity of the proposed optimum design of the orifice for the axial-flow fan in the outdoor unit.

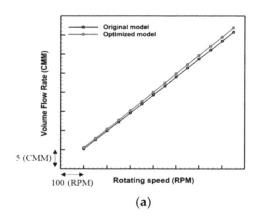
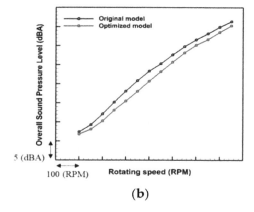

(a)   (b)

Figure 25. Comparison of performances between optimized and original systems at the various rotating speeds of fan: (**a**) volume flow rate and (**b**) overall sound pressure levels.

8. Conclusions

In this study, the aerodynamic and aeroacoustics performances of the axial-flow fan system in the outdoor unit of the split-type air conditioner was investigated using the numerical and experimental methods. First, the virtual fan-performance tester (VFT) using the incompressible unsteady RANS solver was devised to investigate the detailed flow field driven by the axial-flow cooling fan in the outdoor unit of the air-conditioner. Its validity was confirmed by comparing the predicted P–Q curve with that measured using the fan-performance tester. To improve the flow and noise performance of the outdoor unit of the air-conditioner without changing the fan blade shape, the orifice shape is optimized. The geometry of orifice was divided into three sections: inlet, neck, and outlet. According to the role of each section, the curvature radius of the inlet, the height of the neck, and the angle of the outlet were selected as the independent variables for the optimization. The response surface method using these variables was carried out for the maximum flow rate. It was found that the axial-flow fan using the optimal shape of the orifice drove the airflow by 4.6% more than the existing one. The detailed analysis of the flow field showed that this improvement was achieved by reducing the vortex flow between the fan blade tip and the orifice. Finally, the flow and noise experiments using the prototype engineering sample manufactured using the optimized design showed that the flow rate is increased by 2.1%, the sound pressure level is reduced by 2.8 dBA, and the input power is reduced by 4.0% at the same rotational speed.

Author Contributions: C.C. provided the basic idea for this study and the overall numerical strategies. S.-Y.R. carried out the numerical simulations and worked on the analysis of numerical results. S.M.P. and Y.-C.A. Provided the basic concept of optimization. J.W.K. and B.I.P. carried out the experiments. S.K.O. administrated the project.

Acknowledgments: This work was supported by the 'Human Resources Program in Energy Technology' of the Korea Institute of Energy Technology Evaluation and Planning (KETEP), which granted financial resources from the Ministry of Trade, Industry & Energy, Republic of Korea (No. 20164030201230).

References

1. Lee, S.; Heo, S.; Cheong, C. Prediction and reduction of internal blade-passing frequency noise of the centrifugal fan in a refrigerator. *Int. J. Refrig.* **2010**, *33*, 1129–1141. [CrossRef]
2. Heo, S.; Cheong, C.; Kim, T.H. Development of low-noise centrifugal fans for a refrigerator using inclined S-shaped trailing edge. *Int. J. Refrig.* **2011**, *34*, 2076–2091. [CrossRef]
3. Heo, S.; Cheong, C.; Kim, T. Unsteady Fast Random Particle Mesh method for efficient prediction of tonal and broadband noises of a centrifugal fan unit. *AIP Adv.* **2015**, *5*, 97–133. [CrossRef]
4. Heo, S.; Ha, M.; Kim, T.H.; Cheong, C. Development of high-performance and low-noise axial-flow fan units in their local operating region. *J. Mech. Sci. Technol.* **2015**, *29*, 3653–3662. [CrossRef]
5. Zhao, X.; Sun, J.; Zhang, Z. Prediction and measurement of axial flow fan aerodynamic and aeroacoustic performance in a split-type air-conditioner outdoor unit. *Int. J. Refrig.* **2013**, *36*, 1098–1108. [CrossRef]
6. Jiang, C.L.; Chen, J.P.; Chen, Z.J.; Tian, J.; OuYang, H.; Du, Z.H. Experimental and numerical study on aeroacoustic sound of axial flow fan in room air conditioner. *Appl. Acoust.* **2007**, *68*, 458–472. [CrossRef]
7. Wright, T.; Simmons, W.E. Blade sweep for low-speed axial fans. In *ASME 1989 International Gas Turbine and Aeroengine Congress and Exposition*; American Society of Mechanical Engineers: Amsterdam, The Netherlands, 1989.
8. Ye, X.; Li, P.; Li, C.; Ding, X. Numerical investigation of blade tip grooving effect on performance and dynamics of an axial flow fan. *Energy* **2015**, *82*, 556–569. [CrossRef]
9. Wang, H.; Tian, J.; Ouyang, H.; Wu, Y.; Du, Z. Aerodynamic performance improvement of up-flow outdoor unit of air conditioner by redesigning the bell-mouth profile. *Int. J. Refrig.* **2014**, *46*, 173–184. [CrossRef]
10. Hu, J.; Ding, G. Effect of deflecting ring on noise generated by outdoor set of a split-unit air conditioner. *Int. J. Refrig.* **2006**, *29*, 505–513. [CrossRef]
11. Jiang, C.L.; Tian, J.; Ouyang, H.; Chen, J.P.; Chen, Z.J. Investigation of air-flow fields and aeroacoustic noise in outdoor unit for split-type air conditioner. *Noise Control Eng. J.* **2006**, *54*, 146–156. [CrossRef]
12. Korea Standards. *Room air-conditioners, KS C 9306*; Korean Standards Association: Seoul, Korea, 2002.
13. Escue, A.; Cui, J. Comparison of turbulence models in simulating swirling pipe flows. *Appl. Math. Model.* **2010**, *34*, 2840–2849. [CrossRef]
14. Yilmaz, H.; Cam, O. Effect of different turbulence models on combustion and emission characteristics of hydrogen/air flames. *Int. J. Hydrog. Energy* **2017**, *42*, 25744–25755. [CrossRef]
15. Yakhot, V.; Orszag, S.A. Renormalization group analysis of turbulence. I. Basic theory. *J. Sci. Comput.* **1986**, *1*, 3–51. [CrossRef]
16. Galván, S.; Reggio, M.; Guibault, F. Assessment study of K-ε turbulence models and near-wall modeling for steady state swirling flow analysis in draft tube using fluent. *Eng. Appl. Comput. Fluid Mech.* **2011**, *5*, 459–478. [CrossRef]

17. Ffowcs-Williams, J.E.; Hawkings, D.L. Sound Generation by Turbulence and Surfaces in Arbitrary Motion. *Philos. Trans. R. Soc. Lond. Ser. A. Math. Phys. Sci.* **1969**, *264*, 321–342. [CrossRef]

18. Ren, G.; Heo, S.; Kim, T.H.; Cheong, C. Response surface method-based optimization of the shroud of an axial cooling fan for high performance and low noise. *J. Mech. Sci. Technol.* **2013**, *27*, 33–42. [CrossRef]

19. Montgomery, D.C. *Design and Analysis of Experiments*, 6th ed.; John Wiley and Sons: New York, NY, USA, 2005.

20. Khuri, A.I.; Cornell, J.A. *Response Surfaces*, 2nd ed.; Marcel Dekker: New York, NY, USA, 1996.

3

Fractional Order Forced Convection Carbon Nanotube Nanofluid Flow Passing Over a Thin Needle

Taza Gul [1,2], Muhammad Altaf Khan [1], Waqas Noman [1], Ilyas Khan [3,*], Tawfeeq Abdullah Alkanhal [4] and Iskander Tlili [5]

[1]   Department of Mathematics, City University of Science and Information Technology, Peshawar 25000, Pakistan; tazagul@cusit.edu.pk (T.G.); makhan@cusit.edu.pk (M.A.K.); waqarnoman55@yahoo.com (W.N.)

[2]   Department of Mathematics, Govt. Superior Science College Peshawar, Khyber Pakhtunkhwa, Pakistan

[3]   Faculty of Mathematics and Statistics, Ton Duc Thang University, Ho Chi Minh City 72915, Vietnam

[4]   Department of Mechatronics and System Engineering, College of Engineering, Majmaah University, Majmaah 11952, Saudi Arabia; t.alkanhal@mu.edu.sa

[5]   Energy and Thermal Systems Laboratory, National Engineering School of Monastir, Street Ibn El Jazzar, 5019 Monastir, Tunisia; iskander.tlili@enim.rnu.tn

*   Correspondence: ilyaskhan@tdt.edu.vn

Abstract: In the fields of fluid dynamics and mechanical engineering, most nanofluids are generally not linear in character, and the fractional order model is the most suitable model for representing such phenomena rather than other traditional approaches. The forced convection fractional order boundary layer flow comprising single-wall carbon nanotubes (SWCNTs) and multiple-wall carbon nanotubes (MWCNTs) with variable wall temperatures passing over a needle was examined. The numerical solutions for the similarity equations were obtained for the integer and fractional values by applying the Adams-type predictor corrector method. A comparison of the SWCNTs and MWCNTs for the classical and fractional schemes was investigated. The classical and fractional order impact of the physical parameters such as skin fraction and Nusselt number are presented physically and numerically. It was observed that the impact of the physical parameters over the momentum and thermal boundary layers in the classical model were limited; however, while utilizing the fractional model, the impact of the parameters varied at different intervals.

Keywords: SWCNT/MWCNT nanofluid; thin needle; classical and fractional order problems; APCM technique

1. Introduction

This study is concerned with the enhancement of heat transfer through nanofluid, which will play a dynamic role in the field of chemical sciences and the energy sector. The enhancement of heat transfer through nanofluid was studied by many scientists in the field of geometry under diverse conditions. Sparrow and Gregg [1] scrutinized the removal of humidity, using centrifugal force procedures on a cooled rotary disk. The energy obtaining and cooling behavior of the devices mainly depend on the heat transfer liquid used, and the lower thermal efficiency of these fluids can create harsh restrictions for device performance. The limitations and low thermal efficiency of these liquids delay the device performance and compression of heat exchangers. Choi [2] explored the idea of nanofluids by utilizing small nanosized (10–50 nm) particles in base fluids. The anticipated factors influencing the performance of nanofluids during heat transfer were: (i) thermal properties; (ii) chemical stability; (iii) compatibility with the base fluid; (iv) toxicity; (v) accessibility; and (vi) cost. Possible nanomaterials include metals, metal oxides, and carbon materials. Carbon materials play a significant role in enhancing the thermal efficiency of base fluids, and carbon nanotubes (CNTs) are the renowned family of carbons that have been used for thermal and cooling applications in recent studies. Carbon nanotubes are further divided into two classes: single-walled carbon nanotubes (SWCNTs) and multiple-walled carbon nanotubes (MWCNTs). Single-walled carbon nanotubes are created by packaging a layer of carbon one-atom thick, while MWCNTs contain multiple rolled layers of carbon. Carbon nanotubes nanofluids have many important applications in industries such as aerospace, electronics, optics, and energy conservation, as reported by Volder et al. [3] and Terrones [4]. The higher thermal conductivity (2000–6000 W/mK) of carbon nanotubes make them more valuable for the augmentation of heat transfer devices. Ellahi et al. [5] investigated CNTs' nanofluid flow along a vertical cone under the influence of a variable wall temperature, and a comparison between SWCNTs and MWCNTs was made in their study. Gohar et al. [6] have studied SWCNTs/MWCNTs' nanofluid flow over a non-linear stretching disc. The high thermal efficiency of CNTs increased the heat flux and thermal efficiency of the base liquids as the heat fluxed, compared to other nanofluids as reported by Murshed et al. [7,8]. Various thermal conductivity models have been proposed by researchers for nanofluid flow problems. The appropriate and frequently-used thermal conductivity models for CNTs were reported by Xue [9]. The flow problem which passes over a thin needle under the effect of convection has been considered by many scholars Narain and Uberoi [10,11] and Chen [12]. Wang [13] and Grosan and Pop [14] have deliberated the mixed convection boundary layer flows over an upright thin needle including an intense heat source at the tip of the needle.

This study was carried out considering water-based CNT nanofluid flow over a thin needle. Further, the variable surface temperature with forced convection comprising single walled carbon nanotubes (SWCNT) and multi walled carbon nanotube (MWCNT) water-based nanofluid past over a thin needle was investigated in classical and fractional models, respectively.

The integer order derivatives or the classical model of fluid dynamics investigate the flow behavior at the integer steps, while the fractional order derivatives of the same fluid flow explore the natural phenomena to expose the internal behavior of the fluid flow by taking the fractional values among the integers. However, the idea of fractional calculus has been conventional for approximately three hundred years [15–17].

In fluid mechanics, most fluids are not generally linear in characteristic and the fractional order model is more appropriate for the illustration of such a kind of spectacle, rather than traditional methods. Caputo [18] introduced the idea of fractional derivatives from the modified Darcy's law using the concept of unsteadiness. This idea was further modified by other researchers [19–21] through the introduction of a variety of new fractional derivatives and their applications. Agarwal et al. [22] studied neural network models using ynchronization and impulsive Caputo fractional differential equations. Khan et al. [23] examined the fractional order solution of the Phi-4 equations using the GO/G expansion technique. Hameed et al. [24] examined the fractional order second grade fluid peristaltic transport in a vertical cylinder. A variety of numerical techniques have been used to find solutions to the classical models [25–30], and these techniques have been further combined to find solutions for fractional order problems.

The aim of this study is to analyze the force convectional CNT nanofluid flow passing over a thin needle including the elastic heat flux. The FDE-12 method was used for the solution for the fractional order non-linear differential equations. It is the execution of the predictor corrector method of the Adams–Bash Forth–Moulton technique derived by Diethelm and Freed [31]. Diethelm et al. [32] found the convergence and validity of this method for the solution of fractional order differential equations. The solution for classical and fractional order mathematical models containing (SWCNT/MWCNT) water-based nanofluids was obtained through the solution for fractional order systems, which was solved by the Adams-type predictor corrector method as used in References [33,34]. The range of the parameters in this study were selected as per the investigation by Gul et al. [35] using the BVP 2.0 package and the Optimal Homotopy Analysis Method OHAM technique. They used the 20th order approximation for the selected range of the parameters and obtained the minimum square residual error. The important outcomes were presented physically and numerically.

2. Mathematical Formulation

The axisymmetric boundary layer comprising SWCNT and MWCNT nanofluids' flow over the surface of a thin needle, including position-dependent wall temperature at the ambient temperature T_∞ was considered. The radius of the thin needle is defined as. The surface temperature, T_w, of the thin needle is considered heavier than Ambient temperature T_∞, $(T_w > T_\infty)$. The external flow velocity of the nanofluid is considered to be $u_e(x)$. The momentum and thermal boundary layer equations were derived in the axial and radial coordinates and all the assumption are imposed as [14]:

$$\frac{\partial \tilde{r}\tilde{u}}{\partial \tilde{x}} + \frac{\partial \tilde{r}\tilde{v}}{\partial \tilde{r}} = 0, \tag{1}$$

$$\left(\tilde{u}\frac{\partial \tilde{u}}{\partial \tilde{x}} + \tilde{v}\frac{\partial \tilde{u}}{\partial \tilde{r}} \right) = \tilde{u}_e \frac{d\tilde{u}_e}{d\tilde{x}} + v_{nf}\frac{1}{\tilde{r}}\frac{\partial}{\partial \tilde{r}}\left(\tilde{r}\frac{\partial \tilde{u}}{\partial \tilde{r}} \right), \tag{2}$$

$$\left(\tilde{u}\frac{\partial \tilde{T}}{\partial \tilde{x}} + \tilde{v}\frac{\partial \tilde{T}}{\partial \tilde{r}} \right) = \alpha_{nf}\frac{1}{\tilde{r}}\frac{\partial}{\partial \tilde{r}}\left(\tilde{r}\frac{\partial \tilde{T}}{\partial \tilde{r}} \right). \tag{3}$$

The physical conditions satisfy [14] and are defined as:

$$\tilde{u} = 0, \tilde{v} = 0, \tilde{T} = T_w \text{ at } \tilde{r} = R(\tilde{x}),$$
$$\tilde{u} = \tilde{u}_e(\tilde{x}), \tilde{T} = T_\infty, \text{ at } \tilde{r} \to \infty. \tag{4}$$

The velocity components are represented by $\tilde{u}, \tilde{v}$ towards the axial and radial $(\tilde{x}, \tilde{r})$ directions, respectively.

ρ_{nf} is the density of the nanofluids, μ_{nf} is the dynamic viscosity of the nanofluids such that $v_{nf} = \frac{\mu_{nf}}{\rho_{nf}}$ is the kinematic viscosity of the nanofluid, ϕ is the solid particle volume fraction, k_{nf} is the thermal conductivity, and $(\rho C_p)_{nf}$ is the specific heat capacity of the nanofluids such that $\alpha_{nf} = \frac{k_{nf}}{(\rho C_p)_{nf}}$. The thermophysical properties for the CNT nanofluids were presented and satisfy Xue [9]:

$$\rho_{nf} = \rho_f - \phi\rho_f + \phi\rho_s, \mu_{nf} = \mu_f(1-\phi)^{-2.5}, (\rho C_p)_{nf} = (\rho C_p)_f - \phi(\rho C_p)_f + \phi(\rho C_p)_s$$

$$\frac{k_{nf}}{k_f} = \frac{1-\phi+2\left(\frac{k_{CNT}}{k_{CNT}-k_f} \ln \frac{k_{CNT}+k_f}{2\,k_f} \right)\phi}{1-\phi+2\left(\frac{k_f}{k_{CNT}-k_f} \ln \frac{k_{CNT}+k_f}{2\,k_f} \right)\phi}. \tag{5}$$

To bring the basic Equations (1)–(3) into a dimensionless form, under boundary limitations, as per Equation (4), we adopted the scaling transformations as [14]:

$$x = \tilde{x}/L, \ r = (\tilde{r}/L)\text{Re}^{\frac{1}{2}}, \ R(x) = (\tilde{R}(\tilde{x})/L)\text{Re}^{\frac{1}{2}}, \ u = \tilde{u}/U_\infty,$$
$$v = (\tilde{v}/U_\infty)\text{Re}^{\frac{1}{2}}, \ u_e(x) = \tilde{u}_e(\tilde{x})/U_\infty, \ T = (\tilde{T} - T_\infty)/\Delta T. \tag{6}$$

Here, $\text{Re} = \frac{U_\infty L}{v_f}$ is the Reynolds number, L is the characteristic length of the needle, $R(x)$ is the dimensionless radial coordinate, r is the dimensionless radius of the needle, U_∞ is the characteristic velocity, ΔT is the characteristic temperature, and x is the dimensionless axial coordinate. Bringing Equation (6) into the basic Equations (1)–(4) cuts into the following non-linear differential form as:

$$\frac{\partial ru}{\partial x} + \frac{\partial rv}{\partial r} = 0, \tag{7}$$

$$\left(u\frac{\partial u}{\partial x} + v\frac{\partial u}{\partial r} \right) = u_e\frac{du_e}{dx} + \frac{v_{nf}}{v_f}\frac{1}{r}\frac{\partial}{\partial r}\left(r\frac{\partial u}{\partial r} \right), \tag{8}$$

$$\Pr\left(u\frac{\partial T}{\partial x} + v\frac{\partial T}{\partial r} \right) = \frac{\alpha_{nf}}{\alpha_f}\frac{1}{r}\frac{\partial}{\partial r}\left(r\frac{\partial T}{\partial r} \right). \tag{9}$$

The suitable boundary conditions are:

$$\begin{aligned} u = 0, v = 0, T = T_w(x) \text{ at } r = R(x), \\ u = u_e(x), T = 0, \text{ at } r \to \infty. \end{aligned} \tag{10}$$

Next, the similarity variables are:

$$u_e(x) = x^m, T_w(x) = x^n, \psi = x\, f(\eta), \eta = x^{m-1}r^2, T(x) = x^n\Theta(\eta). \tag{11}$$

Here, $u_e(x)$ is the dimensionless velocity of the external flow, ψ is used to demonstrate the stream function and satisfy the continuity Equation (7). The velocity components derived from the stream function ψ are defined as: $u = \frac{1}{r}\frac{\partial\psi}{\partial r}, v = -\frac{1}{r}\frac{\partial\psi}{\partial x}$. Putting $\eta = a$ into Equation (11) describes the size of the needle: $r = R(x) = \sqrt{ax^{(1-m)}}$, along the surface. Using Equation (11) in the basic Equations (7)–(10), the continuity equation is satisfied characteristically, and the rest of the equations are transformed as:

$$\frac{8}{(1-\phi)^{2.5}\left(1 - \phi + \phi\frac{\rho_{CNT}}{\rho_f} \right)}(\eta f'')' + 4ff'' + m(1 - 4(f')^2) = 0, \tag{12}$$

$$\frac{2\left(\frac{k_{nf}}{k_f} \right)}{\Pr\left(1 - \phi + \phi\frac{(\rho C_p)_{CNT}}{(\rho C_p)_f} \right)}(\eta\Theta')' + f\Theta' - nf'\Theta = 0. \tag{13}$$

The suitable boundary conditions are:

$$\begin{aligned} f(a) = 0, f'(a) = 0, \Theta(a) = 1, \\ f'(\infty) = \tfrac{1}{2}, \Theta(\infty) = 0. \end{aligned} \tag{14}$$

The skin friction coefficient and the local Nusselt number satisfy [14]:

$$\text{Re}_x^{\frac{1}{2}}C_f = 4a^{\frac{1}{2}}(1-\phi)^{-2.5}f''(a), \quad \text{Re}_x^{-\frac{1}{2}}Nu_x = \left[-2a^{\frac{1}{2}}\frac{K_{nf}}{K_f} \right]\Theta'(a). \tag{15}$$

Here, $\text{Re}_x = \frac{u_e(x)\,x}{v_f}$, is the local Reynolds number.

3. Preliminaries on the Caputo Fractional Derivatives

The useful definition of Caputo fractional order derivatives and their properties are presented below.

Definition 1. *Let $a > 0$, $t > a$; $a, \alpha, t \in \Re$. The Caputo fractional derivative of order α of the function $f \in C^n$ is given by:*

$$_a^C D_t^\alpha f(t) = \frac{1}{\Gamma(n-\alpha)}\int_a^t \frac{f^{(n)}(\xi)}{(t-\xi)^{\alpha+1-n}}d\xi, \quad n - 1 < \alpha < n \in N. \tag{16}$$

Property 1. *Let* $f(t)$, $g(t) : [a, b] \rightarrow \Re$ *be such that* ${}_{a}^{C}D_{t}^{\alpha}f(t)$ *and* ${}_{a}^{C}D_{t}^{\alpha}g(t)$ *exist almost everywhere, and let* c_1, $c_2 \in \Re$. *Then* ${}_{a}^{C}D_{t}^{\alpha}\{c_1\,f(t) + c_2\,g(t)\}$ *exists almost everywhere and*

$${}_{a}^{C}D_{t}^{\alpha}\{c_1\,f(t) + c_2\,g(t)\} = c_1\,{}_{a}^{C}D_{t}^{\alpha}f(t) + c_2\,{}_{a}^{C}D_{t}^{\alpha}g(t). \tag{17}$$

Property 2. *The function* $f(t) \equiv c$ *is constant and therefore, the fractional derivative is zero:* ${}_{a}^{C}D_{t}^{\alpha}\,c = 0$. *The general description of the fractional differential equation was assumed including the Caputo concept:*

$${}_{a}^{C}D_{t}^{\alpha}\,x(t) = f(t, x(t)), \quad \alpha \in (0, 1). \tag{18}$$

With the initial conditions $x_0 = x(t_0)$.

4. Solution Methodology

The following variables were selected for the momentum and thermal boundary layer (12, 13) to reduce the system into the first order differential equations as:

$$y_1 = \eta, y_2 = f, y_3 = f', y_4 = f'', y_5 = \Theta, y_6 = \Theta'. \tag{19}$$

The Caputo fractional order derivative applied to the first order ODE system was obtained from (12, 13) with the efforts of the proposed variables given in Equation (19).

The fractional order system was obtained from Reference [32]:

$$\begin{pmatrix} D_{\eta}^{\alpha}y_1 \\ D_{\eta}^{\alpha}y_2 \\ D_{\eta}^{\alpha}y_3 \\ D_{\eta}^{\alpha}y_4 \\ D_{\eta}^{\alpha}y_5 \\ D_{\eta}^{\alpha}y_6 \end{pmatrix} = \begin{pmatrix} 1 \\ y_3 \\ y_4 \\ \dfrac{-(1-\phi)^{2.5}\left(1-\phi+\phi\frac{\rho_{CNT}}{\rho_f}\right)}{8y_1}\left(y_4 + 4y_1y_3 + m(1 - 4(y_3)^2)\right) \\ y_6 \\ \dfrac{-\Pr\left(1-\phi+\phi\frac{(\rho c_p)_{CNT}}{(\rho c_p)_f}\right)}{2\left(\frac{k_{nf}}{k_f}\right)}(y_6 + y_2y_6 - ny_3y_5) \end{pmatrix} \begin{pmatrix} y_1 \\ y_2 \\ y_3 \\ y_4 \\ y_5 \\ y_6 \end{pmatrix} = \begin{pmatrix} 0 \\ 0 \\ 0 \\ u_1 \\ 1 \\ u_2 \end{pmatrix}. \tag{20}$$

Equation (20) represents a matrix system of fractional order equations of an initial value problem. Considering $(\alpha = 1)$, we have an integer order model or a classical model.

5. Results and Discussion

The two-dimensional forced conventional boundary layer SWCNT/MWCNT nanofluid flow for the enhancement of heat transmission over a thin needle was examined. A comparison of the influence of the physical constraints was studied for the integer and fractional order values

The fractional order system was solved numerically through the Adams-type predictor corrector method.

The geometry of the problem is displayed in Figure 1. The influence of the constant m versus velocity field $f'(\eta)$ is shown in Figures 2 and 3, for the classical and fractional order values, respectively. The larger values of the parameter m cause lower velocity. Physically, the rising values of m enhance the non-linearity to generate a friction force to decline the radial velocity. This decline is comparatively fast in the fractional order scheme. Due to the high thermophysical properties, the decline effect is comparatively rapid using the SWCNTs. The impact of ϕ over the $f'(\eta)$ for the integer and fractional order values is displayed in Figures 4 and 5, respectively. The larger value of ϕ causes a decrease in the velocity, and this effect is clearly larger when using the SWCNTs when compared to the MWCNTs. In fact, the larger amount of ϕ enhances the efficiency of the frictional force, and as a result, the viscous forces become strong enough to stop the fluid motion. Again, the decline effect is stronger using the fractional values. Figures 6 and 7 indicate the influence of the various values of the nanoparticle volume fraction versus the temperature field. The larger value of ϕ raises the temperature profile, and this effect is comparatively strong by means of the SWCNTs. In fact, the thermal conductivity of SWCNTs is high and provides rapid thermal efficiency to enhance the temperature field.

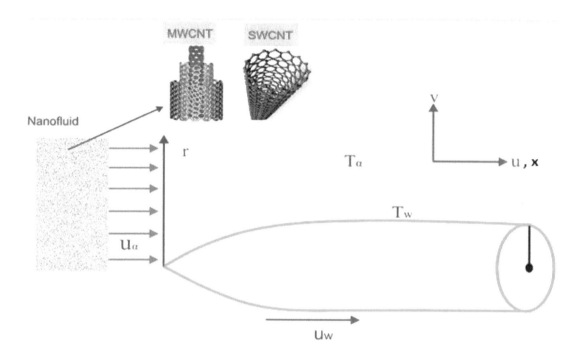

Figure 1. The geometry of the problem.

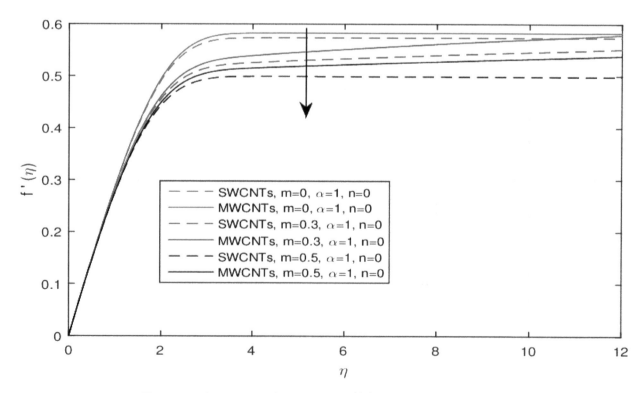

Figure 2. The impact of m over the $f'(\eta)$ for the integer values.

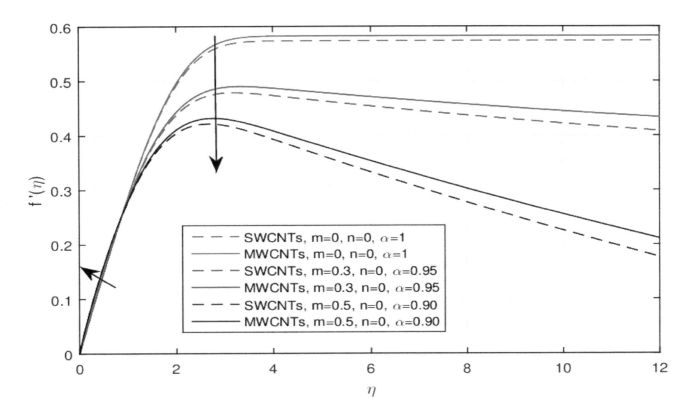

Figure 3. The impact of m over the $f'(\eta)$ for the fractional values.

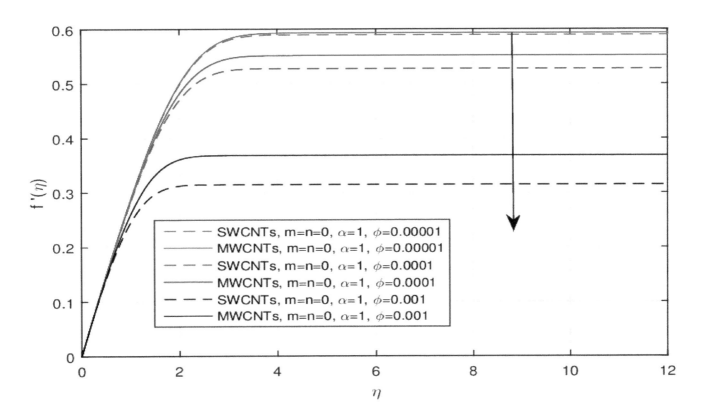

Figure 4. The impact of ϕ over the $f'(\eta)$ for the integer values.

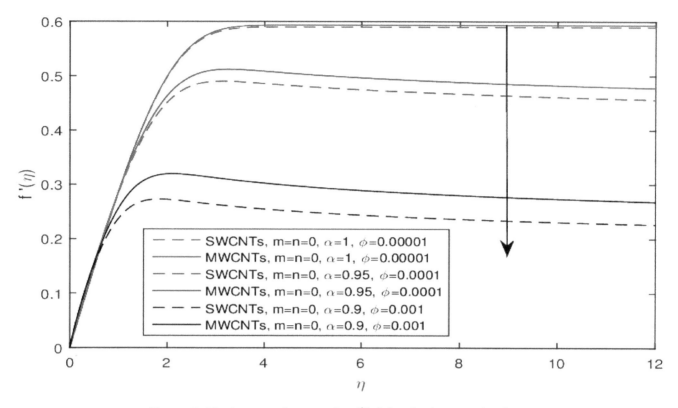

Figure 5. The impact of ϕ over the $f'(\eta)$ for the fractional values.

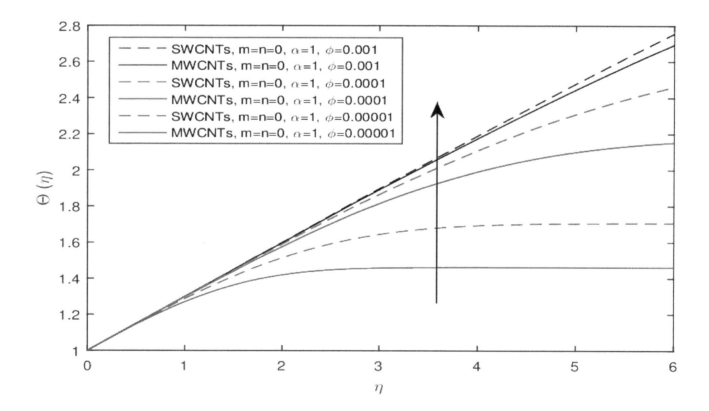

Figure 6. The impact of ϕ over the $\Theta(\eta)$ for the integer values.

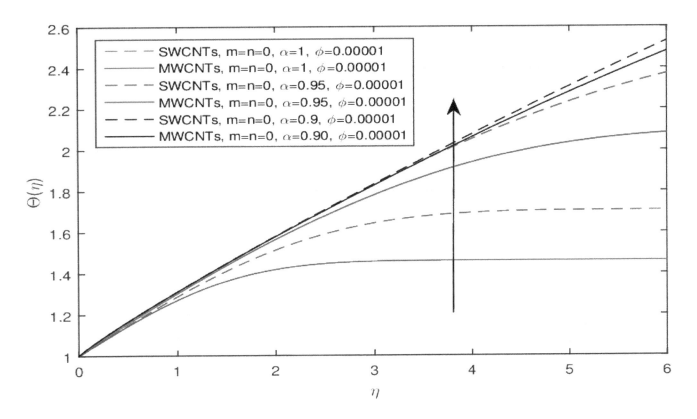

Figure 7. The impact of ϕ over the $\Theta(\eta)$ for the fractional values.

The impact of the wall temperature profile parameter n over the $\Theta(\eta)$ for the integer and fractional values are shown in Figures 8 and 9, respectively. The smaller values of n are enhancing the cooling effect, and as a result, the temperature field declines for the integer values and this effect is reversed for the fractional order values. The performance of the parameter n decreases the temperature field near the surface of the needle for the fractional values $\alpha = 1, 0.95, 0.90$, and this effect changes to increase the temperature profile after the critical point, as shown in Figure 9. The impact of the Prandtl number Pr over the temperature profile $\Theta(\eta)$ for the integer and fractional values is displayed in Figures 10 and 11, respectively. The rising values of Pr causes lower values compared to the classical model, as usually shown in the literature, but using the fractional model for the same values as the Prandtl number, the temperature profile near the needle surface increases and declines after the point of inflection.

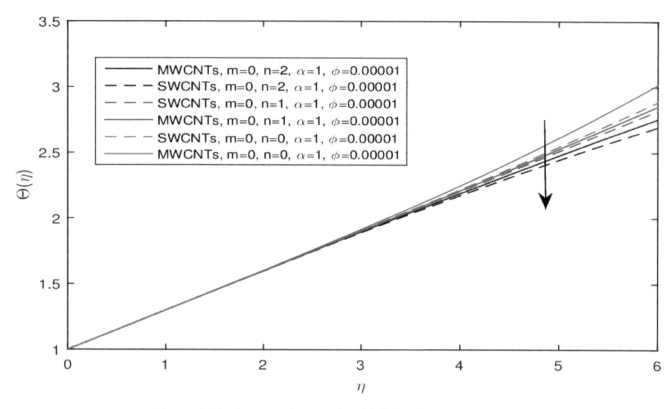

Figure 8. The impact of n over the $\Theta(\eta)$ for the integer values.

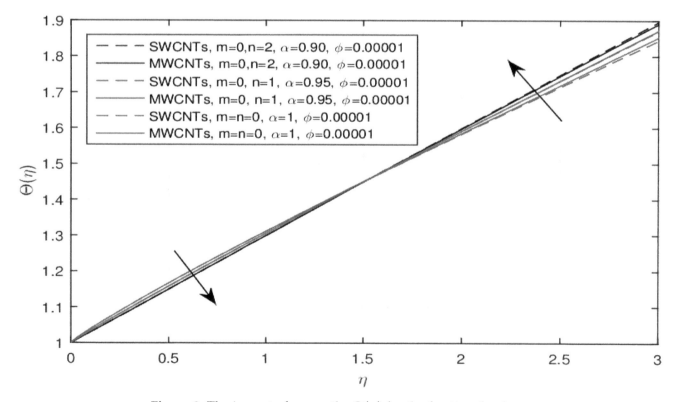

Figure 9. The impact of n over the $\Theta(\eta)$ for the fractional values.

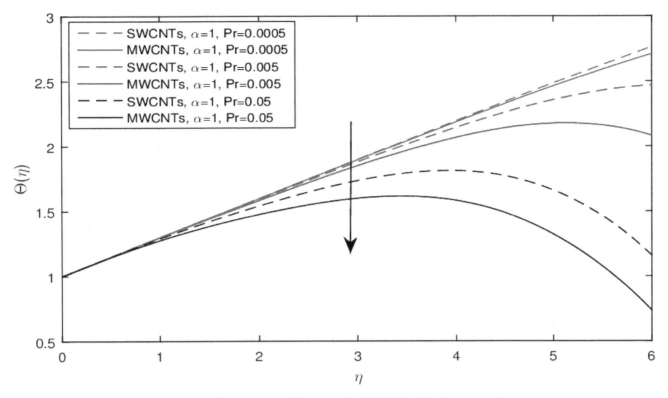

Figure 10. The impact of Pr over the $\Theta(\eta)$ for the integer values.

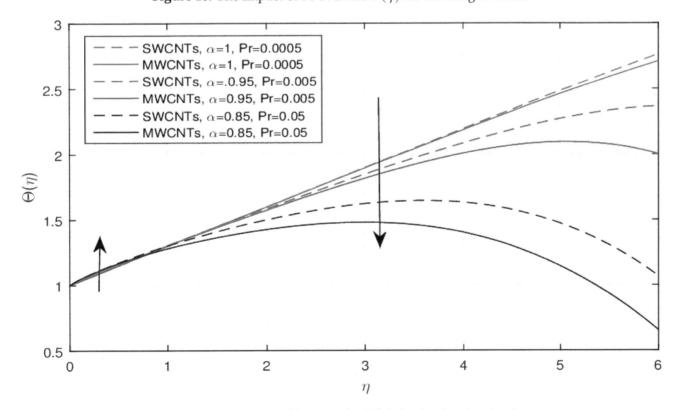

Figure 11. The impact of Pr over the $\Theta(\eta)$ for the fractional values.

The thermophysical properties of the base fluid and SWCNTs/MWCNTs are shown in Table 1. Skin friction and the Nusselt number are the physical parameters of interest under the influence of classical and fractional values. The classical and fractional model outputs for the skin friction and Nusselt number are displayed in Tables 2 and 3, respectively. Both tables specify the decline in the numerical values using the fractional order model. The two types of CNTs were compared in these tables for the fractional order values, and it was observed that the impact of the SWCNTs and MWCNTs varies using the fractional model, which is completely different from the classical model, where identical outputs occur in all cases.

Table 1. The thermo physical properties of carbon nanotubes (CNTs) and the base fluid water.

Physical Properties		Density $\rho(Kg/m^3)$	Thermal Conduct $k(Wm^{-1}/k^{-1})$	Specific Heat $c_p(Kg^{-1}/k^{-1})$
Base fluid Water		997	0.613	4179
Nanoparticles	SWCNT	2600	6600	425
	MWCNT	1600	3000	796

Table 2. The classical and fractional order comparison for the skin fraction comprising (SWCNTs/MWCNTs). When $\alpha = 1, 0.95, 0.90, \mathrm{Pr} = 0.0005, \phi = 0.0001, m = 0.01, n = 0$.

$\alpha=1, \eta.$	$f''(a)$ SWCNTs	$f''(a)$ MWCNTs	$\alpha=0.95, \eta.$	$f''(a)$ SWCNTs	$f''(a)$ MWCNTs	$\alpha=0.90, \eta.$	$f''(a)$ SWCNTs	$f''(a)$ MWCNTs
0.1	0.0988	0.0988	0.1	0.0985	0.0985	0.1	0.0982	0.0982
0.2	0.0967	0.0967	0.2	0.0960	0.0961	0.2	0.0953	0.0954
0.3	0.0938	0.0939	0.3	0.0927	0.0928	0.3	0.0915	0.0916
0.4	0.0901	0.0903	0.4	0.0886	0.0888	0.4	0.0869	0.0871
0.5	0.0856	0.0859	0.5	0.0837	0.0839	0.5	0.0815	0.0818
0.6	0.0804	0.0807	0.6	0.0781	0.0784	0.6	0.0755	0.0758
0.7	0.0745	0.0749	0.7	0.0717	0.0722	0.7	0.0688	0.0693
0.8	0.0679	0.0683	0.8	0.0648	0.0653	0.8	0.0615	0.0621
0.9	0.0606	0.0612	0.9	0.0572	0.0579	0.9	0.0538	0.0545
1.0	0.0527	0.0534	1.0	0.0492	0.0499	1.0	0.0456	0.0464

Table 3. The classical and fractional order comparison for the Nusselt number (SWCNTs/MWCNTs). When $\alpha = 1, 0.95, 0.90, \mathrm{Pr} = 0.0005, \phi = 0.0001, m = 0.0, n = 1$.

$\alpha=1, \eta.$	$\Theta'(a)$ SWCNTs	$\Theta'(a)$ MWCNTs	$\alpha=0.95, \eta.$	$\Theta'(a)$ SWCNTs	$\Theta'(a)$ MWCNTs	$\alpha=0.90, \eta.$	$\Theta'(a)$ SWCNTs	$\Theta'(a)$ MWCNTs
0.1	0.3000	0.3000	0.1	0.3000	0.3000	0.1	0.3000	0.3000
0.2	0.3000	0.2999	0.2	0.3000	0.2999	0.2	0.3000	0.2999
0.3	0.3000	0.2999	0.3	0.3000	0.2999	0.3	0.2999	0.2999
0.4	0.2999	0.2998	0.4	0.2999	0.2998	0.4	0.2999	0.2998
0.5	0.2999	0.2998	0.5	0.2999	0.2997	0.5	0.2999	0.2997
0.6	0.2999	0.2997	0.6	0.2999	0.2997	0.6	0.2998	0.2996
0.7	0.2998	0.2996	0.7	0.2998	0.2996	0.7	0.2998	0.2995
0.8	0.2998	0.2995	0.8	0.2998	0.2995	0.8	0.2998	0.2994
0.9	0.2998	0.2994	0.9	0.2997	0.2994	0.9	0.2997	0.2993
1.0	0.2997	0.2993	1.0	0.2997	0.2992	1.0	0.2997	0.2992

6. Conclusions

The SWCNT and MWCNT water-based nanofluids' flow over a thin needle was analyzed for the enhancement of temperature. Classical and fractional models were used to investigate the impact of the physical parameters and for similar values for the boundary conditions. The non-linear system was solved through the FDE-12 method. The classical and fractional results were obtained for $\alpha = 1$, and $\alpha = 0.95, 0.90$, respectively. The impact of the physical parameters over the velocity and temperature profiles in the classical model were limited, but utilizing the fractional model, the impact of the parameters varied for different intervals. It was observed that the fractional order model specifies the accuracy of the physical parameters more precisely considering the small interval of the derivative between 0 and 1, which have important applications, such as for a fractional order PID controller which may provide a more effective way to improve the system control routine; similarly, non-Fickian transport and anomalous diffusion in porous media, polymer flows, or very high gradients of concentration or heat are important application areas of the fractional order derivative in the field of engineering.

The main findings of this study are:

- Greater values of Pr cause decreases in the thickness of the thermal boundary layer when using the classical model, but by means of the fractional model for the same values of the Prandtl number, the thermal boundary layer near the needle surface increases and decreases after the critical point.

- Lower values of n lead to a decrease in the temperature profile using the classical model values, and this effect is upturned for the fractional order values $\alpha = 0.95, 0.90$ near the wall and change to an upsurge in the thermal boundary layer after the point of inflection.

Author Contributions: T.G.; conceptualization, M.A.K. and W.N.; methodology, I.K.; software, T.A.A.; and I.T.; validation, T.G., M.A.K. and W.N.; formal analysis, I.K.; investigation, T.A.A.; I.K.; and I.T.; writing—original draft preparation, T.G.; M.A.K. and W.N.; writing—review and editing.

Acknowledgments: Authors acknowledge anonymous referees for their valuable suggestions.

References

1. Sparrow, E.M.; Gregg, J.L. A theory of rotating condensation. *J. Heat Transf.* **1959**, *81*, 113–120.

2. Choi, S.U.S. Enhancing thermal conductivity of fluids with nanoparticles, developments and applications of non-Newtonian flows. *FED-231lMD* **1995**, *66*, 99–105.

3. Volder, M.D.; Tawfick, S.; Baughman, R.; Hart, A. Carbon nanotubes: Present and future commercial applications. *Science* **2013**, *339*, 535–539. [CrossRef] [PubMed]

4. Terrones, M. Science and technology of the twenty-first century: Synthesis, properties, and applications of carbon nanotubes. *Annu. Rev. Mater. Res.* **2003**, *33*, 491–501. [CrossRef]

5. Ellahi, R.; Hassan, M.; Zeeshan, A. Study of Natural Convection MHD Nanofluid by Means of Single and Multi-Walled Carbon Nanotubes Suspended in a Salt-Water Solution. *IEEE Trans. Nanotechnol.* **2015**, *14*, 726–734. [CrossRef]

6. Taza, G.; Waris, K.; Muhammad, S.; Muhammad, A.K.; Ebenezer, B. MWCNTs/SWCNTs Nanofluid Thin Film Flow over a Nonlinear Extending Disc: OHAM Solution. *J. Therm. Sci.* **2018**. [CrossRef]

7. Murshed, S.M.S.; Nieto de Castro, C.A.; Loureno, M.J.V.; Lopes, M.L.M.; Santos, F.J.V. A review of boiling and convective heat transfer with nanofluids. *Renew. Sustain. Energy Rev.* **2011**, *15*, 2342–2354. [CrossRef]

8. Murshed, S.M.S.; Leong, K.C.; Yang, C. Thermophysical and electro kinetic properties of nanofluids a critical review. *Appl. Therm. Eng.* **2008**, *28*, 2109–2125. [CrossRef]

9. Xue, Q. Model for thermal conductivity of carbon nanotube-based composites. *Phys. B Condens. Matter.* **2005**, *368*, 302–307. [CrossRef]

10. Narain, J.P.; Uberoi, M.S. Combined Forced and Free-Convection Heat Transfer From Vertical Thin Needles in a Uniform Stream. *Phys. Fluids* **1972**, *15*, 1879–1882. [CrossRef]

11. Narain, J.P.; Uberoi, M.S. Combined Forced and Free- Convection Over Thin Needles. *Int. J. Heat Mass Transf.* **1973**, *16*, 1505–1512. [CrossRef]

12. Chen, J.L.S. Mixed Convection Flow About Slender Bodies of Revolution. *ASME J. Heat Transf.* **1987**, *109*, 1033–1036. [CrossRef]

13. Wang, C.Y. Mixed convection on a vertical needle with heated tip. *Phys. Fluids A* **1990**, *2*, 622–625. [CrossRef]

14. Grosan, T.; Pop, I. Forced convection boundary layer flow past nonisothermal thin needles in nanofluids. *J. Heat Transf.* **2011**, *133*, 1–4. [CrossRef]

15. Oldham, K.B.; Spanier, J. *The Fractional Calculus*; Academic Press: New York, NY, USA, 1974.

16. Benson, D.; Wheatcraft, S.W.; Meerschaert, M.M. The fractional-order governing equation of Lévy motion. *Water Resour. Res.* **2000**, *36*, 1413–1423. [CrossRef]

17. Benson, D.; Wheatcraft, S.W.; Meerschaert, M.M. Application of a fractional advection–dispersion equation. *Water Resour. Res.* **2000**, *36*, 1403–1412. [CrossRef]

18. Caputo, M. Models of flux in porous media with memory. *Water Resour. Res.* **2000**, *36*, 693–705. [CrossRef]

19. El Amin, M.F.; Radwan, A.G.; Sun, S. Analytical solution for fractional derivative gas-flow equation in porous media. *Results Phys.* **2017**, *7*, 2432–2438. [CrossRef]

20. Atangana, A.; Alqahtani, R.T. Numerical approximation of the space-time Caputo-Fabrizio fractional derivative and application to groundwater pollution equation. *Adv. Differ. Equ.* **2016**, *156*, 1–13. [CrossRef]

21. Alkahtani, B.S.T.; Koca, I.; Atangan, A. A novel approach of variable order derivative: Theory and Methods. *J. Nonlinear Sci. Appl.* **2016**, *9*, 4867–4876. [CrossRef]

22. Agarwal, R.; Hristova, S.; O'Regan, D. Global Mittag-Leffler Synchronization for Neural Networks Modeled by Impulsive Caputo Fractional Differential Equations with Distributed Delays. *Symmetry* **2018**, *10*, 473. [CrossRef]

23. Khan, U.; Ellahi, R.; Khan, R.; Mohyud-Din, S.T. extracting new solitary wave solutions of Benny–Luke equation and Phi-4 equation of fractional order by using (G0/G)-expansion method. *Opt. Quant. Electron.* **2017**, *49*, 362–376. [CrossRef]

24. Hameed, M.; Ambreen, A.K.; Ellahi, R.; Raza, M. Study of magnetic and heat transfer on the peristaltic transport of a fractional second grade fluid in a vertical tube. *Eng. Sci. Technol. Int. J.* **2015**, *18*, 496–502. [CrossRef]

25. Shirvan, K.M.; Ellahi, R.; Sheikholeslami, T.F.; Behzadmehr, A. Numerical investigation of heat and mass transfer flow under the influence of silicon carbide by means of plasmaenhanced chemical vapor deposition vertical reactor. *Neural Comput. Appl.* **2018**, *30*, 3721–3731. [CrossRef]

26. Barikbin, Z.; Ellahi, R.; Abbasbandy, S. The Ritz-Galerkin method for MHD Couette ow of non-Newtonian fluid. *Int. J. Ind. Math.* **2014**, *6*, 235–243.

27. Hayat, T.; Saif, R.S.; Ellahi, R.; Muhammad, T.; Ahmad, B. Numerical study of boundary-layer flow due to a nonlinear curved stretching sheet with convective heat and mass conditions. *Results Phys.* **2017**, *7*, 2601–2606. [CrossRef]

28. Hayat, T.; Saif, R.S.; Ellahi, R.; Muhammad, T.; Ahmad, B. Numerical study for Darcy-Forchheimer flow due to a curved stretching surface with Cattaneo-Christov heat flux and homogeneous heterogeneous reactions. *Results Phys.* **2017**, *7*, 2886–2892. [CrossRef]

29. Javeed, S.; Baleanu, D.; Waheed, A.; Khan, M.S.; Affan, H. Analysis of Homotopy Perturbation Method for Solving Fractional Order Differential Equations. *Mathematics* **2019**, *7*, 40. [CrossRef]

30. Srivastava, H.M.; El-Sayed, A.M.A.; Gaafar, F.M. A Class of Nonlinear Boundary Value Problems for an Arbitrary Fractional-Order Differential Equation with the Riemann-Stieltjes Functional Integral and Infinite-Point Boundary Conditions. *Symmetry* **2018**, *10*, 508. [CrossRef]

31. Diethelm, K.; Freed, A.D. The Frac PECE subroutine for the numerical solution of differential equations of fractional order. In *Forschung und Wissenschaftliches Rechnen*; Heinzel, S., Plesser, T., Eds.; Gessellschaft fur Wissenschaftliche Datenverarbeitung: Gottingen, Germany, 1999; pp. 57–71.

32. Diethelm, K.; Ford, N.J.; Freed, A.D. Detailed error analysis for a fractional Adams method. *Numer. Algorithms* **2004**, *36*, 31–52. [CrossRef]

33. Saifullah Khan, M.A.; Farooq, M. A fractional model for the dynamics of TB virus. *Chaos Solitons Fractals* **2018**, *116*, 63–71.

34. Gul, T.; Khan, M.A.; Khan, A.; Shuaib, M. Fractional-order three-dimensional thin-film nanofluid flow on an inclined rotating disk. *Eur. Phys. J. Plus* **2018**, *133*, 500–5011. [CrossRef]

35. Gul, T.; Haleem, I.; Ullah, I.; Khan, M.A.; Bonyah, E.; Khan, I.; Shuaib, M. The study of the entropy generation in a thin film flow with variable fluid properties past over a stretching sheet. *Adv. Mech. Eng.* **2018**, *10*, 1–15. [CrossRef]

4

Influence of Cattaneo–Christov Heat Flux on MHD Jeffrey, Maxwell and Oldroyd-B Nanofluids with Homogeneous-Heterogeneous Reaction

Anwar Saeed [1], Saeed Islam [1], Abdullah Dawar [2], Zahir Shah [1], Poom Kumam [3,4,5,*] and Waris Khan [6]

[1] Department of Mathematics, Abdul Wali Khan University Mardan, Mardan 23200, Pakistan; anwarsaeed769@gmail.com (A.S.); saeedislam@awkum.edu.pk (S.I.); zahir1987@yahoo.com (Z.S.)

[2] Department of Mathematics, Qurtuba University of Science and Information Technology, Peshawar 25000, Pakistan; abdullah.mathematician@gmail.com

[3] KMUTTFixed Point Research Laboratory, Room SCL 802 Fixed Point Laboratory, Science Laboratory Building, Department of Mathematics, Faculty of Science, King Mongkut's University of Technology Thonburi (KMUTT), Bangkok 10140, Thailand

[4] KMUTT-Fixed Point Theory and Applications Research Group, Theoretical and Computational Science Center (TaCS), Science Laboratory Building, Faculty of Science, King Mongkut's University of Technology Thonburi (KMUTT), Bangkok 10140, Thailand

[5] Department of Medical Research, China Medical University Hospital, China Medical University, Taichung 40402, Taiwan

[6] Department of Mathematics, Kohat University of Science and Technology, Kohat 26000, Pakistan; wariskhan758@yahoo.com

* Correspondence: poom.kum@kmutt.ac.th

Abstract: This research article deals with the determination of magnetohydrodynamic steady flow of three combile nanofluids (Jefferey, Maxwell, and Oldroyd-B) over a stretched surface. The surface is considered to be linear. The Cattaneo–Christov heat flux model was considered necessary to study the relaxation properties of the fluid flow. The influence of homogeneous-heterogeneous reactions (active for auto catalysts and reactants) has been taken in account. The modeled problem is solved analytically. The impressions of the magnetic field, Prandtl number, thermal relaxation time, Schmidt number, homogeneous–heterogeneous reactions strength are considered through graphs. The velocity field diminished with an increasing magnetic field. The temperature field diminished with an increasing Prandtl number and thermal relaxation time. The concentration field upsurged with the increasing Schmidt number which decreased with increasing homogeneous-heterogeneous reactions strength. Furthermore, the impact of these parameters on skin fraction, Nusselt number, and Sherwood number were also accessible through tables. A comparison between analytical and numerical methods has been presented both graphically and numerically.

Keywords: Magnetohydrodynamic (MHD); Jefferey, Maxwell and Oldroyd-B fluids; Cattaneo–Christov heat flux; homogeneous–heterogeneous reactions; analytical technique; Numerical technique

1. Introduction

A fluid composed of nanoparticles is called nanofluid. Nanoparticles of materials such as metallic oxides, carbide ceramics, nitride metals, ceramics, semiconductors, single, double or multi walled carbon nanotubes, alloyed, nanoparticles, etc. have been used for the preparation of nanofluids. Nanofluids have many characteristics in heat transfer, including microelectronics, local refrigerator, cooler, machining, and heat exchanger. The idea of nanofluid was introduced by Choi [1]. The Fourier's [2] recommended law of heat conduction normally works for heat transmission features from the time it was presented in the literature. By including the relaxation time parameter Cattaneo [3] it has improved this law and this term overwhelms the paradox of heat conduction. Christov [4] has named this theory Cattaneo-Christov heat flux theory, by further modifying the Cattaneo theory by exchanging the time derivative with Oldroyd-B upper convicted derivative. Mustafa [5] scrutinized the model [4] for heat transmission in a rotating Maxwell nanofluid flow. Chen [6] probed the influence of heat transfer and viscous dissipation of nanofluid flow over a stretching sheet. Sheikholeslami et al. [7–11] deliberated the three-dimensional magnetohydrodynamics (MHD) nanofluid flow in parallel rotating plates. Sheikholeslami [11–15] analytically and numerically deliberated the applications of nanofluids with different properties, behavior, and influences. Dawar et al. [16] examined the flow Williamson nanofluid over a stretching surface. Shah et al. [17] examined the micropolar nanofluid flow in rotating parallel plates with Hall current impact. Maleki et al. [18] scrutinized the non-Newtonian nanofluids flow and heat transfer over a porous surface. Nasiri et al. [19] deliberated the smoothed particle by a hydrodynamics approach for numerical simulation of nanofluid flows. Rashidi et al. [20] used the nanofluids in a circular tube heat exchanger and examined the entropy generation. Safaei et al. [21] studied numerically and experimentally the nanofluids convective heat transfer in closed conduits. Mahian et al. [22,23] presented the advances in the simulation and modeling of the flows of nanofluids.

Due to its relaxation properties, Jeffrey, Maxwell, and Oldroyd-B nanofluids have significant applications in the area of fluid mechanics. Ahmad et al. [24] scrutinized the flow of Jeffrey nanofluid with Magnetohydrodynamic impact. Ahmad and Ishak [25] deliberated the flow of Jeffrey nanofluid with MHD and transverse magnetic field impacts in a porous medium. Hayat et al. [26] probed the Oldroyd-B nanofluid flow with heat transfer and thermal radiation impacts. Raju et al. [27] deliberated the flow of Jeffrey nanofluid with a homogenous-heterogeneous reaction and non-linear thermal radiation impacts. The articles that are related to Jeffrey nanofluid can be found in [28–32]. Hayat et al. [33] inspected the MHD Maxwell nanofluid flow using suction/injection. Raju et al. [34] presented the heat and mass transmission in three-dimensional non-Newtonian nanofluid and Ferrofluid. Sandeep and Sulochana [35] investigated the mixed convection micropolar nanofluid flow over a stretching sheet. Raju et al. [36] deliberated the impacts of an inclined magnetic field, thermal radiation and cross diffusion on the two-dimensional flow. Nadeem et al. [37] presented the heat and mass transfer in Jeffrey nanofluid. Makinde et al. [38] deliberated the unsteady fluid flow with convective boundary conditions. Sheikholeslami [39] discussed the hydro-thermal behavior of nanofluids flow because of its external heated plates. Shah et al. [40] presented the Darcy-Forchheimer flow of radiative carbon nanotubes in a rotating frame. Chai et al. [41] presented a review of the heat transfer and hydrodynamic characteristics of nano/microencapsulated phase. Shah et al. [42] examined the electro-magneto micropoler Casson Ferrofluid over a stretching/shrinking sheet. Dawar et al. [43] analyzed the MHD CNTs Casson nanofluid in rotating channels. Khan et al. [44] deliberated the Williamson nanofluid flow over a linear stretching surface. Imtiaz et al. [45] examined the unsteady MHD flow due to a curved stretchable surface with homogeneous–heterogeneous reactions. Hayat et al. [46] deliberated the flow of nanofluids with homogeneous–heterogeneous reaction impacts over a non-linear stretched sheet with variable thickness. The recent study about nanofluid with application can be seen [47–50].

The present work is based on an analysis of MHD flow of three combine nanofluids (Maxwell, Oldroyd-B, and Jeffrey) over a linear stretching surface. The present model composed of Cattaneo–Christov heat flux. The impact of homogeneous-heterogeneous reactions were taken in this model. A boundary layer methodology was used in the mathematical expansion. The impact of dimensionless parameters on the fluid flow have been presented through graphs and tables.

2. Mathematical Modeling and Formulation

The incompressible electrically conducted three combined nanofluids (Jeffrey, Maxwell, and Oldroyd-B) were confined by a linear stretched surface. The fluid flow was taken in a two-dimensional steady state with stable surface temperature. x-axis was considered parallel to the surface, while y-axis was orthogonal to x-axis in the chosen coordinate system. The stretching velocity in x-axes direction was defined as $U_w(x) = \zeta x$. The conclusion of homogeneous-heterogeneous reactions on the fluid flows of two chemical species I and J were taken in account. In the y-axis direction a uniform magnetic field B_0 was acting. The heat transmission procedure was applied through Cattaneo–Christov heat flux theory.

In case of cubic autocatalysis, the Homogeneous reaction is [45,46]

$$I + 2J \rightarrow 3J, rate = k_c ij^2, \tag{1}$$

While on the catalyst surface, the heterogeneous reaction has been defined by

$$I \rightarrow J, rate = k_s i, \tag{2}$$

where k_c, k_s, I, J, i, j are the rate constants, chemical species, and concentrations of chemical species, respectively.

In the absence of viscous dissipation and thermal radiation, the boundary layer equations leading to the flow of viscoelastic fluids can be written as follows [47]:

$$u_x + v_y = 0, \tag{3}$$

$$uu_x + vu_y = -\lambda_1 \left(u^2 u_{xx} + 2uv u_{xy} + v^2 u_{yy} \right) +$$
$$\frac{v_f}{1+\lambda_2} \left\{ u_{yy} + \lambda_3 \left(uu_{xyy} + u_y u_{xy} + vu_{yyy} - u_x u_{yy} \right) \right\} - \frac{\sigma B_0^2}{\rho_f} u, \tag{4}$$

$$\rho c_p \left(uT_x + vT_y \right) = -\nabla.q, \tag{5}$$

$$ui_x + vi_y = D_I i_{yy} - k_c ij^2, \tag{6}$$

$$uj_x + vj_y = D_J j_{yy} + k_c ij^2. \tag{7}$$

Here u, v, μ, ρ_f, v_f are velocity components in their respective directions, dynamic viscosity, density, and kinematic viscosity respectively. $\lambda_1, \lambda_2, \lambda_3$ are the relaxation time, a proportion of the relaxation to retardation times, respectively. T, σ_f, B_0 indicated the temperature, electrical conductivity and the transverse magnetic field.

The problem is studied based on the following conditions:

i   Oldroyd-B nanofluid when $\lambda_1 \neq 0, \lambda_2 = 0$ and $\lambda_3 \neq 0$.
ii   Maxwell nanofluid when $\lambda_1 \neq 0, \lambda_2 = 0$ and $\lambda_3 = 0$.
iii Jeffrey nanofluid when $\lambda_1 = 0, \lambda_2 \neq 0$ and $\lambda_3 \neq 0$.

The heat flux theory which was presented by Cattaneo–Christov:

$$q + \lambda_1 \left\{ \frac{\partial q}{\partial t} + V \cdot \nabla q - q \cdot \nabla V + (\nabla \cdot V)q + q_t \right\} = -k\nabla T, \tag{8}$$

where k, q represented thermal conductivity and heat flux. Classical Fourier's law was assumed by setting $\lambda_1 = 0$ in Equation (8). By assuming the condition $(\nabla.V = 0)$ and steady flow with $(q_t = 0)$, Equation (8) became:

$$q + \lambda_1 (V.\nabla q - q.\nabla V) = -k\nabla T. \tag{9}$$

The heat transfer equation proceeded as:

$$u T_x + v T_y + \lambda_1 \Phi_E = \alpha \left(T_{yy} \right), \tag{10}$$

where Φ_E is given as:

$$\Phi_E = u u_x T_x + v v_y T_y + u v_x T_y + v u_y T_x + 2 u v T_{xy} + u^2 T_{xx} + v^2 T_{yy}. \tag{11}$$

The accompanying boundary conditions were:

$$u = U_w(x) = \zeta x, v = 0, T = T_w, D_I i_y = k_s i, D_J j_y = -k_s i \text{ at } y = 0, \\ u \to 0, \ T \to T_\infty, i \to i_0, j \to 0 \text{ at } y \to \infty, \tag{12}$$

where $\alpha = \frac{k}{\rho c_p}$ indicated the thermal diffusivity, D_I and D_J indicated the diffusion coefficients, T_w denoted the temperature at the surface, T_∞ for the surrounding fluid temperature and ζ for non-negative stretching rate constant with T^{-1} as the dimension.

$$u = \zeta x F'(\eta), v = -(\zeta v)^{\frac{1}{2}} F(\eta), \eta = \left(\frac{\zeta}{v} \right)^{\frac{1}{2}} y, \\ G(\eta) = \frac{T - T_\infty}{T_w - T_\infty}, i = i_0 \phi(\eta), j = i_0 h(\eta). \tag{13}$$

Apparently the equation of continuity is satisfied and Equations (4)–(13) become:

$$F''' + \kappa_2 \left(F''^2 - F F'''' \right) - M(1 + \lambda_2) F' - (1 + \lambda_2) \left\{ F'^2 - F F'' + \kappa_1 \left(F^2 F''' - 2 F F' F'' \right) \right\} = 0, \tag{14}$$

$$G'' + \Pr \left\{ F G' - \Omega \left(F F' G' + F^2 G'' \right) \right\} = 0, \tag{15}$$

$$\phi'' + Sc \left(F \phi' - K \phi h^2 \right) = 0, \tag{16}$$

$$\delta h'' + Sc \left(F h' + K \phi h^2 \right) = 0, \tag{17}$$

with boundary conditions:

$$F = 0, F' = 1, G = 1, \phi' = K_s \phi, \delta h' = -K_s \phi \text{ at } \omega = 0, \\ F' \to 0, G \to 0, \phi \to 1, h \to 0 \text{ at } \eta \to \infty, \tag{18}$$

In the above equations, $M = \frac{\sigma_{nf} B_0^2}{\rho_f \zeta}$ indicated the magnetic field, $\kappa_1 = \lambda_1 \zeta$ and $\kappa_2 = \lambda_3 \zeta$ were the Debora numbers with respect to relaxation and retardation time, $\Pr = \frac{v_f}{\alpha}$ represented the Prandtl number, $\Omega = \zeta \lambda_1$ indicated the thermal relaxation time, $Sc = \frac{v_f}{D_I}$ is the Schmidt number, $K = \frac{k_c i_0^2}{U_w}$ indicating the homogeneous reaction strength, $K_s = \frac{k_s}{D_I i_0} \sqrt{\frac{\zeta}{v_f}}$ represented the heterogeneous reaction strength, and $\delta = \frac{D_J}{D_I}$ indicated the diffusion coefficient, When $D_I = D_J$ then $\delta = 1$ and as a result:

$$\phi(\eta) + h(\eta) = 1. \tag{19}$$

Now Equations (16) and (17) yield:

$$\phi'' + Sc \left\{ F \phi' - K \phi (1 - \phi)^2 \right\} = 0. \tag{20}$$

The subjected boundary conditions are:

$$\phi'(0) = K_s \phi(0), \phi(\infty) \to 1. \tag{21}$$

Skin friction coefficient through the dimensionless scale is:

$$\mathrm{Re}_x^{\frac{1}{2}} Cf_x = \left(\frac{1+\kappa_1}{1+\kappa_2}\right) F''(0). \tag{22}$$

where Re_x is called the local Reynolds number.

The dimensionless form of Nu_x and Sh_x were found as:

$$Nu_x = -G'(0), \quad Sh_x = -\phi'(0). \tag{23}$$

3. Solution by Homtopy Analysis Method (HAM)

To evaluate the Equations (14), (15) and (20) with boundary conditions (18) and (21) using HAM with the following procedure.

The initial assumptions were picked as below:

$$\overline{F}_0(\eta) = 1 - e^{-\eta}, \ \overline{G}_0(\eta) = e^{-\eta}, \overline{\phi}_0(\eta) = e^{-\eta}. \tag{24}$$

The linear operators were taken as $L_{\overline{F}}, L_{\overline{G}}$ and $L_{\overline{\phi}}$:

$$L_{\overline{F}}(\overline{F}) = \overline{F}''' - \overline{F}', L_{\overline{G}}(\overline{G}) = \overline{G}'' - \overline{G}, L_{\overline{\phi}}(\overline{\phi}) = \overline{\phi}'' - \overline{\phi}, \tag{25}$$

With the following properties:

$$L_{\overline{F}}(r_1 + r_2 e^{-\eta} + r_3 e^{\eta}) = 0, \ L_{\overline{G}}(r_4 e^{-\eta} + r_5 e^{\eta}) = 0, L_{\overline{\phi}}(r_6 e^{-\eta} + r_7 e^{\eta}) = 0, \tag{26}$$

where $r_i (i = 1 - 7)$ were the constants:

The resulting non-linear operators $N_{\overline{F}}, N_{\overline{G}}$ and $N_{\overline{\phi}}$ were specified as:

$$N_{\overline{F}}[\overline{F}(\eta;\tau)] = \frac{\partial^3 \overline{F}(\eta;\tau)}{\partial \eta^3} + \kappa_2 \left\{ \left(\frac{\partial^2 \overline{F}(\eta;\tau)}{\partial \eta^2}\right)^2 - \frac{\partial \overline{F}(\eta;\tau)}{\partial \eta} \frac{\partial^4 \overline{F}(\eta;\tau)}{\partial^4 \eta}\right\}$$
$$-M(1+\lambda_2)\left\{ \begin{array}{c} \left(\frac{\partial \overline{F}(\eta;\tau)}{\partial \eta}\right)^2 - \overline{F}(\eta;\tau)\frac{\partial^2 \overline{F}(\eta;\tau)}{\partial \eta^2} + \\ \kappa_1\left(\overline{F}^2(\eta;\tau)\frac{\partial^3 \overline{F}(\eta;\tau)}{\partial \eta^3} - 2\overline{F}(\eta;\tau)\frac{\partial \overline{F}(\eta;\tau)}{\partial \eta}\frac{\partial^2 \overline{F}(\eta;\tau)}{\partial \eta^2}\right) \end{array}\right\}, \tag{27}$$

$$N_{\overline{G}}\left[\overline{F}(\eta;\tau), \overline{G}(\eta;\tau)\right] = \frac{\partial^2 \overline{G}(\eta;\tau)}{\partial \eta^2} +$$
$$\mathrm{Pr}\left\{\overline{F}(\eta;\tau)\frac{\partial \overline{G}(\eta;\tau)}{\partial \eta} - \Omega\left(\overline{F}(\eta;\tau)\frac{\partial \overline{F}(\eta;\tau)}{\partial \eta}\frac{\partial \overline{G}(\eta;\tau)}{\partial \eta} + \overline{F}^2(\eta;\tau)\frac{\partial^2 \overline{G}(\eta;\tau)}{\partial \eta^2}\right)\right\}, \tag{28}$$

$$N_{\overline{\phi}}[\overline{F}(\eta;\tau), \overline{\phi}(\eta;\tau)] = \frac{\partial^2 \overline{\phi}(\eta;\tau)}{\partial \eta^2} + Sc\left\{\overline{F}(\eta;\tau)\frac{\partial \overline{\phi}(\eta;\tau)}{\partial \eta} - K\left(\overline{\phi}^3(\eta;\tau) - 2\overline{\phi}^2(\eta;\tau) + \overline{\phi}(\eta;\tau)\right)\right\}, \tag{29}$$

The zeroth-order problem for Equations (14), (15) and (20) were:

$$(1-\tau)L_{\overline{F}}[\overline{F}(\eta;\tau) - \overline{F}_0(\eta)] = \tau \hbar_{\overline{F}} N_{\overline{F}}[\overline{F}(\eta;\tau)], \tag{30}$$

$$(1-\tau)L_{\overline{G}}[\overline{G}(\eta;\tau) - \overline{G}_0(\eta)] = \tau \hbar_{\overline{G}} N_{\overline{G}}[\overline{F}(\eta;\tau), \overline{G}(\eta;\tau)], \tag{31}$$

$$(1-\tau)L_{\overline{\phi}}[\overline{\phi}(\eta;\tau) - \overline{\phi}_0(\eta)] = \tau \hbar_{\overline{\phi}} N_{\overline{\phi}}[\overline{F}(\eta;\tau), \overline{\phi}(\eta;\tau)]. \tag{32}$$

The related boundary conditions where:

$$\overline{F}(\eta;\tau)|_{\eta=0} = 0, \ \frac{\partial \overline{F}(\eta;\tau)}{\partial \eta}\bigg|_{\eta=0} = 1, \ \frac{\partial \overline{F}(\eta;\tau)}{\partial \eta}\bigg|_{\eta\to\infty} = 0,$$
$$\overline{G}(\eta;\tau)|_{\eta=0} = 1, \ \overline{G}(\eta;\tau)|_{\eta\to\infty} = 0, \tag{33}$$
$$\frac{\partial \overline{\phi}(\eta;\tau)}{\partial \eta}\bigg|_{\eta=0} = K_s \overline{\phi}(\eta;\tau)|_{\eta=0}, \ \overline{\phi}(\eta;\tau)|_{\eta\to\infty} = 1,$$

where $\tau \in [0,1]$ is the embedding parameter, $\hbar_{\overline{F}}$, $\hbar_{\overline{G}}, \hbar_{\overline{\phi}}$ that were used to control the solution convergence. When $\tau = 0$ and $\tau = 1$ we have:

$$\overline{F}(\eta;1) = \overline{F}(\eta),\ G(\eta;1) = \overline{G}(\eta) \text{ and } \overline{\phi}(\eta;1) = \overline{\phi}(\eta), \tag{34}$$

Expanding $\overline{F}(\eta;\tau)$, $G(\eta;\tau)$ and $\overline{\phi}(\eta;\tau)$ by Taylor's series:

$$\begin{aligned}
\overline{F}(\eta;\tau) &= \overline{F}_0(\eta) + \sum_{q=1}^{\infty} \overline{F}_q(\eta)\tau^q, \\
\overline{G}(\eta;\tau) &= \overline{G}_0(\eta) + \sum_{q=1}^{\infty} \overline{G}_q(\eta)\tau^q, \\
\overline{\phi}(\eta;\tau) &= \overline{\phi}_0(\eta) + \sum_{q=1}^{\infty} \overline{\phi}_q(\eta)\tau^q.
\end{aligned} \tag{35}$$

where:

$$\overline{F}_q(\eta) = \frac{1}{q!}\frac{\partial\overline{F}(\eta;\tau)}{\partial\eta}\bigg|_{\tau=0},\ G_q(\eta) = \frac{1}{q!}\frac{\partial\overline{G}(\eta;\tau)}{\partial\eta}\bigg|_{\tau=0} \text{ and } \overline{\phi}_q(\eta) = \frac{1}{q!}\frac{\partial\overline{\phi}(\eta;\tau)}{\partial\eta}\bigg|_{\tau=0}. \tag{36}$$

The $\hbar_{\overline{F}}, \hbar_{\overline{G}}$ and $\hbar_{\overline{\phi}}$ are taken in such a way that the series (35) converges at $\tau = 1$, we have:

$$\begin{aligned}
\overline{F}(\eta) &= \overline{F}_0(\eta) + \sum_{q=1}^{\infty} \overline{F}_q(\eta), \\
\overline{G}(\eta) &= \overline{G}_0(\eta) + \sum_{q=1}^{\infty} \overline{G}_q(\eta), \\
\overline{\phi}(\eta) &= \overline{\phi}_0(\eta) + \sum_{q=1}^{\infty} \overline{\phi}_q(\eta).
\end{aligned} \tag{37}$$

The following are satisfied by the qth-order problem.

$$\begin{aligned}
L_{\overline{F}}\left[\overline{F}_q(\eta) - \chi_q\overline{F}_{q-1}(\eta)\right] &= \hbar_{\overline{F}}U_q^{\overline{F}}(\eta), \\
L_{\overline{G}}\left[\overline{G}_q(\eta) - \chi_q\overline{G}_{q-1}(\eta)\right] &= \hbar_{\overline{G}}U_q^{\overline{G}}(\eta), \\
L_{\overline{\phi}}\left[\overline{\phi}_q(\eta) - \chi_q\overline{\phi}_{q-1}(\eta)\right] &= \hbar_{\overline{\phi}}U_q^{\overline{\phi}}(\eta).
\end{aligned} \tag{38}$$

Which have the following boundary conditions:

$$\begin{aligned}
\overline{F}_q(0) = \overline{F}'_q(0) = \overline{F}'_q(\infty) &= 0, \\
\overline{G}_q(0) = \overline{G}_q(\infty) &= 0, \\
\overline{\phi}'_q(0) - K_s\overline{\phi}_q(0) = \overline{\phi}_q(\infty) &= 0.
\end{aligned} \tag{39}$$

Here

$$U_q^{\overline{F}}(\eta) = \overline{F}'''_{q-1} + \kappa_2\left(\sum_{k=0}^{q-1}\overline{F}_{q-1-k}\overline{F}''_k - \sum_{k=0}^{q-1}\overline{F}_{q-1-k}\overline{F}_k^{iv}\right) - M(1+\lambda_2)\overline{F}'_{q-1} -$$

$$(1+\lambda_2)\left\{\sum_{k=0}^{q-1}\overline{F}'_{q-1-k}\overline{F}' - \sum_{k=0}^{q-1}\overline{F}_{q-1-k}\overline{F}''_k + \kappa_1\left(\sum_{k=0}^{q-1}\overline{F}_{q-1-k}\sum_{j=0}^{k}\overline{F}_{k-j}\overline{F}'''_j - 2\sum_{k=0}^{q-1}\overline{F}_{q-1-k}\sum_{j=0}^{k}\overline{F}'_{k-j}\overline{F}''_j\right)\right\}, \tag{40}$$

$$U_q^{\overline{G}}(\eta) = \overline{G}''_{q-1} + Pr\left\{\sum_{k=0}^{q-1}\overline{F}_{q-1-k}\overline{G}'_k - \Omega\left(\sum_{k=0}^{q-1}\overline{F}_{q-1-k}\sum_{j=0}^{k}\overline{F}'_{k-j}\overline{G}'_j + \sum_{k=0}^{q-1}\overline{F}_{q-1-k}\sum_{j=0}^{k}\overline{F}_{k-j}\overline{G}''_j\right)\right\}, \tag{41}$$

$$U_q^{\overline{\phi}}(\eta) = \overline{\phi}''_{q-1} - KSc\left(\sum_{k=0}^{q-1}\overline{\phi}_{q-1-k}\sum_{j=0}^{k}\overline{\phi}_{k-j}\overline{\phi}_j - 2\sum_{k=0}^{q-1}\overline{\phi}_{q-1-k}\overline{\phi}_k + \overline{\phi}_{q-1}\right), \tag{42}$$

where:

$$\chi_q = \begin{cases} 0, & \text{if } \tau \leq 1 \\ 1, & \text{if } \tau > 1 \end{cases} \tag{43}$$

4. HAM Solution Convergences

In this segment we graphically discussed the superior effect of the concerned parameters. The convergence of Equation (36) was subjected entirely through the auxiliary constraints $\hbar_F, \hbar_G, \hbar_\phi$. This is a collection in a way that it controls and converges the series solutions. The optional division of $\hbar$, was plotted through $\hbar$-curves $F''(0), G'(0), \phi'(0)$ for the 2nd ordered approximated solution of HAM. The operational region of $\hbar$ is $-2.2 < \hbar_F < 0.2, -2.1 < \hbar_G < -0.1, -2.4 < \hbar_\phi < 0.1$. The convergence of HAM through the $\hbar$-curve on velocity profile $F''(0)$, temperature profile $G'(0)$ and concentration profile $\phi'(0)$ is presented in Figure 1.

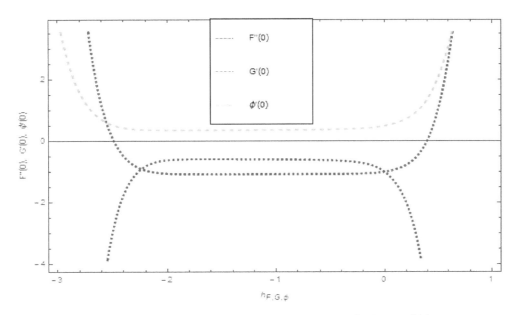

Figure 1. The combined $\hbar$-curves for $F''(0), G'(0)$ and $\phi'(0)$.

5. Results and Discussion

In this segment the impact of emerging parameters on velocity function $F'(\eta)$, temperature function $G(\eta)$ and concentration function $\phi(\eta)$ within the defined domain have been discussed. The impact of M on $F'(\eta)$ is deliberated in Figure 2. The Lorentz force theory deliberated that the magnetic field grows at a reversed force to the fluids flow. This force reduced the momentum boundary layer while it improved the thickness of the boundary layer. Therefore, with the escalating magnetic field M the velocity profile $F'(\eta)$ declined. From here we concluded that Jeffrey nanofluid was greatly subjected by the magnetic field compared to the other two. In Figures 3 and 4 the impact of Pr and Ω on $G(\eta)$ were presented respectively. In Figure 3 we perceived that $G(\eta)$ diminished with the rise in Pr. Physically the thickness of the boundary layer increased with the reduction in thermal diffusion. In addition, it can also be seen from the figure that Pr is more effective on Jeffrey and Maxwell nanofluids compared to the Oldroyd-B nanofluid. In Figure 4 the effect of thermal relaxation parameter Ω on $G(\eta)$ has been described. From here we saw that $G(\eta)$ reduced with the escalation in Ω. This was attributable to the fact that as we escalate Ω, the material particles need more time for heat transmission to its nearest particles. In addition, it can be stated that this material shows a non-conducting behavior with higher values of Ω which results in a reduction in $G(\eta)$. The impact of Sc, K and K_s on $\phi(\eta)$ are schemed in Figures 5–7 respectively. In Figure 5 the effect of Sc on $\phi(\eta)$ has been described. Schmidt number is the ratio of momentum diffusivity to mass diffusivity. Physically, the Schmidt number is related to hydrodynamic layer's thickness and boundary layer. The escalating

Sc intensifies the momentum of the boundary layer flow which results in an increase in concentration profile. It is clear from the figure that $\phi(\eta)$ upsurges with the rise in Sc. In Figure 6 the impact of K on $\phi(\eta)$ has been described. From here we concluded that larger K results in a reduction in $\phi(\eta)$. This may have been caused by the fact that the reaction rates controlled the diffusion coefficients. To a certain extent similar results are displayed in Figure 7. In Figure 7 the impact of K_s on $\phi(\eta)$ has been described. From this figure we have concluded that the growing values of K_s showed a drop in behavior in $\phi(\eta)$. This results from an agreement with the general physical behavior of the homogeneous reaction K and the heterogeneous reaction K_s. In Figures 8 and 9 the impact of M on Cf_x and Nu_x for Jeffrey, Maxwell and Oldroyd-B nanofluids have been described. It is clear from the figures that the growing values of M were decreasing for both Cf_x and Nu_x. The magnetic field was applied perpendicular to the flow of fluids and had an inverse variation with the skin friction of the fluid flow. This is the reason why the increasing magnetic field reduced the skin friction of the fluids flow. Similarly, the behavior of the heat transfer rate was due to the growing magnetic force on the fluids flow phenomena, with the fluid particles requiring more time to transfer the heat to the nearest particle. This was because the heat transfer rate reduced with the escalating magnetic field. The impact of Pr and Ω on Nu_x for the nanofluids flow has been described in Figures 10 and 11. From here we have concluded that the escalation in Pr increased the heat transfer rate while the increased Ω reduced the heat transfer rate for the nanofluids flow. Figure 12 shows the Total Residual error for the three types of nanofluid flow.

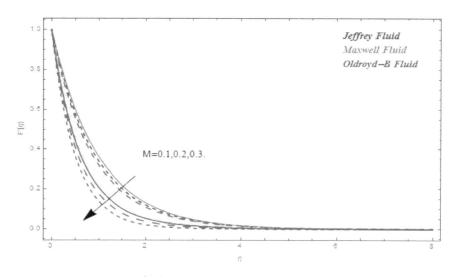

Figure 2. Impact of M on $F'(\eta)$, when $\Omega = 0.5, Sc = 0.6, K = 0.8, Pr = 0.7, K_s = 0.9$.

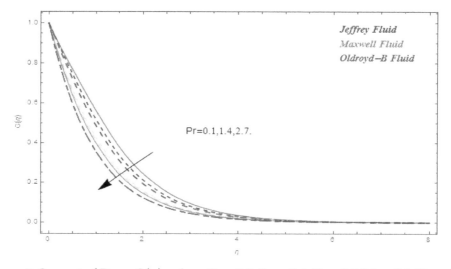

Figure 3. Impact of Pr on $G(\eta)$, when $\Omega = 0.5, Sc = 0.6, K = 0.8, M = 0.1, K_s = 0.9$.

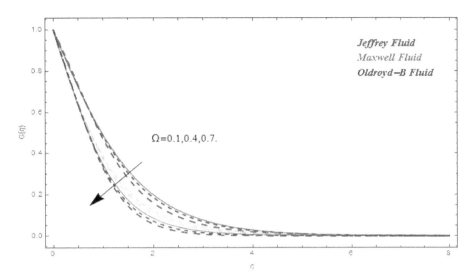

Figure 4. Impact of Ω on $G(\eta)$, when $Sc = 0.6, \mathrm{Pr} = 0.7, \mathrm{K} = 0.8, \mathrm{M} = 0.1, \mathrm{K}_s = 0.9.$

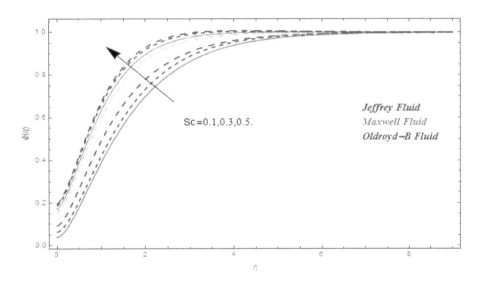

Figure 5. Impact of Sc on $\phi(\eta)$, when $\Omega = 0.1, \mathrm{Pr} = 0.7, \mathrm{K} = 0.8, \mathrm{M} = 0.1, \mathrm{K}_s = 0.9.$

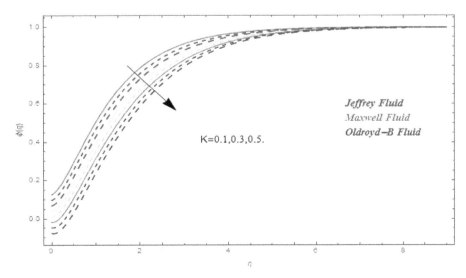

Figure 6. Impact of K on $\phi(\eta)$, when $\Omega = 0.1, \mathrm{Pr} = 0.7, Sc = 0.6, \mathrm{M} = 0.1, \mathrm{K}_s = 0.9.$

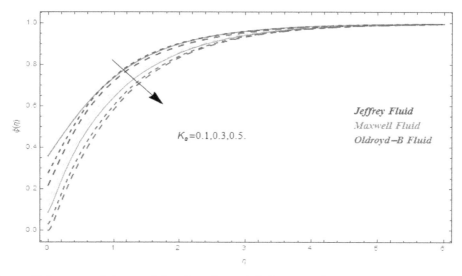

Figure 7. Impact of K_s on $\phi(\eta)$, when $\mathrm{Pr}=0.7, \Omega=0.1, Sc=0.6, \mathrm{M}=0.1, \mathrm{K}=0.8$.

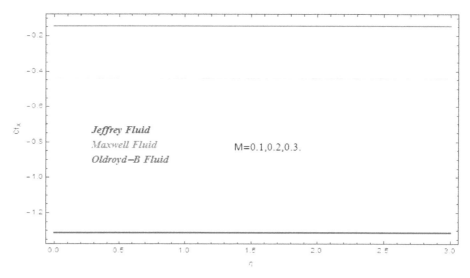

Figure 8. Impact of M on Cf_x.

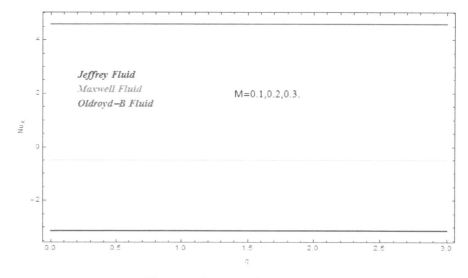

Figure 9. Impact of M on Nu_x.

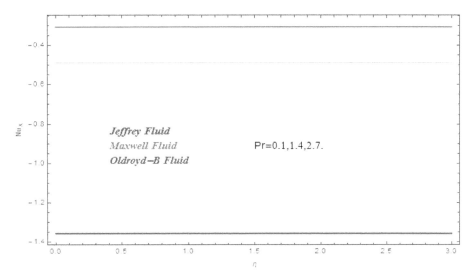

Figure 10. Impact of Pr on Nu_x.

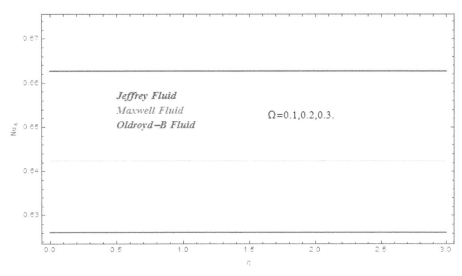

Figure 11. Impact of Ω on Nu_x.

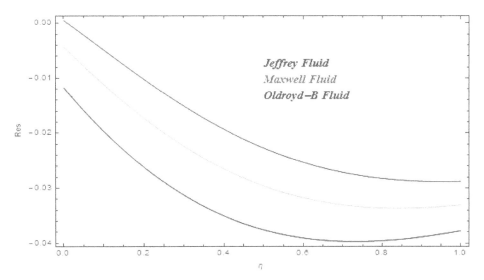

Figure 12. Total residual error for Jeffery, Maxwell and Oldroyd-B nanofluids.

Tables Discussion

In this section we have demonstrated the effect of emerging dimensionless parameters on the presented model of nanofluids. Table 1 displayed the conclusions associated with emerging parameters of skin fraction coefficients. This shows the impression of magenatic field parameter M on skin fraction coefficients. The magenatic field parameter shows a reduction in the skin fraction coefficient. Table 2 demonstrated the conclusion of incipient parameters on local Nusselt numbers. The heat transfer rate decreases with the rise in thermal relaxation parameter Ω while escalates with the increase in Prandtl number Pr. Table 2 shows that the thermal relaxation parameter has more effect on Jefferey nanofluids in comparison to Maxwell and Oldroyd-B nanofluids. Table 3 demonstrated the conclusion of an emerging parameter on the Sherwood number. The Sherwood number reduces with its rise, which upsurges with the escalation of the strength of homogeneous reaction K and the strength of heterogeneous reaction K_s.

Table 1. Distinction in $-Cf_x$ for different M.

M	Ref. [47]	Present Results for Jeffrey Nanofluid	Ref. [47]	Present Results for Maxwell Nanofluid	Ref. [47]	Present Results for Oldroyd-B Nanofluid
1.0	1.210458	0.210462	1.504151	1.504153	1.071019	1.071022
2.0	1.431584	1.431587	1.804788	1.804791	1.248081	1.248084

Table 2. Distinction in Nu_x for different Ω and Pr.

Ω	Pr	Ref. [47]	Jeffrey Nanofluid	Ref. [47]	Maxwell Nanofluid	Ref. [47]	Oldroyd-B Nanofluid
1.0		———	0.610394	———	0.595298	———	0.610846
1.2		———	0.607503	———	0.593311	———	0.607993
	6.0	0.418081	0.513786	0.421167	0.511247	0.426476	0.5154367
	7.0	0.439695	0.626865	0.441919	0.548966	0.447670	0.5477974

Table 3. Distinction in Sh_x for different Sc, K and K_s.

Sc	K	K_s	Jeffrey	Maxwell	Oldroyd-B
1.2			−0.096477	−0.095593	−0.096771
1.5			−0.096782	−0.095890	−0.097081
	1.5		−0.058030	−0.047238	−0.049135
	1.7		−0.056699	−0.037230	−0.039262
		0.5	−0.018933	−0.037233	0.012399
		0.8	−0.160028	0.046603	−0.125205

6. Comparison of Analytical Solutions and Numerical Solutions

An analytical solution means an exact solution. To study the behavior of systems, an analytical solution can be used with varying properties. Regrettably there are many practical systems that lead to an analytical solution, and analytical solutions are often of limited use. This is why we use a numerical approach to generate answers that are closer to practical results. These solutions which cannot be used as complete mathematical expressions are numerical solutions. In the natural worldthere are almost no problems that are exactly solvable, which makes the problem more difficult than all the exactly solvable problems. There are three or four of them in nature that have already been solved, unfortunately even numerical methods cannot always give an exact solution. Numerical techniques can handle any completed physical geometries which are often impossible to solve analytically. In this article both analytical and numerical approaches are tested to solve the modeled problem. A comparison of HAM and ND-Solve technique for $F'(\eta)$, $G(\eta)$ and $\phi(\eta)$ are deliberated in Figures 13–15 and Tables 4–6, respectively.

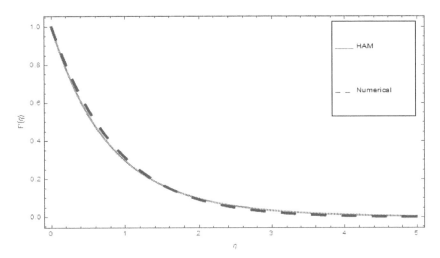

Figure 13. HAM versus numerical comparison for $F'(\eta)$.

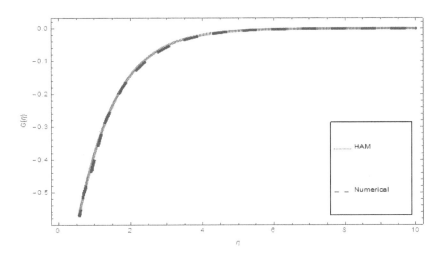

Figure 14. HAM versus numerical comparison for $G(\eta)$.

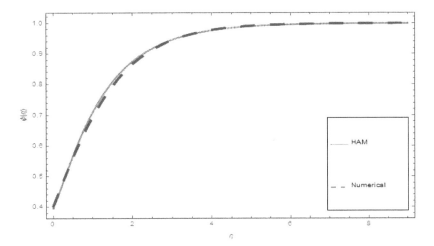

Figure 15. HAM versus numerical comparison for $\phi(\eta)$.

Table 4. Symmetry of HAM versus numerical solutions for $F'(\eta)$, when $Sc = \text{Pr} = K_2 = 1.0$, $\kappa = \kappa_1 = \kappa_2 = \lambda_2 = M = K = 0.1$.

η	HAM $F'(\eta)$	Numerical $F'(\eta)$	Absolute Error AE
0.0	3.33067×10^{-16}	0.000000	3.33067×10^{-16}
0.1	0.0950433	0.093334	0.002876
0.2	0.180827	0.173972	0.009647
0.3	0.258260	0.242791	0.016488
0.4	0.328165	0.300578	0.020276
0.5	0.391281	0.348033	0.019089
0.6	0.448277	0.385777	0.011828
0.7	0.499753	0.414357	0.085396
0.8	0.546251	0.434262	0.111989
0.9	0.588258	0.445923	0.142335
1.0	0.626213	0.449730	0.176483

Table 5. Symmetry of HAM versus numerical solutions for $G(\eta)$, when $Sc = \text{Pr} = K_s = 1.0$, $\kappa = \kappa_1 = \kappa_2 = \lambda_3 = M = K = 0.1$.

η	HAM $G(\eta)$	Numerical $G(\eta)$	Absolute Error AE
0.0	1.000000	1.00000	0.000000
0.1	0.915165	0.887477	0.002876
0.2	0.835725	0.775917	0.009647
0.3	0.761845	0.666195	0.016488
0.4	0.693506	0.559077	0.020276
0.5	0.630563	0.45522	0.019089
0.6	0.572791	0.355165	0.011828
0.7	0.519912	0.259342	0.250570
0.8	0.471618	0.168073	0.303545
0.9	0.427594	0.0815811	0.346013
1.0	0.387518	5.60459×10^{-9}	0.387517

Table 6. Symmetry of HAM versus numerical solutions for $\phi(\eta)$, when $Sc = \text{Pr} = K_s = 1.0$, $\kappa = \kappa_1 = \kappa_2 = \lambda_2 = M = K = 0.1$.

η	HAM $\phi(\eta)$	Numerical $\phi(\eta)$	Absolute Error AE
0.0	0.396762	0.404080	0.007318
0.1	0.429884	0.436425	0.006841
0.2	0.464667	0.468606	0.003939
0.3	0.499830	0.500349	0.000519
0.4	0.534490	0.531418	0.003072
0.5	0.568056	0.561618	0.006438
0.6	0.600146	0.590789	0.009357
0.7	0.630533	0.618810	0.011723
0.8	0.659102	0.645589	0.013513
0.9	0.685811	0.671065	0.014746
1.0	0.710675	0.695202	0.015473

7. Conclusions

In this article the MHD flow of three combined nanofluids (Jefferey, Maxwell, and Oldroyd-B) over a linear stretched surface have been scrutinized. The problem was solved analytically by HAM.

The convergence of HAM has been presented through graphical presentations. The concluding remarks are as follows:

➢   The upsurges in magnetic field diminishes the velocity field.
➢   The upsurges in Prandtl number and thermal relaxation parameters diminish the temperature field.
➢   The upsurges in Schmidt number upsurges the concentration field.
➢   The larger homogeneous reaction and heterogeneous reaction strengths falloff from the concentration field.

Author Contributions: A.S. and Z.S. modeled the problem and wrote the manuscript. S.I., A.D. and P.K. thoroughly checked the mathematical modeling and English corrections. W.K. and A.S. solved the problem using Mathematica software. Z.S., S.I. and P.K. contributed to the results and discussions. All authors finalized the manuscript after its internal evaluation.

Nomenclature

B_0	Magnetic field strength ($\mathrm{NmA^{-1}}$)
Cf_x	Skin friction coefficient
D_I, D_J	Diffusion coefficients
F	Velocity profile
G	Temperature profile
I, J	Chemical species
i, j	Concentration
K	Strength of homogenous reaction
K_s	Strength of heterogeneous reaction
k	Thermal conductivity ($\mathrm{Wm^{-1}K^{-1}}$)
M	Magnetic parameter
Nu_x	Nusselt number
Pr	Prandtl number
q	Heat flux ($\mathrm{Wm^{-2}}$)
Re_x	Local Reynolds number
Sc	Schmidt number
Sh_x	Sherwood number
T	Fluid temperature (K)
T_w	Surface temperature (K)
T_∞	Temperature at infinity (K)
u, v	Velocity components ($\mathrm{ms^{-1}}$)
x, y	Coordinates
α	Thermal diffusivity ($\mathrm{m^2 s^{-1}}$)
η	Similarity variable
μ	Dynamic viscosity (mPa)
v_f	Kinematic viscosity (mPa)
ρ_f	Density ($\mathrm{Kgm^{-3}}$)
λ_1	Relaxation time
λ_2	Relaxation to retardation time
λ_3	Retardation time
ζ	Stretching rate
κ	Deborah number
Ω	Thermal relaxation parameter
σ	Electrical conductivity ($\mathrm{Sm^{-1}}$)
ϕ	Dimensional concentration profile

References

1. Choi, S.U.S. Enhancing thermal conductivity of fluids with nanoparticles. *Int. Mech. Eng. Congr. Expo.* **1995**, *66*, 99–105.

2. Fourier, J.B.J. *TheÂorie Analytique De La Chaleur*; Chez Firmin Didot, père et fils: Paris, France, 1822.

3. Cattaneo, C. Sulla conduzione delcalore. *Atti Semin Mat Fis Univ Modena Reggio Emilia* **1948**, *3*, 83–101.

4. Christov, C.I. On frame indifferent formulation of the Maxwell-Cattaneo model of finite-speed heat conduction. *Mech. Res. Commun.* **2009**, *36*, 481–486. [CrossRef]

5. Mustafa, M. Cattaneo-Christov heat flux model for rotating flow and heat transfer of upper-convected Maxwell fluid. *AIP Adv.* **2015**, *5*, 047109. [CrossRef]

6. Chen, C.H. Effect of viscous dissipation on heat transfer in a non-Newtonian liquid film over an unsteady stretching sheet. *J. Non-Newton. Fluid Mech.* **2005**, *135*, 128–135. [CrossRef]

7. Sheikholeslami, M.; Shah, Z.; Shafi, A.; Khan, I.; Itili, I. Uniform magnetic force impact on water based nanofluid thermal behavior in a porous enclosure with ellipse shaped obstacle. *Sci. Rep.* **2019**, *9*, 1196. [CrossRef] [PubMed]

8. Sheikholeslami, M.; Shah, Z.; Tassaddiq, A.; Shafee, A.; Khan, I. Application of Electric Field for Augmentation of Ferrofluid Heat Transfer in an Enclosure Including Double Moving Walls. *IEEE Access* **2019**, *7*, 21048–21056. [CrossRef]

9. Sheikholeslami, M.; Haq, R.L.; Shafee, A.; Zhixiong, L. Heat transfer behavior of Nanoparticle enhanced PCM solidification through an enclosure with V shaped fins. *Int. J. Heat Mass Transf.* **2019**, *130*, 1322–1342. [CrossRef]

10. Sheikholeslami, M.; Gerdroodbary, M.B.; Moradi, R.; Shafee, A.; Zhixiong, L. Application of Neural Network for estimation of heat transfer treatment of Al_2O_3-H_2O nanofluid through a channel. *Comput. Methods Appl. Mech. Eng.* **2019**, *344*, 1–12. [CrossRef]

11. Sheikholeslami, M. Numerical study of heat transfer enhancement in a pipe filled with porous media by axisymmetric TLB model based on GPU. *Int. J. Heat Mass. Trans.* **2014**, *70*, 1040–1049.

12. Shah, Z.; Islam, S.; Gul, T.; Bonyah, E.; Altaf Khan, M. The Elcerical MHD And Hall Current Impact on Micropolar Nanofluid Flow Between Rotating Parallel Plates. *Results Phys.* **2018**. [CrossRef]

13. Shah, Z.; Bonyah, E.; Islam, S.; Gul, T. Impact of thermal radiation on electrical mhd rotating flow of carbon nanotubes over a stretching sheet. *AIP Adv.* **2019**, *9*, 015115. [CrossRef]

14. Shah, Z.; Tassaddiq, A.; Islam, S.; Alklaibi, A.; Khan, I. Cattaneo–Christov Heat Flux Model for Three-Dimensional Rotating Flow of SWCNT and MWCNT Nanofluid with Darcy–Forchheimer Porous Medium Induced by a Linearly Stretchable Surface. *Symmetry* **2019**, *11*, 331. [CrossRef]

15. Shah, Z.; Bonyah, E.; Islam, S.; Khan, W.; Ishaq., M. Radiative MHD thin film flow of Williamson fluid over an unsteady permeable stretching. *Heliyon* **2018**, *4*, e00825. [CrossRef] [PubMed]

16. Dawar, A.; Shah, Z.; Khan, W.; Islam, S.; Idrees, M. An optimal analysis for Darcy-Forchheimer 3-D Williamson Nanofluid Flow over a stretching surface with convective conditions. *Adv. Mech. Eng.* **2019**, *11*, 1–15. [CrossRef]

17. Shah, Z.; Islam, S.; Ayaz, H.; Khan, S. Radiative Heat and Mass Transfer Analysis of Micropolar Nanofluid Flow of Casson Fluid between Two Rotating Parallel Plates with Effects of Hall Current. *ASME J. Heat Transf.* **2019**. [CrossRef]

18. Maleki, H.; Safaei, M.R.; Alrashed, A.A.A.A.; Kasaeian, A. Flow and heat transfer in non-Newtonian nanofluids over porous surfaces. *J. Anal. Calorim.* **2019**, *135*, 1655. [CrossRef]

19. Nasiri, H.; Abdollahzadeh Jamalabadi, M.Y.; Sadeghi, R.; Safaei, M.R.; Nguyen, T.K.; Shadloo, M.S. A smoothed particle hydrodynamics approach for numerical simulation of nano-fluid flows. *J. Anal. Calorim.* **2019**, *135*, 1733. [CrossRef]

20. Rashidi, M.M.; Nasiri, M.; Shadloo, M.S.; Yang, Z. Entropy Generation in a Circular Tube Heat Exchanger Using Nanofluids: Effects of Different Modeling Approaches. *Heat Transf. Eng.* **2017**, *38*, 853–866. [CrossRef]

21. Safaei, M.R.; Shadloo, M.S.; Goodarzi, M.S.; Hadjadj, A.; Goshayeshi, H.R.; Afrand, M.; Kazi, S.N. A survey on experimental and numerical studies of convection heat transfer of nanofluids inside closed conduits. *Adv. Mech. Eng.* **2016**. [CrossRef]

22. Mahian, O.; Kolsi, L.; Amani, M.; Estelle, P.; Ahmadi, G.; Kleinstreuer, C.; Marshall, J.S.; Siavashi, M.; Taylor, R.A.; Niazmand, H.; et al. Recent advances in modeling and simulation of nanofluid flows-Part I: Fundamentals and theory. *Phys. Rep.* **2019**, *790*, 1–48. [CrossRef]

23. Mahian, O.; Kolsi, L.; Amani, M.; Estelle, P.; Ahmadi, G.; Kleinstreuer, C.; Marshall, J.S.; Taylor, R.A.; Abu-Nada, E.; Rashidi, S.; et al. Recent advances in modeling and simulation of nanofluid flows-Part II: Applications. *Phys. Rep.* **2019**, *791*, 1–59. [CrossRef]

24. Kartini, A.; Zahir, H.; Anuar, I. Mixed convection Jeffrey fluid flow over an exponentially stretching sheet with magnetohydrodynamic effect. *AIP Adv.* **2016**, *6*, 035024.

25. Kartini, A.; Anuar, I. Magnetohydrodynamic Jeffrey fluid over a stretching vertical surface in a porous medium. *Propuls. Power Res.* **2017**, *6*, 269–276.

26. Hayat, T.; Hussain, T.; Shehzad, S.A.; Alsaedi, A. Flow of Oldroyd-B fluid with nano particles and thermal radiation. *Appl. Math. Mech. Engl. Ed.* **2015**, *36*, 69–80. [CrossRef]

27. Raju, C.S.K.; Sandeep, N.; Gnaneswar, R.M. Effect of nonlinear thermal radiation on 3D Jeffrey fluid flow in the presence of homogeneous–heterogeneous reactions. *Int. J. Eng. Res. Afr.* **2016**, *21*, 52–68. [CrossRef]

28. Raju, C.S.K.; Sandeep, N. Heat and mass transfer in MHD non-Newtonian bio-convection flow over a rotating cone/plate withcross diffusion. *J. Mol. Liq.* **2016**, *215*, 115–126. [CrossRef]

29. Alla, A.M.A.; Dahab, S.M.A. Magnetic field and rotation effects on peristaltic transport on Jeffrey fluid in an asymmetric channel. *J. Mag. Magn. Mater.* **2015**, *374*, 680–689. [CrossRef]

30. Khan, A.A.; Ellahi, R.; Usman, M. Effects of variable viscosity on the flow of non-Newtonian fluid through a porous medium in an inclined channel with slip conditions. *J. Porous Media* **2013**, *16*, 59–67. [CrossRef]

31. Ellahi, R.; Bhatti, M.M.; Riaz, A.; Sheikholeslami, M. Effects of magnetohydrodynamics on peristaltic flow of Jeffrey fluid in a rectangular duct through a porous medium. *J. Porous Media* **2014**, *17*, 143–147. [CrossRef]

32. Hayat, T.; Ali, N. A mathematical description of peristaltic hydromagnetic flow in a tube. *Appl. Math. Comput.* **2007**, *188*, 1491–1502. [CrossRef]

33. Hayat, T.; Abbas, Z.; Sajid, M. Series solution for the upper convected Maxwell fluid over a porous stretching plate. *Phys. Lett. A* **2006**, *358*, 396–403. [CrossRef]

34. Raju, C.S.K.; Sandeep, N. Heat and mass transfer in 3D non-Newtonian nano and Ferro fluids over a bidirectional stretching surface. *Int. J. Eng. Res. Afr.* **2016**, *21*, 33–51. [CrossRef]

35. Sandeep, N.; Sulochana, C. Dual solutions for unsteady mixed convection flow of MHD micropolar fluid over a stretching/ shrinking sheet with non-uniform heat source/sink. *Eng. Sci. Technol. Int. J.* **2015**, *18*, 738–745. [CrossRef]

36. Raju, C.S.K.; Sandeep, N.; Sulochana, C.; Sugunamma, V.; Jayachandra, B.M. Radiation, inclined magnetic field and cross diffusion effects on flow over a stretching surface. *J. Niger. Math. Soc.* **2015**, *34*, 169–180. [CrossRef]

37. Nadeem, S.; Akbar, N.S. Influence of haet and mass transfer on a peristaltic motion of a Jeffrey-six constant fluid in an annulus. *Heat Mass Transf.* **2010**, *46*, 485–493. [CrossRef]

38. Makinde, O.D.; Chinyoka, T.; Rundora, L. Unsteady flow of a reactive variable viscosity non-Newtonian fluid through a porous saturated medium with asymmetric convective boundary conditions. *Comput. Math. Appl.* **2011**, *62*, 3343–3352. [CrossRef]

39. Sheikholeslami, M. KKL correlation for simulation of nanofluid flow and heat transfer in a permeable channel. *Phys. Lett. A* **2014**, *378*, 3331–3339. [CrossRef]

40. Shah, Z.; Dawar, A.; Islam, S.; Khan, I.; Ching, D.L.C. Darcy-Forchheimer Flow of Radiative Carbon Nanotubes with Microstructure and Inertial Characteristics in the Rotating Frame. *Case Stud. Therm. Eng.* **2018**, *12*, 823–832. [CrossRef]

41. Chai, L.R.; Shaukat, L.; Wang, H.; Wang, S. A review on heat transfer and hydrodynamic characteristics of nano/microencapsulated phase change slurry (N/MPCS) in mini/microchannel heat sinks. *Appl. Therm. Eng.* **2018**. [CrossRef]

42. Shah, Z.; Dawar, A.; Islam, S.; Khan, I.; Ching, D.L.C.; Khan, A.Z. Cattaneo-Christov model for Electrical Magnetite Micropoler Casson Ferrofluid over a stretching/shrinking sheet using effective thermal conductivity model. *Case Stud. Therm. Eng.* **2018**. [CrossRef]

43. Dawar, A.; Shah, Z.; Islam, S.; Idress, M.; Khan, W. Magnetohydrodynamic CNTs Casson Nanofluid and Radiative heat transfer in a Rotating Channels. *J. Phys. Res. Appl.* **2018**, *1*, 017–032.

44. Khan, A.S.; Nie, Y.; Shah, Z.; Dawar, A.; Khan, W.; Islam, S. Three-Dimensional Nanofluid Flow with Heat and Mass Transfer Analysis over a Linear Stretching Surface with Convective Boundary Conditions. *Appl. Sci.* **2018**, *8*, 2244. [CrossRef]

45. Imtiaz, M.; Hayat, T.; Alsaedi, A.; Hobiny, A. Homogeneous-heterogeneous reactions in MHD flow due to an unsteady curved stretching surface. *J. Mol. Liq.* **2016**, *221*, 245–253. [CrossRef]

46. Hayat, T.; Hussain, Z.; Muhammad, T.; Alsaedi, A. Effects of homogeneous and heterogeneous reactions in flow of nanofluids over a nonlinear stretching surface with variable surface thickness. *J. Mol. Liq.* **2016**, *221*, 1121–1127. [CrossRef]

47. Nasir, S.; Shah, Z.; Islam, S.; Khan, W.; Khan, S.N. Radiative flow of magneto hydrodynamics single-walled carbon nanotube over a convectively heated stretchable rotating disk with velocity slip effect. *Adv. Mech. Eng.* **2019**, *11*, 1–11.

48. Nasir, S.; Shah, Z.; Islam, S.; Khan, W.; Bonyah, E.; Ayaz, M.; Khan, A. Three dimensional Darcy-Forchheimer radiated flow of single and multiwall carbon nanotubes over a rotating stretchable disk with convective heat generation and absorption. *AIP Adv.* **2019**, *9*, 035031. [CrossRef]

49. Hammed, K.; Haneef, M.; Shah, Z.; Islam, I.; Khan, W.; Asif, S.M. The Combined Magneto hydrodynamic and electric field effect on an unsteady Maxwell nanofluid Flow over a Stretching Surface under the Influence of Variable Heat and Thermal Radiation. *Appl. Sci.* **2018**, *8*, 160. [CrossRef]

50. Ullah, A.; Alzahrani, E.O.; Shah, Z.; Ayaz, M.; Islam, S. Nanofluids Thin Film Flow of Reiner-Philippoff Fluid over an Unstable Stretching Surface with Brownian Motion and Thermophoresis Effects. *Coatings* **2019**, *9*, 21. [CrossRef]

5

MHD Boundary Layer Flow of Carreau Fluid over a Convectively Heated Bidirectional Sheet with Non-Fourier Heat Flux and Variable Thermal Conductivity

Dianchen Lu [1], Mutaz Mohammad [2], Muhammad Ramzan [3,4,*], Muhammad Bilal [5], Fares Howari [6] and Muhammad Suleman [1,7]

[1] Department of Mathematics, Faculty of Science, Jiangsu University, Zhenjiang 212013, China; dclu@ujs.edu.cn (D.L.); suleman@ujs.edu.cn (M.S.)
[2] Department of Mathematics & Statistics, College of Natural and Health Sciences, Zayed University, 144543 Abu Dhabi, UAE; Mutaz.Mohammad@zu.ac.ae
[3] Department of Computer Science, Bahria University, Islamabad Campus, Islamabad 44000, Pakistan
[4] Department of Mechanical Engineering, Sejong University, Seoul 143-747, Korea
[5] Department of Mathematics, University of Lahore, Chenab Campus, Gujrat 50700, Pakistan; me.bilal.786@outlook.com
[6] College of Natural and Health Sciences, Zayed University, 144543 Abu Dhabi, UAE; Fares.Howari@zu.ac.ae
[7] Department of Mathematics, COMSATS University, Islamabad 45550, Pakistan
* Correspondence: mramzan@bahria.edu.pk

Abstract: In the present exploration, instead of the more customary parabolic Fourier law, we have adopted the hyperbolic Cattaneo–Christov (C–C) heat flux model to jump over the major hurdle of "parabolic energy equation". The more realistic three-dimensional Carreau fluid flow analysis is conducted in attendance of temperature-dependent thermal conductivity. The other salient impacts affecting the considered model are the homogeneous-heterogeneous (h-h) reactions and magnetohydrodynamic (MHD). The boundary conditions supporting the problem are convective heat and of h-h reactions. The considered boundary layer problem is addressed via similarity transformations to obtain the system of coupled differential equations. The numerical solutions are attained by undertaking the MATLAB built-in function bvp4c. To comprehend the consequences of assorted parameters on involved distributions, different graphs are plotted and are accompanied by requisite discussions in the light of their physical significance. To substantiate the presented results, a comparison to the already conducted problem is also given. It is envisaged that there is a close correlation between the two results. This shows that dependable results are being submitted. It is noticed that h-h reactions depict an opposite behavior versus concentration profile. Moreover, the temperature of the fluid augments for higher values of thermal conductivity parameters.

Keywords: Carreau fluid; Cattaneo–Christov heat flux model; convective heat boundary condition; temperature dependent thermal conductivity; homogeneous-heterogeneous reactions

1. Introduction

Non-Newtonian fluids have gained substantial attention of researchers and scientists owing to their widespread applications. A number of examples like apple sauce, chyme, emulsions, mud, soaps, shampoos and blood at low shear stress may be quoted as non-Newtonian fluids. Existing literature does not facilitate us to identify a single relation that exhibits numerous physiognomies of non-Newtonian fluids. This is why a variety of mathematical models have been suggested, as deemed appropriate, to the requirement. The viscosity of the fluid plays a vital role in the chemical engineering industry. In case of generalized Newtonian fluids, viscosity is dependent on shear stress. In some fluids, a change up to two to three orders in magnitude may not make a visible effect in some fluids, but its impact can't be ignored particularly in polymer industry and lubrication processes. Bird et al. [1] presented the idea of generalized Newtonian fluids with the idea that the viscosity fluctuates with the shear rate. Fluid flows over a solid surface have been frequently studied and the reports revealed that surface forces become significant on a micro level and lead to the enhanced fluid viscosity due to fluid layering [2–5]. The major shortcoming of the Power-law model is that it does not properly address the viscosity in case of very low or high shear rates. To overcome this hurdle, the Carreau fluid model is introduced [6]. Contrary to the Power law model, the viscosity remains finite as the shear rate vanishes. This is why the constitutive relation for the Carreau fluid model is more appropriate in case of free surface flows. Owing to such important characteristics, the Carreau fluid model has attracted researchers for many years. Chhabra and Uhlherr [7] deliberated the Carreau fluid flow over the spheres and this concept was extended by Bush and Phan-Thein [8]. The squeezing Carreau fluid flow past sphere is examined by Uddin et al. [9]. Tshehla [10] deliberated the flow of the Carreau fluid over an inclined plane. The nonlinear radiation impact on the 3D Carreau fluid flow was deliberated by Khan et al. [11]. Khan et al. [12] obtained the solution of the Carreau nanofluid flow analytically with entropy generation. Similar explorations discussing Carreau fluid flow may be found at [13–15] and many therein.

Flows under the influence of magnetohydrodynamics (MHD) have a wide range of applications including thermal insulators, blood flow measurements, petroleum and polymer technologies, nuclear reactors and MHD generators. Taking into account all such applications, many researchers have examined the flows stimulated by magnetohydrodynamics. Waqas et al. [16] conversed micropolar fluid's flow with the effect of convective boundary condition and magnetohydrodynamics. He also considered effects of viscous dissipation and mixed convection. Ramzan et al. [17] deliberated the series solution of micropolar fluid flow in attendance of MHD, partial slip and convective boundary condition over a porous stretching sheet. Besthapu et al. [18] calculated the numerical solution of double stratification nanofluid flow with MHD and viscous dissipation using the finite element method past an exponentially stretching sheet. Khan and Azam [19] explored the flow of Carreau nanofluid under the influence of magnetohydrodynamic using a numerical technique named bvp4c. Turkyilmazoglu [20] examined the exact solution of micropolar fluid flow with the mixed convection and magnetohydrodynamic past a permeable heated/cooled deformable plate. Hayat et al. [21] premeditated the flow of Oldroyd-B nanofluid in attendance of MHD and heat generation/absorption using Optimal Homotopy analysis method HAM. Some recent attempts highlighting effects of magnetohydrodynamic may be found at [22–25].

The importance of heat transfer is fundamental in many engineering processes like nuclear reactors, fuel cells, and microelectronics. Thermal conductivity is considered to be constant in all such procedures. Nevertheless, the requirement of variable characteristics is fundamental. A variation from $0°F$ to $400°F$ [26] in temperature is observed in such cases. Fourier law of heat conduction [27] has been a customary gauge for years in heat transfer applications. However, a major drawback of parabolic energy equation experiences a disruption in the beginning which prevails throughout the entire process, which forces the researchers to look for some modification to Fourier's law. Cattaneo [28] proposed an improved Fourier's law by instituting a thermal relaxation term. Later, Oldroyd's upper-convected derivatives [29] are considered as an alternative to the thermal relaxation time in Cattaneo's model. Recent attempts in various scenarios with an emphasis on C–C flux model may encompass a study by Ramzan et al. [30], highlighting effects of the 2D third grade-fluid flow accompanying Cattaneo–Christov heat flux and magnetohydrodynamics. Flow analysis is done in the presence of h-h reactions and convective boundary condition. Hayat et al. [31] found an analytical solution of Jeffrey fluid flow past a stretched cylinder with the effect of C–C heat flux and thermal stratification. Sui et al. [32] studied upper-convected Maxwell nanofluid flow with C–C heat flux and slip boundary condition past a linearly stretched sheet. Liu et al. [33] discussed a fractional C–C flux model numerically where the fractional derivative is represented by a weight coefficient.

Many chemical reactions necessitate the presence of h-h reactions. Fewer of these reactions act at a slow pace, whereas some absolutely not except in the attendance of the catalyst. These reactions are involved in many scenarios like fibrous insulations, production of polymers and ceramics, and air and water pollution. The heterogeneous reactions cover the complete phase evenly and is found in solid phase. Nevertheless, the homogeneous reactions are covered by catalysis and combustion. The homogeneous catalyst occurs in liquid and gaseous states but heterogeneous catalyst exists in solid form. The latest research discussing the h-h reactions effects may comprise the study by Kumar et al. [34] who examined the irreversibility process with h-h reactions of the flow of carbon nanotubes based nanofluid past a bi-directional stretched surface. The flow with h-h reactions of Blasius nanofluid is pondered by Xu [35]. Sithole et al. [36] used a Bivariate spectral local linearization method to investigate the effects of h-h reactions on the flow of time dependent micropolar nanofluid past a stretched surface. The numerical simulations are conducted for h-h reactions and nonlinear thermal radiation past a 3D crossfluid flow with MHD by Khan et al. [37]. In a gravity driven nanofluid film flow, the effects of h-h reactions with mixed convection are deliberated by Rasees et al. [38]. Ramzan et al. [39,40] highlighted the time dependent nanofluid squeezing flow with carbon nanotubes under the influence of h-h reactions and C–C heat flux, and in Micropolar nanofluid flow with thermal radiation past a nonlinear stretched surface and many therein [40–45].

In all the aforementioned literature surveys, it is observed that either the effect of only C–C heat flux or h-h reactions have been discussed in various geometries. Even if the simultaneous effects of C–C heat flux and h-h reactions have been discussed, it is in the two-dimensional case. However, much less literature is available featuring effects of both C–C and h-h reactions in 3D models. The present study discusses the 3D Carreau fluid model in attendance of temperature-dependent thermal conductivity, C–C heat flux and h-h reactions accompanied by the impact of convective heat with h-h boundary conditions. A MATLAB built-in bvp4c routine is betrothed to obtain series solutions. Graphs are drawn depicting effects of pertinent parameters on involved distributions. Validation of presented results in the limiting case is also an additional feature of this exploration.

2. Mathematical Formulation

Let us presume a 3D flow of Carreau fluid in x- and y-directions with respective velocities $u = u_w(x) = cx$ and $v = v_w(y) = dy$ occupying the region $z = 0$ under the influence of C–C heat flux and variable thermal conductivity past a bidirectional stretching surface as shown in Figure 1. Flow analysis is performed subject to h-h reactions with magnetohydrodynamics. Temperature at the surface T_w is considered to be more than the temperature away from the surface T_∞. A magnetic field with strength B_o is introduced along the z-axis. Electric and Hall effects are ignored. Small Reynolds number's assumption needs to omit an induced magnetic field. For two chemical species A and B, analysis is performed in the presence of h-h reactions. For homogeneous reaction, the cubic autocatalysis is epitomized by the following expression [46]:

$$A + 2B \rightarrow 3B, rate = k_c ab^2. \tag{1}$$

However, on the catalyst surface, the first order isothermal reaction is given by:

$$A \rightarrow B, rate = k_s a. \tag{2}$$

For both the h-h reaction processes, it is assumed that temperature is constant. Governing equations that abide by the above mentioned assumptions are given below:

$$u_x + v_y + w_z = 0, \tag{3}$$

$$uu_x + vu_y + wu_z = vu_{zz} \left[\beta^* + (1 - \beta^*) \left\{ 1 + \Gamma^2 (u_z)^2 \right\}^{\frac{n-1}{2}} \right] - \frac{\sigma B_o^2}{\rho} u$$
$$+v (n-1) (1 - \beta^*) \Gamma^2 (u_{zz}) (u_z)^2 \left\{ 1 + \Gamma^2 (u_z)^2 \right\}^{\frac{n-3}{2}}, \tag{4}$$

$$uv_x + vv_y + wv_z = vv_{zz} \left[\beta^* + (1 - \beta^*) \left\{ 1 + \Gamma^2 (v_z)^2 \right\}^{\frac{n-1}{2}} \right] - \frac{\sigma B_o^2}{\rho} v$$
$$+v (n-1) (1 - \beta^*) \Gamma^2 (v_{zz}) (v_z)^2 \left\{ 1 + \Gamma^2 (v_z)^2 \right\}^{\frac{n-3}{2}}, \tag{5}$$

$$\rho C_P \mathbf{V}.\nabla T = -\nabla.\mathbf{q}, \tag{6}$$

$$ua_x + va_y + wa_z = D_A a_{zz} - k_c ab^2, \tag{7}$$

$$ub_x + vb_y + wb_z = D_B b_{zz} + k_c ab^2, \tag{8}$$

with **q** being the heat flux satisfying the relation

$$\mathbf{q} + K_1 \left(\mathbf{q}_t + \mathbf{V}.\nabla \mathbf{q} - \mathbf{q}.\nabla \mathbf{V} + (\nabla.\mathbf{V}) \mathbf{q} \right) = -\nabla \left(\alpha T \right). \tag{9}$$

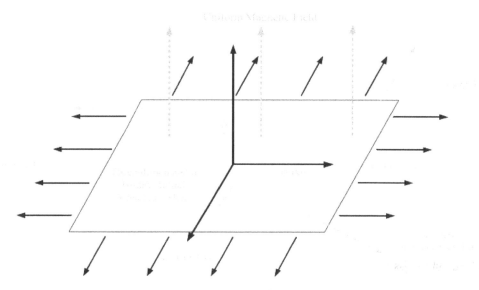

Figure 1. Geometry of the problem.

Using the fluid's incompressibility condition and Christov [29], Equations (6) and (9) take the following form after omission of **q**:

$$uT_x + vT_y + +wT_z = \frac{1}{\rho C_P} (\alpha T_z)_z,$$

$$- K_1 \left(\begin{array}{c} u^2 T_{xx} + v^2 T_{yy} + w^2 T_{zz} + 2uv T_{xy} \\ +2vw T_{yz} + 2uw T_{xz} + (uu_x + vu_y + wu_z) T_x + \\ (uv_x + vv_y + wv_z) T_y + (uw_x + vw_y + ww_z) T_z \end{array} \right). \tag{10}$$

The supporting boundary conditions to the given model are

$$u = u_w (x) = cx, \ v = v_w (y) = dy, \ w = 0,$$

$$-k_h T_z = h_f (T_w - T), \ D_A a_z = k_s a, \ D_B b_z = -k_s a, \ at z = 0,$$

$$u \to 0, \ v \to 0, \ a \to a_o, \ b \to 0, T \to T_\infty as z \to \infty, \tag{11}$$

considering temperature dependent thermal conductivity $\epsilon = \frac{k_w - k_\infty}{k_\infty}$ as defined in [47].

Taking into account the transformations

$$u = cxf'(\eta), \; v = cyg'(\eta), \; w = -\sqrt{cv}(f(\eta) + g(\eta)), \; \alpha = \alpha_\infty(1 + \epsilon\theta),$$
$$\theta(\eta) = \frac{T - T_\infty}{T_w - T_\infty}, \; \eta = \sqrt{\frac{c}{v}}z, \; a = a_0\phi(\eta), \; b = a_0h(\eta). \tag{12}$$

The requirement of Equation (3) is met inevitably, whereas Equations (4), (5), (7), (8), (10) and (11) take the following form:

$$\left[\beta^* + (1 - \beta^*)\left\{1 + We_1^2(f'')^2\right\}^{\frac{n-3}{2}}\left\{1 + nWe_1^2(f'')^2\right\}\right]f''' + (f + g)f'' - f'^2 - M^2f' = 0, \tag{13}$$

$$\left[\beta^* + (1 - \beta^*)\left\{1 + We_2^2(g'')^2\right\}^{\frac{n-3}{2}}\left\{1 + nWe_2^2(g'')^2\right\}\right]g''' + (f + g)g'' - g'^2 - M^2g' = 0, \tag{14}$$

$$(1 + \epsilon\theta)\theta'' + \epsilon\theta'^2 + Pr(f + g)\theta' - Pr\lambda_1\left((f + g)^2\theta'' + (f + g)(f' + g')\theta'\right) = 0, \tag{15}$$

$$\phi'' + Sc(f + g)\phi' - Sc\gamma_1\phi h^2 = 0, \tag{16}$$

$$\zeta h'' + Sc(f + g)h' + Sc\gamma_1\phi h^2 = 0, \tag{17}$$

$$f(0) = 0, f'(0) = 1, \; g(0) = 0, g'(0) = \lambda,$$
$$\phi'(0) = \gamma_2\,\phi(0), \theta'(0) = -\delta(1 - \theta(0)),$$
$$\zeta h'(0) = -\gamma_2\,\phi(0), \; f'(\infty) \to 0, \; f''(\infty) \to 0, \; g'(\infty) \to 0, \tag{18}$$
$$g''(\infty) \to 0, \; \theta(\infty) \to 0, \; \phi(\infty) \to 1.$$

Different parameters used in the above equations are defined as follows:

$$\gamma_1 = \frac{k_c a_0^2}{c}, \; \gamma_2 = \frac{k}{D_A a_0}\sqrt{\frac{v}{c}}, \; Sc = \frac{v}{D_A}, \; Pr = \frac{\mu C_p}{k}, \; \lambda = \frac{d}{c},$$
$$We_1 = \sqrt{\frac{c\Gamma^2 U_w^2}{v}}, We_2 = \sqrt{\frac{c\Gamma^2 V_w^2}{v}}, \; \lambda_1 = K_1 c, \; \zeta = \frac{D_B}{D_A}, \; \delta = \frac{h_f}{k}\sqrt{\frac{v}{c}}, M^2 = \frac{\sigma B_0^2}{c\rho}. \tag{19}$$

The expectation as in most applications that coefficients of chemical species A and B are of equivalent magnitude lead us to make a supplementary presumption that diffusion coefficients D_A and D_B are equivalent i.e., $\zeta = 1$, [46]. Thus, we get:

$$\phi(\eta) + h(\eta) = 1. \tag{20}$$

Now, Equations (16) and (17) take the following form:

$$\phi'' + Sc(f + g)\phi' - Sc\gamma_1\phi(1 - \phi)^2 = 0, \tag{21}$$

with boundary conditions

$$\phi'(0) = \gamma_2\phi(0), \; \phi(\infty) = 1. \tag{22}$$

The skin friction coefficient in dimensional form is

$$C_{fx} = \frac{\tau_{xz}}{\rho U_w^2(x)}, C_{fy} = \frac{\tau_{yz}}{\rho u_w^2(y)}. \tag{23}$$

Dimensionless forms of Skin friction coefficient is

$$C_{fx}Re_x^{1/2} = f''(0)\left[\beta^* + (1 - \beta^*)\left\{1 + We_1^2(f''(0))^2\right\}^{\frac{n-3}{2}}\right], \tag{24}$$

$$C_{fy}Re_x^{1/2} = g''(0)\left[\beta^* + (1 - \beta^*)\left\{1 + We_2^2(g''(0))^2\right\}^{\frac{n-3}{2}}\right]. \tag{25}$$

3. Numerical Solutions

The software MATLAB with built-in bvp4c function is engaged to solve the system of differential equations. It requires converting the differential equation with higher order to the first order along with their respective boundary conditions. We have considered f as y_1, g as y_4, θ as y_7 and ϕ as y_9 during the conversion as

$$
\begin{aligned}
y_1' &= y_2, \quad y_2' = y_3, \\
y_3' &= \frac{y_2^2 + My_2 - (y_1 + y_4)\,y_3}{\beta^* + (1-\beta^*)\left(1 + We_1^2 y_3^2\right)^{\frac{n-3}{2}}\left(1 + nWe_1^2 y_3^2\right)}, \\
y_4' &= y_5, \quad y_5' = y_6, \\
y_6' &= \frac{y_5^2 + My_5 - (y_1 + y_4)\,y_6}{\beta^* + (1-\beta^*)\left(1 + We_2^2 y_6^2\right)^{\frac{n-3}{2}}\left(1 + nWe_2^2 y_6^2\right)}, \\
y_7' &= y_8, \\
y_8' &= \frac{\Pr K_2 (y_1 + y_4)(y_2 + y_5)y_8 - \Pr(y_1 + y_4)y_8}{1 + \epsilon y_7 - \Pr K_2 (y_1 + y_4)^2}, \\
y_9' &= y_{10}, \\
y_{10}' &= Sc\gamma_1 y_9 (1 - y_9)^2 - Sc(y_1 + y_4)y_{10},
\end{aligned}
$$

accompanying the conditions

$$
\begin{aligned}
y_1(0) &= 0, y_2(0) = 1, y_4(0) = 0, y_5(0) = \lambda, y_2(\infty) = 0, y_5(\infty) = 0, \\
y_8(0) &= -\delta(1 - y_7(0)), y_7(\infty) = 0, y_{10}(0) = \gamma_2 y_9(0), y_9(\infty) = 1.
\end{aligned}
$$

This MATLAB built-in routine is verified by drawing Table 1, in which the results are compared with the previously published article in a limiting case. Previously, Khan et al. [11] have used the same bvp4c technique to tackle the 3D Carreau fluid model. In Table 1, the Skin friction coefficients for varied values of λ is calculated. It is found that all obtained values are in total alignment to [11].

Table 1. Comparison of $-f''(0)$ varied estimates of λ when $n = 3, We_1 = We_2 = 0$.

λ	Khan et al. [11]	Present (bvp4c)
0.1	1.020264	1.020264
0.2	1.039497	1.039497
0.3	1.057956	1.057956
0.4	1.075788	1.075788
0.5	1.093095	1.093095
0.6	1.109946	1.109946
0.7	1.126397	1.126397
0.8	1.142488	1.142488
0.9	1.158253	1.158253
1.0	1.173720	1.173720

In Table 2, a comparison is tabulated for various magnetic parameters and stretching ratio parameter values against the skin friction coefficient along vertical and horizontal directions. It is noted that the skin friction along the x-direction is gradually increasing for the mounting values of λ.

Table 2. Comparison of $-f''(0)$ and $-g''(0)$ for various values of M and λ.

M	$\lambda = 0$		$\lambda = 0.5$		$\lambda = 0.5$		$\lambda = 1.0$	
	$-f''(0)$		$-f''(0)$		$-g''(0)$		$-g''(0)$	
	[48]	Present	[48]	Present	[48]	Present	[48]	Present
0.0	1.0042	1.0045	1.0932	1.0930	0.4653	0.4652	1.1748	1.1742
10	3.3165	3.3149	3.3420	3.3137	1.6459	1.6440	3.3667	3.3654
100	10.0498	10.0427	10.0582	10.0531	5.0208	5.0201	10.0663	10.0654

4. Results and Discussion

This segment is dedicated to highlight the impacts of prominent parameters on all involved profiles. In all the figures, the solid lines show the effect of shear thickening ($n > 1$) fluid while the dashed lines show the shear thinning ($n < 1$) fluid properties. Figures 2 and 3 are illustrated to distinguish the impact of local Weissenberg numbers We_1 and We_2 on the velocity components $f'(\eta)$ and $g'(\eta)$ used for shear thickening and shear thinning fluids respectively. From these figures, it is noted that, for the augmented estimates of We_1, the velocity declines in the case of shear thickening phenomena, while, for the shear thinning phenomena, the velocity increases. Physically, We_1 denotes the proportion between the relaxation time of fluid and increment of viscosity growth of the liquid. For the shear thinning case, the fluid viscosity decreases; consequently, the velocity of the fluid increases. Moreover, for the shear thickening phenomenon, the thickness of the boundary layer escalates for higher values of We_1. In Figure 3, we observed the contradictory behavior for the velocity component $g'(\eta)$. Consequently, it is also found that shear thickening fluid increases the values of We_2, which results in increasing the velocity of fluid and thickness of its related boundary layer. In Figure 4, the effect of viscosity ratio parameter β^* on the velocity is profile $f'(\eta)$ is discussed for the case of shear thinning and shear thickening and keeping all other parameters fixed. An inverse relation is observed, in the case of the shear thinning fluid and for shear thickening fluid velocity of the fluid augmented with escalating values of the viscosity ratio parameter. Moreover, it has been observed that the corresponding boundary layer thickness is less in the case of shear thinning fluid as compared to the shear thickening fluid. Figures 5 and 6 depict declines in velocity profile against the mounting values of magnetic field strength M. A decline in velocity profile is being observed because of the fact that larger values of M enhance the Lorentz force, which increases the resistance for the fluid motion. This decrease in the thickness of the boundary layer is more vigorous for the shear thinning of fluids. Figures 7 and 8 exhibit the effect of stretching ratio parameter λ on $f'(\eta)$ and $g'(\eta)$ velocity profiles, respectively. In Figure 7, the mounting values of stretching ratio parameter resist the fluid flow along the x-axis and this decline is more prominent in shear thinning fluid. Figure 8 shows the opposite trend for the large values of λ on the velocity profile, as the velocity increases for both shear thinning and thickening of fluids. Stretching ratio parameter is the ratio of velocity components along the y-axis to the x-axis. An increase in λ implies the increment in the y-component of the velocity. The effect of Prandtl number Pr on temperature field is shown in Figure 9. Temperature profile decreases for higher values of Pr. The Prandtl number represents the fraction of momentum diffusivity to thermal diffusivity. Thus, an increase in Pr deteriorates the thermal conductivity; ultimately, it decreases the temperature distribution. The effect of thermal conductivity parameter ϵ on the temperature field is being displayed in Figure 10. It is observed that an increase in values of ϵ boosts the temperature distribution. It is an accepted truth that liquids with larger thermal conductivity possess higher temperature. The impact of Schmidt number Sc on concentration profile is being displayed in Figure 11. The augmented values of Sc number boosts the concentration profile and thickness of boundary layer for both the thinning and shear thickening fluids, respectively. The Schmidt number represents the ratio of the molecular diffusion to the viscous diffusion, and the viscous diffusion decreases upon increasing the Sc, which enhances the mass transfer in fluid flow. In Figure 12, the thermal relaxation time parameter λ_1 is portrayed against the temperature profile. It is witnessed that, for the increasing values of λ_1, the temperature profile and thickness of the thermal boundary layer decrease. Figure 13 portrays the effect of Biot number δ on temperature profile. It is noticed that larger values of Biot number escalate the temperature field. A direct relation of heat transfer coefficient with Biot number implies an increase in temperature profile for increasing values of δ. The strength of homogeneous and heterogeneous reactions γ_1 and γ_2 against concentration profile is shown in Figures 14 and 15 as reactants expend in homogeneous reactions. Thus, a reduction in concentration profile is seen for mounting values of γ_1. This fact is shown in Figure 14. An opposite behavior for concentration distribution is observed in Figure 15. Escalating values of heterogeneous reactions decrease diffusion and thereby decrement in concentration is perceived for less diffused particles.

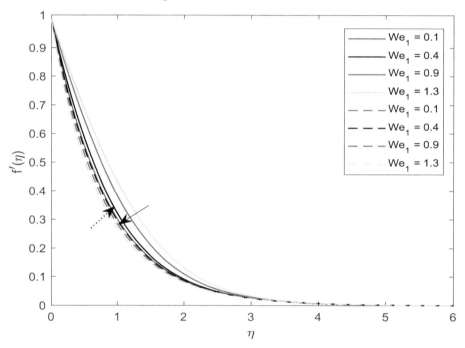

Figure 2. Illustration of We_1 versus $f'(\eta)$.

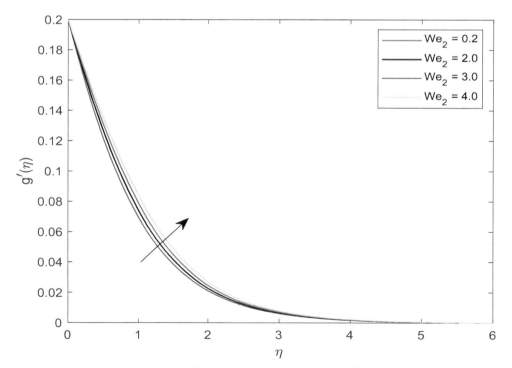

Figure 3. Illustration of We_2 versus $g'(\eta)$.

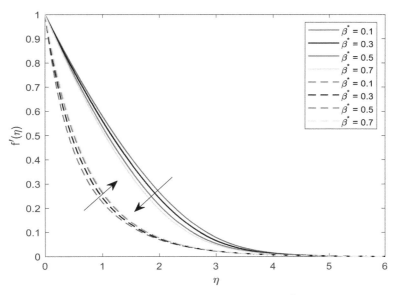

Figure 4. Illustration of β^* versus $f'(\eta)$.

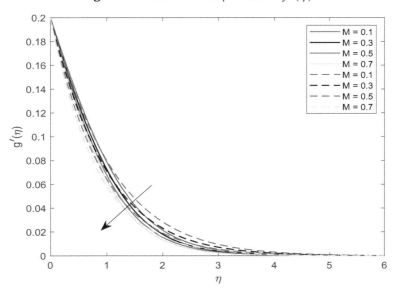

Figure 5. Illustration of M versus $g'(\eta)$.

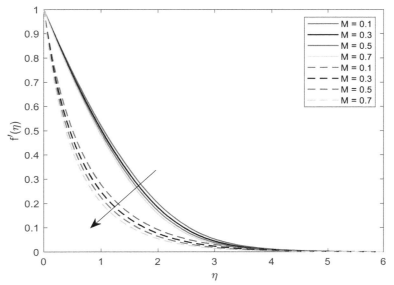

Figure 6. Illustration of M versus $f'(\eta)$.

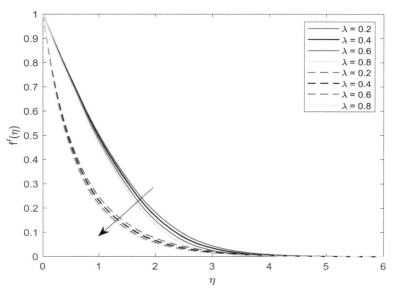

Figure 7. Illustration of λ versus $f'(\eta)$.

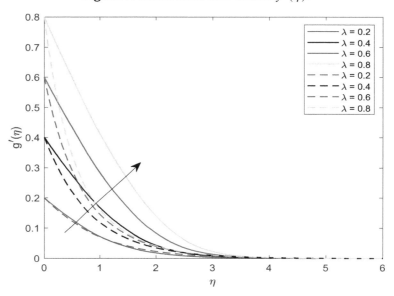

Figure 8. Influence of λ on $g'(\eta)$.

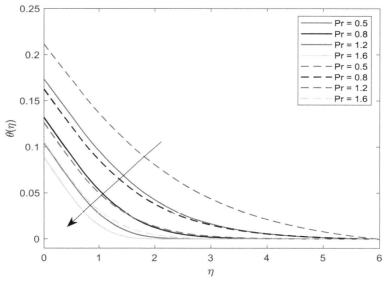

Figure 9. Illustration of Pr versus $\theta(\eta)$.

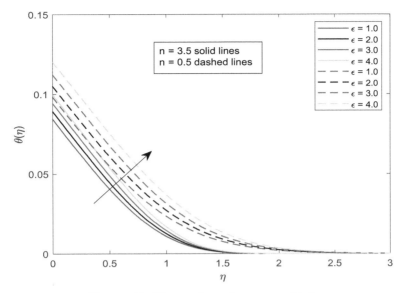

Figure 10. Illustration of ϵ versus $\theta(\eta)$.

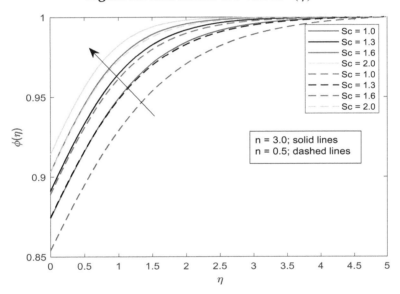

Figure 11. Illustration of Sc versus $\phi(\eta)$.

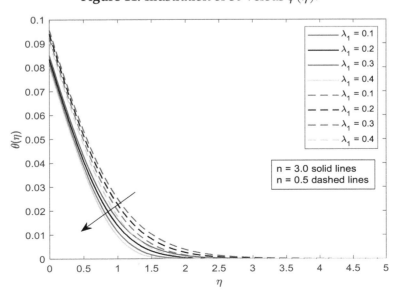

Figure 12. Illustration of λ_1 versus $\theta(\eta)$.

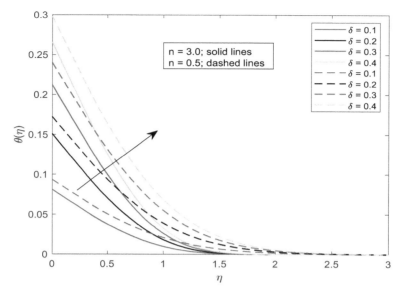

Figure 13. Illustration of δ versus $\theta(\eta)$.

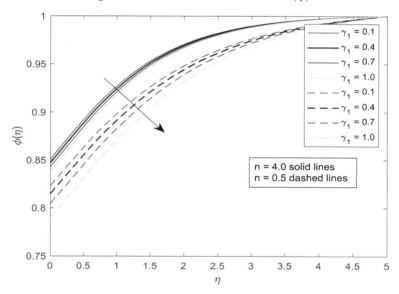

Figure 14. Illustration of γ_1 versus $\phi(\eta)$.

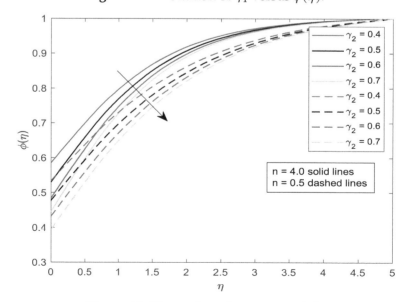

Figure 15. Illustration of γ_2 versus $\phi(\eta)$.

5. Conclusions

In this exploration, impacts of h-h reactions on three-dimensional Carreau fluid flow is witnessed with the presence of temperature dependent thermal conductivity and magneto-hydrodynamic past a bidirectional stretched surface. Furthermore, the impact of C–C heat flux accompanying convective boundary condition is also witnessed. A numerical method is betrothed to find the solution. The notable features of the present study are appended below:

- Strength of homogeneous and heterogeneous reactions show the same decreasing trend on concentration distribution.
- Effects of Prandtl number and Biot number on temperature field are also conflicting.
- The velocity of the fluid is in decline for a stronger magnetic effect.
- Velocity escalates for growing estimates of ratios of stretching rate.
- With an increase in the value of the Schmidt number, the concentration of the fluid is enhanced.

Author Contributions: Formal Analysis, M.R.; Funding Acquisition, M.M., F.H.; Investigation, M.R.; Methodology, D.L.; Project Administration, M.R., M.B.; Software, M.M., M.S.; Validation, M.B.; Writing—Original Draft, M.B.; Writing—Review and Editing, M.S.

Acknowledgments: The authors are highly thankful for exceptional support raised by the Zayed University, Abu Dhabi, UAE, Jiangsu University, China, and Bahria University, Islamabad, Pakistan.

Abbreviations

a, b	concentrations of chemical species
A, B	chemical species
a_0	positive dimensional constants
B_0	Magnetic field strength [kg s^{-2} A^{-1}]
C_p	Specific heat [J/kg K]
c, d	stretching constants
Cf_x	Skin friction coefficient
D_A	diffusion coefficient of species A
D_B	diffusion coefficient of species B
f', g'	Dimensionless velocities
h_f	Heat transfer coefficient
h	dimensionless concentration due to heterogeneous reaction
K_1	thermal relaxation time
k_∞	ambient thermal conductivity
k_c	rate constant of chemical species A
k_s	rate constant of chemical species B
k_w	thermal conductivity at wall
M	Magnetic parameter
n	power law index
Pr	Prandtl number
$\mathbf{q}$	heat flux
Sc	Schmidth number
t	time
T_∞	Ambient temperature [K]
T	Temperature of fluid [K]
T_w	Wall temperature [K]
u_w	sheet velocity along $x-$axis [m/s]
v_w	sheet velocity along y-axis [m/s]

$\mathbf{V}$	Velocity vector
(u, v, w)	Velocity components [m/s]
$u_w(x)$	Stretching velocity along x-axis [m/s]
(x, y, z)	Rectangular coordinate axis [m]
We_1	Weissenberg number
We_2	Weissenberg number
α	variable thermal diffusivity
β^*	ratio of viscosities
γ_1	Thermal Biot number
γ_2	Concentration Biot number
λ	ratio of stretching rates
λ_1	thermal relaxation time coefficient
ν	Kinematic viscosity [m^2/s]
θ	Dimensionless temperature
σ	Electrical conductivity [m^{-3} kg^{-1} s^3 A^2]
μ	Dynamic viscosity [kg/m/s]
η	Similarity variable
ρ	Density of fluid [kg/m^3]
δ	Deborah number
ϕ	dimensionless concentration
ξ	ratio of diffusion coefficients
∇	nibla operator
Γ	material parameter
ϵ	variable thermal conductivity

References

1. Bird, R.; Byron, R.; Armstrong, R.C.; Hassager, O. *Dynamics of Polymeric Liquids. Vol. 1: Fluid Mechanics*; Wiley: London, UK, 1987.
2. Quoc, V.T.; Kim B. Transport phenomena of water in molecular fluidic channels. *Sci. Rep.* **2016**, *6*, 33881.
3. Quoc, V.T.; Park, B.; Park, C.; Kim, B. Nano-scale liquid film sheared between strong wetting surfaces: Effects of interface region on the flow. *J. Mech. Sci. Technol.* **2015**, *29*, 1681–1688.
4. Ghorbanian, J.; Beskok, A. Scale effects in nanochannel liquid Flows. *Microfluid. Nanofluid.* **2016**, *20*, 121. [CrossRef]
5. Ghorbanian, J.; Celebi, A.T.; Beskok, A. A phenomenological continuum model for force-driven nano-channel liquid flows. *J. Chem. Phys.* **2016**, *145*, 184109. [CrossRef]
6. Carreau, P.J. Rheological equations from molecular network Theories. *Trans. Soc. Rheol.* **1972**, *16*, 99–127. [CrossRef]
7. Chhabra, R.P.; Uhlherr P.H.T. Creeping motion of spheres through shear-thinning elastic fluids described by the Carreau viscosity equation. *Rheol. Acta* **1980**, *19*, 187–195. [CrossRef]
8. Bush, M.B.; Phan-Thien, N. Drag force on a sphere in creeping motion throug a Carreau model fluid. *J. Non-Newton. Fluid Mech.* **1984**, *16*, 303–313. [CrossRef]
9. Uddin, J.; Marston, J.O.; Thoroddsen, S.T. Squeeze flow of a Carreau fluid during sphere impact. *Phys. Fluids* **2012**, *24*, 073104. [CrossRef]
10. Tshehla, M.S. The flow of a Carreau fluid down an incline with a free surface. *Int. J. Phys. Sci.* **2011**, *6*, 3896–3910.
11. Khan, M.; Irfan, M.; Khan, W.A.; Alshomrani, A.S. A new modeling for 3D Carreau fluid flow considering nonlinear thermal radiation. *Results Phys.* **2017**, *7*, 2692–2704. [CrossRef]
12. Khan, M.; Ijaz, M.; Kumar, A.; Hayat, T.; Waqas, M.; Singh, R. Entropy generation in flow of Carreau nanofluid. *J. Mol. Liq.* **2019**, *278*, 677–687. [CrossRef]
13. Khan, M.; Irfan, M.; Khan, W.A. Thermophysical properties of unsteady 3D flow of magneto Carreau fluid in the presence of chemical species: A numerical approach. *J. Braz. Soc. Mech. Sci. Eng.* **2018**, *40*, 108. [CrossRef]

14. Irfan, M.; Khan, W.A.; Khan, M.; Gulzar, M. Influence of Arrhenius activation energy in chemically reactive radiative flow of 3D Carreau nanofluid with nonlinear mixed convection. *J. Phys. Chem. Solids* **2019**, *125*, 141–152. [CrossRef]

15. Vasu, B.; Ray, A.K. Numerical study of Carreau nanofluid flow past vertical plate with the Cattaneo–Christov heat flux model. *Int. J. Numer. Methods Heat Fluid Flow* **2019**, *29*, 702–723.

16. Waqas, M.; Farooq, M.; Khan, M.I.; Alsaedi, A.; Hayat, T.; Yasmeen, T. Magnetohydrodynamic (MHD) mixed convection flow of micropolar liquid due to nonlinear stretched sheet with convective condition. *Int. J. Heat Mass Transf.* **2016**, *102*, 766–772. [CrossRef]

17. Ramzan, M.; Farooq, M.; Hayat, T.; Chung, J.D. Radiative and Joule heating effects in the MHD flow of a micropolar fluid with partial slip and convective boundary condition. *J. Mol. Liq.* **2016**, *221*, 394–400. [CrossRef]

18. Besthapu, P.; Haq, R.U.; Bandari, S.; Al-Mdallal, Q.M. Mixed convection flow of thermally stratified MHD nanofluid over an exponentially stretching surface with viscous dissipation effect. *J. Taiwan Inst. Chem. Eng.* **2017**, *71*, 307–314. [CrossRef]

19. Khan, M.; Azam, M. Unsteady heat and mass transfer mechanisms in MHD Carreau nanofluid flow. *J. Mol. Liq.* **2017**, *225*, 554–562. [CrossRef]

20. Turkyilmazoglu, T. Mixed convection flow of magnetohydrodynamic micropolar fluid due to a porous heated/cooled deformable plate: Exact solutions. *Int. J. Heat Mass Transf.* **2017**, *106*, 127–134. [CrossRef]

21. Hayat, T.; Muhammad, T.; Shehzad, S.A.; Alsaedi, A. An analytical solution for magnetohydrodynamic Oldroyd-B nanofluid flow induced by a stretching sheet with heat generation/absorption. *Int. J. Therm. Sci.* **2017**, *111*, 274e288. [CrossRef]

22. Khan, M.I.; Waqas, M.; Hayat, T.; Alsaedi, A. A comparative study of Casson fluid with homogeneous-heterogeneous reactions. *J. Colloid Interface Sci.* **2017**, *498*, 85–90. [CrossRef]

23. Ramzan, M.; Bilal, M.; Chung, J.D. MHD stagnation point Cattaneo–Christov heat flux in Williamson fluid flow with homogeneous–heterogeneous reactions and convective boundary condition—A numerical approach. *J. Mol. Liq.* **2017**, *225*, 856–862. [CrossRef]

24. Ramzan, M.; Farooq, M.; Hayat, T.; Alsaedi, A.; Cao, J. MHD stagnation point flow by a permeable stretching cylinder with Soret-Dufour effects. *J. Cent. South Univ.* **2015**, *22*, 707–716. [CrossRef]

25. Su, X.; Zheng, L.; Zhang, X.; Zhang, J. MHD mixed convective heat transfer over a permeable stretching wedge with thermal radiation and ohmic heating. *Chem. Eng. Sci.* **2012**, *78*, 1–8. [CrossRef]

26. Pal, D.; Chatterjee, S. Soret and Dufour effects on MHD convective heat and mass transfer of a power-law fluid over an inclined plate with variable thermal conductivity in a porous medium. *Appl. Math. Comput.* **2013**, *219*, 7556–7574. [CrossRef]

27. Fourier, J.B.J. *Théorie Analytique de la Chaleur*; Chez Firmin Didot: Paris, France, 1822.

28. Cattaneo, C. Sulla conduzione del calore, Attidel Seminario Matematico e Fisico Dell. *Modena Reggio Emilia* **1948**, *3*, 83–101.

29. Christov, C.I. On frame indifferent formulation of the Maxwell–Cattaneo model of finite-speed heat conduction. *Mech. Res. Commun.* **2009**, *36*, 481–486. [CrossRef]

30. Ramzan, M.; Bilal, M.; Chung, J.D. Effects of MHD homogeneous-heterogeneous reactions on third grade fluid flow with Cattaneo–Christov heat flux. *J. Mol. Liq.* **2016**, *223*, 1284–1290. [CrossRef]

31. Hayat, T.; Khan, M.I.; Farooq, M.; Alsaedi, A.; Khan, M.I. Thermally stratified stretching flow with Cattaneo–Christov heat flux. *Int. J. Heat Mass Transf.* **2017**, *106*, 289–294. [CrossRef]

32. Sui, J.; Zheng, L.; Zhang, X. Boundary layer heat and mass transfer with Cattaneo–Christov double-diffusion in upper-convected Maxwell nanofluid past a stretching sheet with slip velocity. *Int. J. Therm. Sci.* **2016**, *104*, 461–468. [CrossRef]

33. Liu, L.; Zheng, L.; Liu, F.; Zhang, X. Heat conduction with fractional Cattaneo–Christov upper-convective derivative flux model. *Int. J. Therm. Sci.* **2017**, *112*, 421–426. [CrossRef]

34. Kumar, R.; Kumar, R.; Sheikholeslami, M.; Chamkha, A.J. Irreversibility analysis of the three-dimensional flow of carbon nanotubes due to nonlinear thermal radiation and quartic chemical reactions. *J. Mol. Liq.* **2019**, *274*, 379–392. [CrossRef]

35. Xu, H. Homogeneous–Heterogeneous Reactions of Blasius Flow in a Nanofluid. *J. Heat Transf.* **2019**, *141*, 024501. [CrossRef]

36. Sithole, H.; Mondal, H.; Magagula, V.M.; Sibanda, P.; Motsa, S. Bivariate Spectral Local Linearisation Method (BSLLM) for unsteady MHD Micropolar-nanofluids with Homogeneous–Heterogeneous chemical reactions over a stretching surface. *Int. J. Appl. Comput. Math.* **2019**, *5*, 12. [CrossRef]

37. Khan, W.A.; Ali, M.; Sultan, F.; Shahzad, M.; Khan, M.; Irfan, M. Numerical interpretation of autocatalysis chemical reaction for nonlinear radiative 3D flow of cross magnetofluid. *Pramana* **2019**, *92*, 16. [CrossRef]

38. Raees, A.; Wang, R.Z.; Xu. H. A homogeneous-heterogeneous model for mixed convection in gravity-driven film flow of nanofluids. *Int. Commun. Heat Mass Transf.* **2018**, *95*, 19–24. [CrossRef]

39. Lu, D.; Li, Z.; Ramzan, M.; Shafee, M.; Chung, J.D. Unsteady squeezing carbon nanotubes based nano-liquid flow with Cattaneo–Christov heat flux and homogeneous–heterogeneous reactions. *Appl. Nanosci.* **2019**, *9*, 169–178. [CrossRef]

40. Lu, D.; Ramzan, M.; Ahmad, S.; Chung, J.D.; Farooq, U. A numerical treatment of MHD radiative flow of Micropolar nanofluid with homogeneous-heterogeneous reactions past a nonlinear stretched surface. *Sci. Rep.* **2018**, *8*, 12431. [CrossRef]

41. Lu, D.; Ramzan, M.; Bilal, M.; Chung, J.D.; Farooq, U.; Tahir, S. On three-dimensional MHD Oldroyd-B fluid flow with nonlinear thermal radiation and homogeneous–heterogeneous reaction. *J. Braz. Soc. Mech. Sci. Eng.* **2018**, *40*, 387. [CrossRef]

42. Ramzan, M.; Bilal, M.; Chung, J.D. Influence of homogeneous-heterogeneous reactions on MHD 3D Maxwell fluid flow with Cattaneo–Christov heat flux and convective boundary condition. *J. Mol. Liq.* **2017**, *230*, 415–422. [CrossRef]

43. Nadeem, S.; Muhammad, N. Impact of stratification and Cattaneo–Christov heat flux in the flow saturated with porous medium. *J. Mol. Liq.* **2016**, *224*, 423–430. [CrossRef]

44. Hayat, T.; Rashid, M.; Alsaedi, A. Three dimensional radiative flow of magnetite-nanofluid with homogeneous-heterogeneous reactions. *Results Phys.* **2018**, *8*, 268–275. [CrossRef]

45. Merkin, J.H. A model for isothermal homogeneous–heterogeneous reactions in boundary layer flow. *Math. Comput. Model.* **1996**, *24*, 125–136. [CrossRef]

46. Chaudhary, M.A.; Merkin, J.H. A simple isothermal model for homogeneous-heterogeneous reactions in boundary layer flow: I. Equal diffusivities. *Fluid Dyn. Res.* **1995**, *16*, 311–333. [CrossRef]

47. Zargartalebi, H.; Ghalambaz, M.; Noghrehabadi, A.; Chamkha, A. Stagnation-point heat transfer of nanofluids toward stretching sheets with variable thermo-physical properties. *Adv. Powder Technol.* **2015**, *26*, 819–829. [CrossRef]

48. Ahmad, K.; Nazar, R. Magnetohydrodynamic three dimensional flow and heat transfer over a stretching surface in a viscoelastic fluid. *J. Sci. Technol.* **2010**, *3*, 1–14.

6

Significance of Velocity Slip in Convective Flow of Carbon Nanotubes

Ali Saleh Alshomrani and Malik Zaka Ullah *

Department of Mathematics, Faculty of Science, King Abdulaziz University, Jeddah 21589, Saudi Arabia;
aszalshomrani@kau.edu.sa
* Correspondence: malikzakas@gmail.com

Abstract: The present article inspects velocity slip impacts in three-dimensional flow of water based carbon nanotubes because of a stretchable rotating disk. Nanoparticles like single and multi walled carbon nanotubes (CNTs) are utilized. Graphical outcomes have been acquired for both single-walled carbon nanotubes (SWCNTs) and multi-walled carbon nanotubes (MWCNTs). The heat transport system is examined in the presence of thermal convective condition. Proper variables lead to a strong nonlinear standard differential framework. The associated nonlinear framework has been tackled by an optimal homotopic strategy. Diagrams have been plotted so as to examine how the temperature and velocities are influenced by different physical variables. The coefficients of skin friction and Nusselt number have been exhibited graphically. Our results indicate that the skin friction coefficient and Nusselt number are enhanced for larger values of nanoparticle volume fraction.

Keywords: stretchable rotating disk; CNTs (MWCNTs and SWCNTs); velocity slip; convective boundary condition; OHAM

1. Introduction

The investigation of liquid flow by a rotating disk has various applications in aviation science, pivot of hardware, synthetic enterprises and designing, creating frameworks of warm power, rotor-stator frameworks, medicinal contraption, electronic and PC putting away apparatuses, gem developing wonders, machines of air cleaning, nourishment preparing advances, turbo apparatus and numerous others. Von Karman [1] analyzed flow of thick fluid by a rotating disk. Turkyilmazoglu and Senel [2] explored effects of heat and mass transport in thick liquid flow over a permeable rotating disk. Rashidi et al. [3] dissected MHD flow of viscous liquid because of a turn of disk. Turkyilmazoglu [4] exhibited nanoliquid flow by a rotating plate. Hatami et al. [5] examined laminar flow of a thick nanofluid because of the revolution and constriction of disks. Nanoliquid flow because of an extending disk is considered by Mustafa et al. [6]. Sheikholeslami et al. [7] examined nanoliquid flow by a slanted rotatory plate. Recently Hayat et al. [8] analyzed MHD nanoliquid flow over a rotatory disk with slip impacts.

Carbon nanotubes (CNTs) were first discovered by Lijima in 1991. CNTs have long cylindrical pofiles such as frames of carbon atoms with diameter ranges from 0.70–50 nm. CNTs have individual importance in nano-technology, hardwater, air purification systems, structural composite materials, conductive plastics, extra strong fibres, sensors, flat-panel displays, gas storage, biosensors and many others. Thus Choi et al. [9] examined anamolous enhancement of thermal conductivity in nanotubes suspension. Ramasubramaniam et al. [10] examined homogeneous polymer composites/carbon nanotubes for electrical utilizations. Xue [11] proposed a relation for CNT-based composites. Heat transfer enhancement using carbon nanotubes-based-non-Newtonian nanofluids is discussed by Kamali et al. [12]. Wang et al. [13] illustrated laminar flows of nanofluids containing single-walled carbon nanotubes (SWCNT) and multi-walled carbon nanotubes (MWCNTs). Hammouch et al. [14] analyzed squeezed flow of CNTs between parallel disks. Thermal transfer upgrade in front aligned contracting channel by taking FMWCNT nanoliquids is analyzed by Safaei et al. [15]. MHD flow of carbon nanotubes is portrayed by Ellahi et ai. [16]. Karimipour et al. [17] dissected MHD laminar flow of carbon nanotubes in a microchannel with a uniform warmth transition. Hayat et al. [18] represented homogeneous-heterogeneous responses in nanofluid flows over a non-direct extending surface of variable thickness. Unsteady squeezed flow of CNTs with convective surface was contemplated by Hayat et al. [19]. Hayat et al. [20] likewise talked about Darcy Forchheimer flow of CNTs over a turning plate. Further relevant investigations on nanofluids can be seen through the studies [21–25].

Motivated by the aforementioned applications of rotating flows, the underlying objective of this article is to develop a mathematical model for three-dimensional flow of water-based carbon nanotubes because of a stretchable rotating disk considering velocity slip effects. Thermal conductivity of carbon nanotubes is estimated through the well-known Xue model. Such research work was not carried out in the past even in the absence of a convective heating surface. Researchers also found that dispersion of carbon nanotubes in water elevates the thermal conductivity of the resulting nanofluid by 100% (see Choi et al. [9]). Both single-walled carbon nanotubes (SWCNTs) and multi-walled carbon nanotubes (MWCNTs) are considered. Optimal homotopic strategy (OHAM) [26–35] is utilized for solutions of temperature and velocities. Impacts of different flow variables are examined and investigated. Nusselt number and skin friction have been analyzed graphically. Emphasis is given to the role of the main ingredients of the problem, namely volume fraction of carbon nanotubes and a rotating stretchable disk. The benefits of carbon nanotubes towards heat transfer enhancement are also justified via thorough analysis.

2. Mathematical Formulation

Let us assume three-dimensional flow of water-based carbon nanotubes by a stretchable rotating disk. The disk at $z = 0$ rotates subject to constant angular velocity Ω (see Figure 1). Let us assume CNT nanoparticles: SWCNTs and MWCNTs within base liquid (water). Due to axial symmetry, derivatives of φ are neglected. The surface of the disk has temperature T_f, while ambient fluid temperature is T_∞. The velocity components are (u, v, w) in cylindrical coordinate (r, φ, z) respectively. The resulting boundary-layer expressions are [8,20]:

$$\frac{\partial u}{\partial r} + \frac{u}{r} + \frac{\partial w}{\partial z} = 0, \tag{1}$$

$$u\frac{\partial u}{\partial r} - \frac{v^2}{r} + w\frac{\partial u}{\partial z} = \nu_{nf}\left(\frac{\partial^2 u}{\partial r^2} + \frac{\partial^2 u}{\partial z^2} + \frac{1}{r}\frac{\partial u}{\partial r} - \frac{u}{r^2}\right), \tag{2}$$

$$u\frac{\partial v}{\partial r} + \frac{uv}{r} + w\frac{\partial v}{\partial z} = \nu_{nf}\left(\frac{\partial^2 v}{\partial r^2} + \frac{\partial^2 v}{\partial z^2} + \frac{1}{r}\frac{\partial v}{\partial r} - \frac{v}{r^2}\right), \tag{3}$$

$$u\frac{\partial w}{\partial r} + w\frac{\partial w}{\partial z} = \nu_{nf}\left(\frac{\partial^2 w}{\partial r^2} + \frac{\partial^2 w}{\partial z^2} + \frac{1}{r}\frac{\partial w}{\partial r}\right), \tag{4}$$

$$u\frac{\partial T}{\partial r} + w\frac{\partial T}{\partial z} = \alpha_{nf}\left(\frac{\partial^2 T}{\partial r^2} + \frac{\partial^2 T}{\partial z^2} + \frac{1}{r}\frac{\partial T}{\partial r}\right), \tag{5}$$

with subjected boundary conditions [8]:

$$u = rs + L_1\mu_{nf}\frac{\partial u}{\partial z}, \ v = r\Omega + L_1\mu_{nf}\frac{\partial v}{\partial z}, \ w = 0, \ -k_{nf}\frac{\partial T}{\partial z} = h_f\left(T_f - T\right) \text{ at } z = 0, \tag{6}$$

$$u \to 0, \ v \to 0, \ T \to T_\infty \text{ as } z \to \infty. \tag{7}$$

Here u, v and w depict flow velocities in increasing directions of r, φ and z respectively, while $\nu_{nf} = (\mu_{nf}/\rho_{nf})$ stands for kinematic viscosity, $\alpha_{nf} = k_{nf}/(\rho c_p)_{nf}$ for thermal diffusivity, μ_{nf} for dynamic viscosity, L_1 for wall-slip coefficient, T for fluid temperature, k_{nf} for thermal conductivity of nanofluids, ρ_{nf} for effective density, k_{CNT} for thermal conductivity of CNTs and $(\rho c_p)_{nf}$ for effective heat capacitance of nanoparticle material. Xue [11] proposed a theoratical model which is expressed by

$$
\begin{aligned}
\rho_{nf} &= \rho_f(1-\phi) + \rho_{CNT}\phi, \quad \mu_{nf} = \frac{\mu_f}{(1-\phi)^{2.5}}, \\
(\rho c_p)_{nf} &= (\rho c_p)_f(1-\phi) + (\rho c_p)_{CNT}\phi, \\
\frac{k_{nf}}{k_f} &= \frac{(1-\phi)+2\phi\frac{k_{CNT}}{k_{CNT}-k_f}\ln\frac{k_{CNT}+k_f}{2k_f}}{(1-\phi)+2\phi\frac{k_f}{k_{CNT}-k_f}\ln\frac{k_{CNT}+k_f}{2k_f}},
\end{aligned}
\tag{8}
$$

where ϕ represents solid volume fraction of nanoparticles and nf represents thermophysical properties of nanofluid. Table 1 describes thermo-physical features of water and CNT.

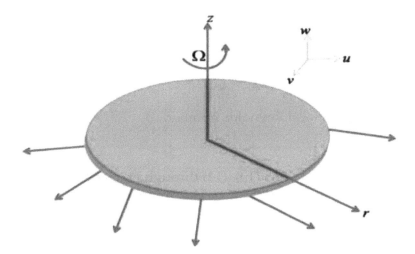

Figure 1. Geometry of the problem.

Table 1. Thermophysical features of water and carbon nanotubes (CNT).

Physical Features	Water	CNT	
		SWCNTs	MWCNTs
$\rho\ (\text{kg/m}^3)$	997.1	2600	1600
$k\ (\text{W/mK})$	0.613	6600	3000
$c_p\ (\text{J/kgK})$	4179	425	796

We now introduce the following transformations:

$$
\begin{aligned}
u &= r\Omega f'(\zeta),\ v = r\Omega g(\zeta),\ w = -\sqrt{2\nu_f\Omega}f(\zeta), \\
\zeta &= z\left(\frac{2\Omega}{\nu_f}\right)^{1/2},\ \theta(\zeta) = \frac{T-T_\infty}{T_f-T_\infty}.
\end{aligned}
\tag{9}
$$

Expression (1) is automatically satisfied while Equations (2)–(8) yield

$$
\frac{1}{(1-\phi)^{5/2}(1-\phi+\frac{\rho_{CNT}}{\rho_f}\phi)}f'''(\zeta) + f(\zeta)f''(\zeta) - \frac{1}{2}f'^2(\zeta) + \frac{1}{2}g^2(\zeta) = 0,
\tag{10}
$$

$$
\frac{1}{(1-\phi)^{5/2}(1-\phi+\frac{\rho_{CNT}}{\rho_f}\phi)}g''(\zeta) + f(\zeta)g'(\zeta) - f'(\zeta)g(\zeta) = 0,
\tag{11}
$$

$$
\frac{1}{\text{Pr}}\frac{k_{nf}}{k_f}\theta''(\zeta) + \left(1-\phi+\phi\frac{(\rho c_p)_{CNT}}{(\rho c_p)_f}\right)f(\zeta)\theta'(\zeta) = 0,
\tag{12}
$$

with the boundary conditions

$$f(0) = 0,\ f'(0) = C + \frac{\alpha}{(1-\phi)^{5/2}}f''(0),\ g(0) = 1 + \frac{\alpha}{(1-\phi)^{5/2}}g'(0),\ \theta'(0) = -\frac{k_f}{k_{nf}}Bi\left(1 - \theta(0)\right), \tag{13}$$

$$f'(\infty) \to 0,\ g(\infty) \to 0,\ \theta(\infty) \to 0. \tag{14}$$

Here C stands for stretching-strength parameter, α for velocity slip number, Pr for Prandtl number and Bi for the Biot number. These numbers are described by:

$$C = \frac{s}{\Omega},\ \alpha = L_1\mu_f\left(\frac{2\Omega}{\nu_f}\right)^{1/2},\ Bi = \frac{h_f}{k_f}\left(\frac{\nu_f}{2\Omega}\right)^{1/2},\ \Pr = \frac{\nu_f\left(\rho c_p\right)_f}{k_f}. \tag{15}$$

Nusselt number and skin friction are defined by

$$\left.\begin{array}{l} \mathrm{Re}_r^{-1/2}Nu_r = -\frac{k_{nf}}{k_f}\theta'(0), \\ \mathrm{Re}_r^{1/2}C_f = \frac{1}{(1-\phi)^{5/2}}\left(f''(0)^2 + g'(0)^2\right)^{1/2}, \end{array}\right\} \tag{16}$$

where $\mathrm{Re}_r = 2\Omega r^2/\nu_f$ depicts the local Reynolds number.

3. Solutions by OHAM

The optimal solutions of expressions (10)–(12) through (13) and (14) have been established by considering optimal homotopic strategy (OHAM). The proper operators and guesses are

$$f_0(\zeta) = \frac{C}{\left(1 + \frac{\alpha}{(1-\phi)^{5/2}}\right)}(1 - e^{-\zeta}),\ g_0(\zeta) = \frac{1}{\left(1 + \frac{\alpha}{(1-\phi)^{5/2}}\right)}e^{-\zeta},\ \theta_0(\zeta) = \frac{Bi}{\left(\frac{k_{nf}}{k_f} + Bi\right)}e^{-\zeta}, \tag{17}$$

$$\mathcal{L}_g = \frac{d^2g}{d\zeta^2} - g,\ \mathcal{L}_\theta = \frac{d^2\theta}{d\zeta^2} - \theta,\ \mathcal{L}_f = \frac{d^3f}{d\zeta^3} - \frac{df}{d\zeta}. \tag{18}$$

The above operators satisfy

$$\mathcal{L}_f\left[F_1^{****} + F_2^{****}e^\zeta + F_3^{****}e^{-\zeta}\right] = 0,\ \mathcal{L}_g\left[F_4^{****}e^\zeta + F_5^{****}e^{-\zeta}\right] = 0,\ \mathcal{L}_\theta\left[F_6^{****}e^\zeta + F_7^{****}e^{-\zeta}\right] = 0, \tag{19}$$

in which F_i^{****} ($i = 1$–7) portrays arbitrary constants. The m-th and zero-th order systems are easily established in view of above operators. By using BVPh2.0 of the software Mathematica, the obtained deformation problems have been computed.

4. Optimal Convergence-Control Parameters

In homotopic solutions, the non zero auxiliary variables $\hbar_f$, $\hbar_g$ and $\hbar_\theta$ determine the convergence portion and also rate of homotopy solution. The idea of minimization has been applied by defining averaged squared residuals errors as proposed by Liao [26].

$$\varepsilon_m^f = \frac{1}{k+1}\sum_{j=0}^{k}\left[\mathcal{N}_f\left(\sum_{i=0}^{m}\hat{f}(\zeta),\sum_{i=0}^{m}\hat{g}(\zeta)\right)_{\zeta=j\delta\zeta}\right]^2, \tag{20}$$

$$\varepsilon_m^g = \frac{1}{k+1}\sum_{j=0}^{k}\left[\mathcal{N}_g\left(\sum_{i=0}^{m}\hat{f}(\zeta),\sum_{i=0}^{m}\hat{g}(\zeta)\right)_{\zeta=j\delta\zeta}\right]^2, \tag{21}$$

$$\varepsilon_m^\theta = \frac{1}{k+1}\sum_{j=0}^{k}\left[\mathcal{N}_\theta\left(\sum_{i=0}^{m}\hat{f}(\zeta),\sum_{i=0}^{m}\hat{g}(\zeta),\sum_{i=0}^{m}\hat{\theta}(\zeta)\right)_{\zeta=j\delta\zeta}\right]^2. \tag{22}$$

Following Liao [26]:

$$\varepsilon_m^t = \varepsilon_m^f + \varepsilon_m^g + \varepsilon_m^\theta, \tag{23}$$

where ε_m^t represents total squared residual error, $\delta\zeta = 0.5$ and $k = 20$. At the second order of deformations, convergence-control parameters for SWCNTs–water have optimal values i.e., $h_f = -0.35923$, $h_g = -0.736096$ and $h_\theta = -0.00105197$ and total averaged squared residuals error is $\varepsilon_m^t = -0.0255367$ while optimal data of convergence-control parameters for MWCNTs–water is $h_f = -0.385385$, $h_g = -0.729057$ and $h_\theta = -0.00232643$ and total averaged squared residuals error is $\varepsilon_m^t = -0.025173$. Figures 2 and 3 display error plots for MWCNTs–water and SWCNTs–water. Tables 2 and 3 show that averaged squared residuals error decreases for higher order deformations.

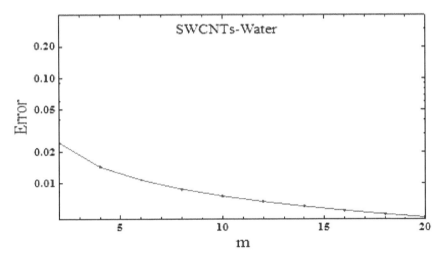

Figure 2. Error sketch for SWCNTs-Water.

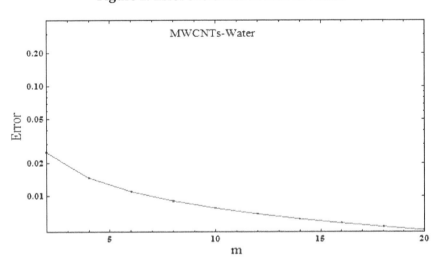

Figure 3. Error sketch for MWCNTs-Water.

Table 2. Individual averaged squared residuals errors for single-walled carbon nanotubes (SWCNTs)–water.

m	ε_m^f	ε_m^g	ε_m^θ
2	9.95225×10^{-5}	2.35341×10^{-2}	7.29447×10^{-7}
6	4.17686×10^{-5}	1.03083×10^{-2}	6.04738×10^{-7}
10	2.95796×10^{-5}	7.24672×10^{-3}	5.69806×10^{-7}
14	2.37325×10^{-5}	5.77429×10^{-3}	5.53122×10^{-7}
18	2.01939×10^{-5}	4.88653×10^{-3}	5.43213×10^{-7}
20	1.88867×10^{-5}	4.55942×10^{-3}	5.39608×10^{-7}

Table 3. Individual averaged squared residuals errors for single-walled carbon nanotubes (MWCNTs)–water.

m	ε_m^f	ε_m^g	ε_m^θ
2	1.0164×10^{-4}	2.40503×10^{-2}	7.29447×10^{-7}
6	4.27165×10^{-5}	1.05447×10^{-2}	6.04522×10^{-7}
10	3.02678×10^{-5}	7.41739×10^{-3}	5.69547×10^{-7}
14	2.42942×10^{-5}	5.91293×10^{-3}	5.52829×10^{-7}
18	2.06785×10^{-5}	5.00567×10^{-3}	5.42892×10^{-7}
20	1.93426×10^{-5}	4.67132×10^{-3}	5.39275×10^{-7}

5. Results and Discussion

The present section presents behaviors of various physical parameters like stretching-strength parameter C, volume fraction ϕ, velocity slip parameter α and Biot number Bi on radial $f'(\zeta)$ and azimuthal $g(\zeta)$ velocities and temperature $\theta(\zeta)$. The results are obtained for both SWCNTs and MWCNTs. Figure 4 shows variation in the radial velocity $f'(\zeta)$ for larger values of α. Radial velocity $f'(\zeta)$ shows reduction for increasing values of α. Figure 5 presents impact of stretching-strength parameter C on radial velocity $f'(\zeta)$. For larger values of C, the radial velocity shows an increasing trend. Figure 6 depicts the effect of nanoparticle volume fraction ϕ on radial velocity $f'(\zeta)$. For higher ϕ, the radial velocity $f'(\zeta)$ is increased. Figure 7 presents that how velocity slip parameter α affects the azimuthal velocity $g(\zeta)$. It is observed that an increment in velocity slip parameter α lead to lower $g(\zeta)$. Figure 8 depicts impact of C on azimuthal velocity $g(\zeta)$. Azimuthal velocity reduces for larger values of streching-strength parameter. Figure 9 depicts the impact of nanoparticles volume fraction ϕ on $g(\zeta)$. The azimuthal velocity $g(\zeta)$ is increased for higher estimations of ϕ. Figure 10 examines that how Biot number Bi affects the temperature profile. For higher values of Bi, the temperature field $\theta(\zeta)$ is enhanced. Higher estimations of Biot number correspond to stronger convection which produces higher temperature field and more associated layer thickness. Figure 11 highlights the impact of stretching-strength parameter C on temperature field $\theta(\zeta)$. Temperature field $\theta(\zeta)$ is reduced for increasing values of C. Figure 12 presents that how volume fraction ϕ affects the temperature field $\theta(\zeta)$. Higher values of ϕ shows an enhancement in temperature $\theta(\zeta)$. Figure 13 shows the effects of volume fraction ϕ and velocity slip parameter α on $\mathrm{Re}_r^{1/2}C_f$. Skin friction $\mathrm{Re}_r^{1/2}C_f$ is increased for higher estimations of ϕ. Figure 14 displays the behavior of the volume fraction ϕ and Biot number Bi on Nusselt number $\mathrm{Re}_r^{-1/2}Nu_r$. The Nusselt number is enhanced for increasing values of ϕ.

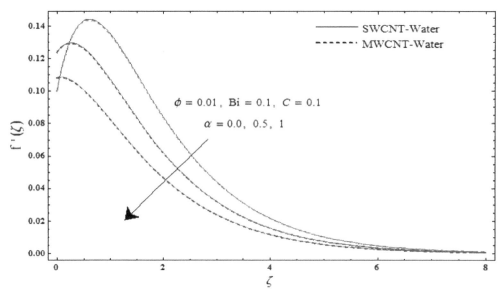

Figure 4. Sketch of $f'(\zeta)$ for α.

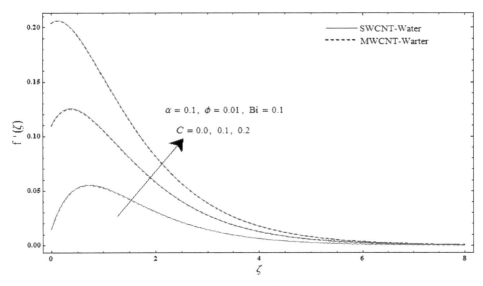

Figure 5. Sketch of $f'(\zeta)$ for C.

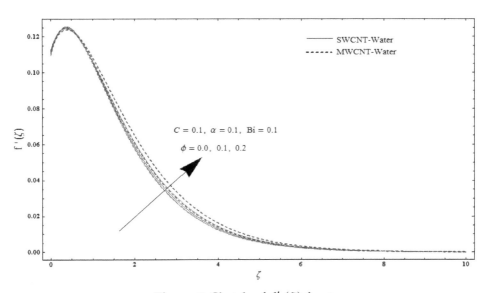

Figure 6. Sketch of $f'(\zeta)$ for ϕ.

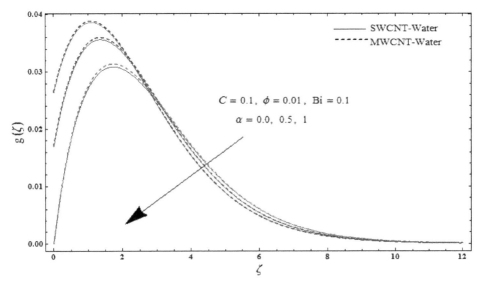

Figure 7. Sketch of $g(\zeta)$ for α.

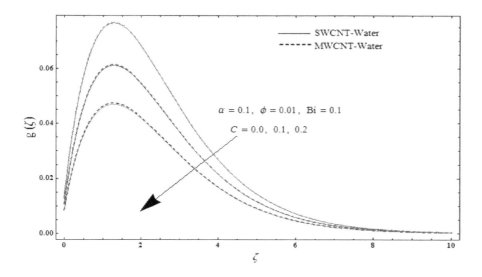

Figure 8. Sketch of $g\left(\zeta\right)$ for C.

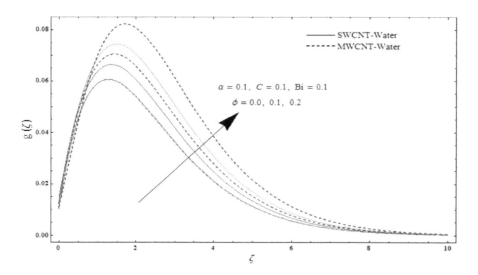

Figure 9. Sketch of $g\left(\zeta\right)$ for ϕ.

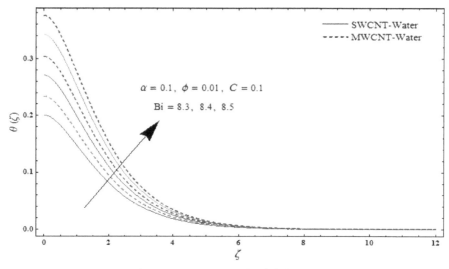

Figure 10. Sketch of $\theta\left(\zeta\right)$ for Bi.

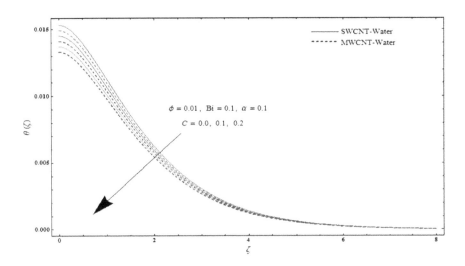

Figure 11. Sketch of $\theta\left(\zeta\right)$ for C.

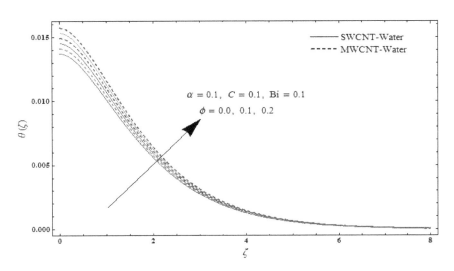

Figure 12. Sketch of $\theta\left(\zeta\right)$ for ϕ.

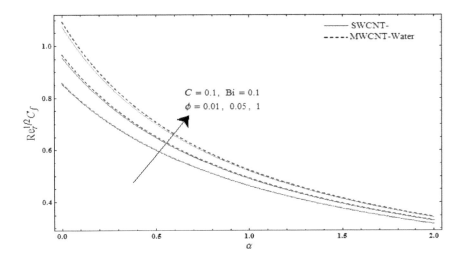

Figure 13. Sketch of $\mathrm{Re}_r^{1/2}C_f$ for ϕ and α.

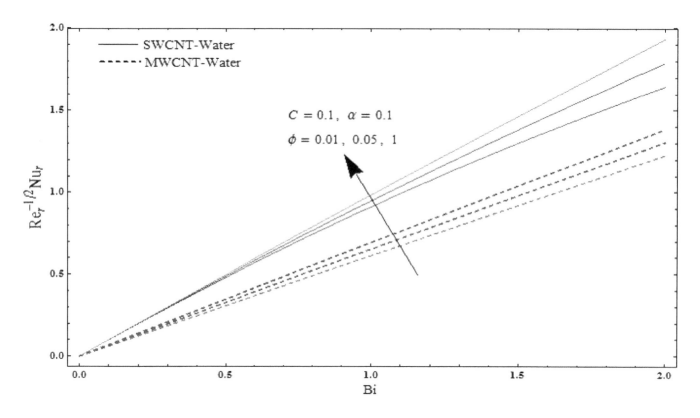

Figure 14. Sketch of $Re_r^{-1/2} Nu_r$ for ϕ and Bi.

6. Conclusions

Three-dimensional flow of carbon nanotubes by a stretchable rotating disk with velocity slip effects is studied. Heat transport is explained by convective heating surface. The key findings of current research are listed below:

- Both velocities $f'(\zeta)$ and $g(\zeta)$ show reduction for higher values of velocity slip parameter α.
- Larger stretching-strength parameter C presents an increase in radial velocity $f'(\zeta)$ while opposite trend is noticed for azimuthal velocity $g(\zeta)$ and temperature $\theta(\zeta)$.
- For higher estimations of the volume fraction ϕ, both the velocity and temperatue field are enhanced.
- Temperature field $\theta(\zeta)$ is enhanced for larger values of the Biot number Bi.
- Nusselt number $Re_r^{-1/2} Nu_r$ is increased for larger values of volume fraction ϕ.
- Coefficient of skin-friction $Re_r^{1/2} C_f$ increases for higher volume fraction ϕ and velocity slip parameter α.
- The used technique for the solution's development has advantages over the other in the sense of the following points:

  a. It is independent of small/large physical parameters.
  b. It provides a simple way to ensure the convergence of series solutions.
  c. It provides a large freedom to choose the base functions and related auxiliary linear operators.

Author Contributions: All the authors contributed equally to the conception of the idea, implementing and analyzing the experimental results, and writing the manuscript.

Nomenclature

r, φ, z	space coordinates [m]	u, v, w	velocity components [m·s^{-1}]
ρ_f	fluid density [kg·m^{-3}]	μ_f	fluid dynamic viscosity [Pa·s]
k_{nf}	nanofluids themal conductivity [W·m^{-1}·K^{-1}]	ν_{nf}	kinematic nanofluid viscosity [m^2·s^{-1}]
k_f	basefluid themal conductivity [W·m^{-1}·K^{-1}]	ν_f	kinematic fluid viscosity [m^2·s^{-1}]
α_f	thermal diffusivity of base fluid [m^2·s^{-1}]	α_{nf}	thermal diffusivity of nanofluid [m^2·s^{-1}]
T_f	hot fluid temperature [K]	T_∞	ambient temperature [K]
C	stretching-strength parameter	k_{CNT}	CNTs thermal conductivity [W·m^{-1}·K^{-1}]
α	velocity slip parameter	ϕ	nanomaterial volume fraction
Bi	Biot number	Pr	Prandtl number
f'	dimensionless velocity	Nu_r	Nusselt number
C_f	skin friction coefficient	ζ	dimensionless variable
Re_r	local Reynolds number	θ	dimensionless temperature
CNTs	carbon nanotubes	F_i^{****}	arbitrary constants

References

1. Von Karman, T. Uber laminare and turbulente Reibung. *ZAMM Z. Angew. Math. Mech.* **1921**, *1*, 233–252. [CrossRef]

2. Turkyilmazoglu, M.; Senel, P. Heat and mass transfer of the flow due to a rotating rough and porous disk. *Int. J. Therm. Sci.* **2013**, *63*, 146–158. [CrossRef]

3. Rashidi, M.M.; Kavyani, N.; Abelman, S. Investigation of entropy generation in MHD and slip flow over rotating porous disk with variable properties. *Int. J. Heat Mass Transf.* **2014**, *70*, 892–917. [CrossRef]

4. Turkyilmazoglu, M. Nanofluid flow and heat transfer due to a rotating disk. *Comput. Fluids* **2014**, *94*, 139–146. [CrossRef]

5. Hatami, M.; Sheikholeslami, M.; Gangi, D.D. Laminar flow and heat transfer of nanofluids between contracting and rotating disks by least square method. *Power Technol.* **2014**, *253*, 769–779. [CrossRef]

6. Mustafa, M.; Khan, J.A.; Hayat, T.; Alsaedi, A. On Bödewadt flow and heat transfer of nanofluids over a stretching stationary disk. *J. Mol. Liq.* **2015**, *211*, 119–125. [CrossRef]

7. Sheikholeslami, M.; Hatami, M.; Ganji, D.D. Numerical investigation of nanofluid spraying on an inclined rotating disk for cooling process. *J. Mol. Liq.* **2015**, *211*, 577–583. [CrossRef]

8. Hayat, T.; Muhammad, T.; Shehzad, S.A.; Alsaedi, A. On magnetohydrodynamic flow of nanofluid due to a rotating disk with slip effect: A numerical study. *Comput. Methods Appl. Mech. Eng.* **2017**, *315*, 467–477. [CrossRef]

9. Choi, S.U.S.; Zhang, Z.G.; Yu, W.; Lockwood, F.E.; Grulke, E.A. Anomalous thermal conductivity enhancement in nanotube suspensions. *Appl. Phys. Lett.* **2001**, *79*, 2252. [CrossRef]

10. Ramasubramaniam, R.; Chen, J.; Liu, H. Homogeneous carbon nanotube/polymer composites for electrical applications. *Appl. Phys. Lett.* **2003**, *83*, 2928. [CrossRef]

11. Xue, Q.Z. Model for thermal conductivity of carbon nanotube-based composites. *Phys. B Condens. Matter* **2005**, *368*, 302–307. [CrossRef]

12. Kamali, R.; Binesh, A. Numerical investigation of heat transfer enhancement using carbon nanotube-based non-Newtonian nanofluids. *Int. Commun. Heat Mass Transf.* **2010**, *37*, 1153–1157. [CrossRef]

13. Wang, J.; Zhu, J.; Zhang, X.; Chen, Y. Heat transfer and pressure drop of nanofluids containing carbon nanotubes in laminar flows. *Exp. Therm. Fluid Sci.* **2013**, *44*, 716–721. [CrossRef]

14. Haq, R.U.; Hammouch, Z.; Khan, W.A. Water-based squeezing flow in the presence of carbon nanotubes between two parallel disks. *Therm. Sci.* **2014**, *20*, 148. [CrossRef]

15. Safaei, M.R.; Togun, H.; Vafai, K.; Kazi, S.N.; Badarudin, A. Investigation of heat transfer enhancement in a forward-facing contracting channel using FMWCNT nanofluids. *Numer. Heat Transf. Part A* **2014**, *66*, 1321–1340. [CrossRef]

16. Ellahi, R.; Hassan, M.; Zeeshan, A. Study of natural convection MHD nanofluid by means of single and multi walled carbon nanotubes suspended in a salt water solutions. *IEEE Trans. Nanotechnol.* **2015**, *14*, 726–734. [CrossRef]

17. Karimipour, A.; Taghipour, A.; Malvandi, A. Developing the laminar MHD forced convection flow of water/FMWNT carbon nanotubes in a microchannel imposed the uniform heat flux. *J. Magn. Magn. Mater.* **2016**, *419*, 420–428. [CrossRef]

18. Hayat, T.; Hussain, Z.; Muhammad, T.; Alsaedi, A. Effects of homogeneous and heterogeneous reactions in flow of nanofluids over a nonlinear stretching surface with variable surface thickness. *J. Mol. Liq.* **2016**, *221*, 1121–1127. [CrossRef]

19. Hayat, T.; Muhammad, K.; Farooq, M.; Alsaedi, A. Unsteady squeezing flow of carbon nanotubes with convective boundary conditions. *PLoS ONE* **2016**, *11*, e0152923. [CrossRef]

20. Hayat, T.; Haider, F.; Muhammad, T.; Alsaedi, A. On Darcy-Forchheimer flow of carbon nanotubes due to a rotating disk. *Int. J. Heat Mass Transf.* **2017**, *112*, 248–254. [CrossRef]

21. Akbar, N.S.; Khan, Z.H.; Nadeem, S. The combined effects of slip and convective boundary conditions on stagnation-point flow of CNT suspended nanofluid over a stretching sheet. *J. Mol. Liq.* **2014**, *196*, 21–25. [CrossRef]

22. Arani, A.A.A.; Akbari, O.A.; Safaei, M.R.; Marzban, A.; Alrashed, A.A.A.; Ahmadi, G.R.; Nguyen, T.K. Heat transfer improvement of water/single-wall carbon nanotubes (SWCNT) nanofluid in a novel design of a truncated double-layered microchannel heat sink. *Int. J. Heat Mass Transf.* **2017**, *113*, 780–795. [CrossRef]

23. Goodarzi, M.; Javid, S.; Sajadifar, A.; Nojoomizadeh, M.; Motaharipour, S.H.; Bach, Q.V.; Karimipour, A. Slip velocity and temperature jump of a non-Newtonian nanofluid, aqueous solution of carboxy-methyl cellulose/aluminum oxide nanoparticles, through a microtube. *Int. J. Numer. Methods Heat Fluid Flow* **2018**. [CrossRef]

24. Ellahi, R.; Zeeshan, A.; Hussain, F.; Asadollahi, A. Peristaltic blood flow of couple stress fluid suspended with nanoparticles under the influence of chemical reaction and activation energy. *Symmetry* **2019**, *11*, 276. [CrossRef]

25. Suleman, M.; Ramzan, M.; Ahmad, S.; Lu, D.; Muhammad, T.; Chung, J.D. A numerical simulation of silver-water nanofluid flow with impacts of Newtonian heating and homogeneous-heterogeneous reactions past a nonlinear stretched cylinder. *Symmetry* **2019**, *11*, 295. [CrossRef]

26. Liao, S.J. An optimal homotopy-analysis approach for strongly nonlinear differential equations. *Commun. Nonlinear. Sci. Numer. Simul.* **2010**, *15*, 2003–2016. [CrossRef]

27. Dehghan, M.; Manafian, J.; Saadatmandi, A. Solving nonlinear fractional partial differential equations using the homotopy analysis method. *Numer. Meth. Part. Diff. Equ.* **2010**, *26*, 448–479. [CrossRef]

28. Malvandi, A.; Hedayati, F.; Domairry, G. Stagnation point flow of a nanofluid toward an exponentially stretching sheet with nonuniform heat generation/absorption. *J. Thermodyn.* **2013**, *2013*, 764827. [CrossRef]

29. Abbasbandy, S.; Hayat, T.; Alsaedi, A.; Rashidi, M.M. Numerical and analytical solutions for Falkner-Skan flow of MHD Oldroyd-B fluid. *Int. J. Numer. Methods Heat Fluid Flow* **2014**, *24*, 390–401. [CrossRef]

30. Sheikholeslami, M.; Hatami, M.; Ganji, D.D. Micropolar fluid flow and heat transfer in a permeable channel using analytic method. *J. Mol. Liq.* **2014**, *194*, 30–36. [CrossRef]

31. Hayat, T.; Muhammad, T.; Alsaedi, A.; Alhuthali, M.S. Magnetohydrodynamic three-dimensional flow of viscoelastic nanofluid in the presence of nonlinear thermal radiation. *J. Magn. Magn. Mater.* **2015**, *385*, 222–229. [CrossRef]

32. Turkyilmazoglu, M. An effective approach for evaluation of the optimal convergence control parameter in the homotopy analysis method. *Filomat* **2016**, *30*, 1633–1650. [CrossRef]

33. Zeeshan, A.; Majeed, A.; Ellahi, R. Effect of magnetic dipole on viscous ferro-fluid past a stretching surface with thermal radiation. *J. Mol. Liq.* **2016**, *215*, 549–554. [CrossRef]

34. Hayat, T.; Abbas, T.; Ayub, M.; Muhammad, T.; Alsaedi, A. On squeezed flow of Jeffrey nanofluid between two parallel disks. *Appl. Sci.* **2016**, *6*, 346. [CrossRef]

35. Muhammad, T.; Alsaedi, A.; Shehzad, S.A.; Hayat, T. A revised model for Darcy-Forchheimer flow of Maxwell nanofluid subject to convective boundary condition. *Chin. J. Phys.* **2017**, *55*, 963–976. [CrossRef]

7

Peristaltic Pumping of Nanofluids through a Tapered Channel in a Porous Environment: Applications in Blood Flow

J. Prakash [1], Dharmendra Tripathi [2,*], Abhishek Kumar Tiwari [3], Sadiq M. Sait [4] and Rahmat Ellahi [5,6]

[1] Department of Mathematics, Avvaiyar Government College for Women, Karaikal 609602, Puducherry-U.T., India
[2] Department of Mathematics, National Institute of Technology, Uttarakhand 246174, India
[3] Department of Applied Mechanics, MNNIT Allahabad, Prayagraj, Uttar Pradesh 211004, India
[4] Center for Communications and IT Research, Research Institute, King Fahd University of Petroleum & Minerals, Dhahran 31261, Saudi Arabia
[5] Center for Modeling & Computer Simulation, Research Institute, King Fahd University of Petroleum & Minerals, Dhahran 31261, Saudi Arabia
[6] Department of Mathematics & Statistics, Faculty of Basic and Applied Sciences (FBAS), International Islamic University (IIUI), Islamabad 44000, Pakistan
* Correspondence: dtripathi@nituk.ac.in

Abstract: In this study, we present an analytical study on blood flow analysis through with a tapered porous channel. The blood flow was driven by the peristaltic pumping. Thermal radiation effects were also taken into account. The convective and slip boundary conditions were also applied in this formulation. These conditions are very helpful to carry out the behavior of particle movement which may be utilized for cardiac surgery. The tapered porous channel had an unvarying wave speed with dissimilar amplitudes and phase. The non-dimensional analysis was utilized for some approximations such as the proposed mathematical modelling equations were modified by using a lubrication approach and the analytical solutions for stream function, nanoparticle temperature and volumetric concentration profiles were obtained. The impacts of various emerging parameters on the thermal characteristics and nanoparticles concentration were analyzed with the help of computational results. The trapping phenomenon was also examined for relevant parameters. It was also observed that the geometric parameters, like amplitudes, non-uniform parameters and phase difference, play an important role in controlling the nanofluids transport phenomena. The outcomes of the present model may be applicable in the smart nanofluid peristaltic pump which may be utilized in hemodialysis.

Keywords: peristaltic transport; tapered channel; porous medium; smart pumping for hemodialysis; thermal radiation

1. Introduction

Peristaltic motion [1–6] is a fundamental physiological mechanism which has many applications in bio-mechanical and engineering sciences where transport phenomena at micro/macro level occur. This mechanism is also applicable in transporting the nanofluids without any contaminations. The nanofluid term was first invented by Choi and Eastman [7] with reference to a conventional heat transfer liquid retention distribution of a nanosized particle. The behavior of nanoliquid in thermal conductivity enhancement has been observed by Masuda et al. [8] and experimental results of nanoparticle into pure fluid may conduct to reduce in heat transfer. The closed form model for convective transport in nanofluids studying the thermophoresis and Brownian diffusion has been studied by Buongiorno and Hu [9] and Buongiorno [10].

In the peristaltic pumping models, nanoliquid was introduced by Akbar and Nadeem [11]. They investigated endoscopic influences on the peristaltic motion of a nanofluid. Akbar [12] further presented the peristaltic transport of a Sisko nanoliquid in an asymmetric channel. It was noticed that enhances in the Sisko nanoliquid parameter axial pressure rise in the peristaltic pumping region. The effect of nanoliquid features on peristaltic heat transfer in a two-dimensional axisymmetric channel was discussed by Tripathi and Beg [13]. They examined that the nanoliquids incline to suppress backflow equated with Newtonian fluids. Akbar et al. [14] discussed the magnetohydrodynamic (MHD) peristaltic motion of a Carreau nanoliquid in an asymmetric channel. Furthermore, Beg and Tripathi [15] introduced the double diffusion process in peristaltic pumping. They discussed the salute and nanoparticle concentrations in their analysis. The effects of nanoparticle geometry on peristaltic motion has been analyzed by Akbar et al. [16]. MHD peristaltic pumping with viscoelastic nanofluids have been studied by Reddy and Makinde [17]. The velocity and slip influences on peristaltic pumping of nanoliquids have been examined by the Akbar et al. [18]. Heat and mass transfer analysis on peristaltic pumping through the rectangular duct was presented by Nadeem et al. [19]. Peristaltic transport of Prandtl nanofluid through the rectangular duct was studied by Ellahi et al. [20] and with magnetic field [21]. Hyperbolic tangent nanofluid with peristaltic pumping was implemented by Kothandapani and Prakash [22] in the presence of a radiation parameter and inclined magnetic field. Peristaltic pumping by eccentric cylinders has been discussed by the Nadeem et al. [23]. In similar directions, many more investigations [24–37] on peristaltic pumping, nanofluids and non-Newtonian nanofluids with various physical constraints and various flow geometries had been described in the literature.

The analysis of fluid flow through porous channels or tubes had gained attention recently because of its several applications in biomedical engineering and many other engineering areas like the flow of blood oxygenators, gall bladder with stones, in small blood vessels, the design of filters, in transpiration cooling boundary layer control, the flow of blood in the capillaries, the dialysis of blood in artificial kidney, gaseous diffusion in the spreading of fatty cholesterol and artery-clogging blood clots in the lumen of a coronary artery [38–47]. The steady laminar incompressible free convective flow of a nanofluid over a permeable upward facing horizontal plate located in a porous medium in an existence of thermal convective boundary condition was considered numerically by Uddin et al. [48]. Chamkha et al. [49] studied the mixed convection boundary layer flow in the existence of laminar and isothermal vertical porous medium. The onset of convection in a horizontal layer of a porous medium by a nanofluid was analytically studied by Kuznetsov and Nield [50]. Akbar [51] investigated the double-diffusive peristaltic transport of Jeffrey nanoliquids in a porous region in the presence of natural convective. Double-diffusive natural convective peristaltic flow of a Jeffrey nanofluid in a porous channel has been analyzed by Nadeem et al. [52] and investigated the peristaltic flow of nanofluid eccentric tubes which comprises a porous medium. Two-phase flow driven by the peristaltic pumping through porous medium was studied by Bhatti et al. [53]. Perturbation solutions have been obtained and it is observed that chemical reaction and Soret numbers oppose the particle concentration. The applications of porosity can be deeply studied by using nanofluid model in [54,55].

Moreover, the no-slip condition is inadequate when a fluid revealing macroscopic wall slip is considered and that, in general, is governed by the relation between the slip velocity and grip. The slip condition plays significant role in shear skin, spurt and hysteresis belongings. The nanofluids that exhibit boundary slip have vital technological purposes such as in shining valves of the artificial heart and internal holes. The proposed mathematical geometry is very similar to the blood vessel models. The blood vessels can be classified into three types: the largest vessels, small vessels and intermediate blood vessels. The largest vessels are identified in the aorta and vena cava and also experience very little heat transfer with the tissue. In addition, there are also more blood vessels that fall into this category. The smallest vessels are noticed in place of arterioles, capillaries and venules which basically experience ideal heat transfer with the blood departure at tissue temperature. The intermediate blood vessels fall in a relatively narrow band with uniformly distributed temperature. These classifications are dependent on the amplitude of vessels and width of channel. Hence, the main purpose of this paper is to study a theoretical analysis of peristaltic transport of a Newtonian nanofluid with slip through a porous medium in the tapered wavy channel subject to convective boundary conditions. The long wavelength and low Reynolds number assumptions are considered. The exact solutions are found in the form of axial velocity from which temperature and volumetric concentration are deduced. Computational results are illustrated and discussed in detail.

2. Mathematical Formulation

Consider an incompressible viscous nanofluid filling the porous space in the tapered wavy channel. The heat transfer between the blood network and living tissues which passes through the channel depends on the geometry of the blood vessel and it is important to understand the behavior of the blood flow and the neighboring tissue nature. Let $\bar{\eta} = \bar{H}_1$ and $\bar{\eta} = \bar{H}_2$ be, correspondingly, the lower and upper blood vessel boundaries of the channel. The sinusoidal waves propagating along the wavy walls of the tapered channel are demonstrated in Figure 1 and mathematically shown as:

$$\bar{H}_2(\bar{\xi},\ t') = \bar{d} + \bar{m}\bar{\xi} + a_2 \sin\left[\frac{\pi}{\lambda}\left(\bar{\xi} - ct'\right)\right]\cos\left[\frac{\pi}{\lambda}\left(\bar{\xi} - ct'\right)\right],$$
$$\bar{H}_1(\bar{\xi}, t') = -\bar{d} - \bar{m}\bar{\xi} - a_1 \sin\left[\frac{\pi}{\lambda}\left(\bar{\xi} - ct'\right) + \phi\right]\cos\left[\frac{\pi}{\lambda}\left(\bar{\xi} - ct'\right) + \phi\right] \tag{1}$$

here $2\bar{d}$, a_1, a_2, $\bar{m}(<< 1)$, λ, c, ϕ, are the width of the channel at the inlet, amplitudes of lower wavy wall, amplitude of upper wavy wall, dimensional non–uniform parameter, wave length, phase speed of the wave and phase difference varies in the range $0 \le \phi \le \pi$, $\phi = 0$ which corresponds to tapered symmetric channel i.e., together walls move towards inward or outward concurrently.

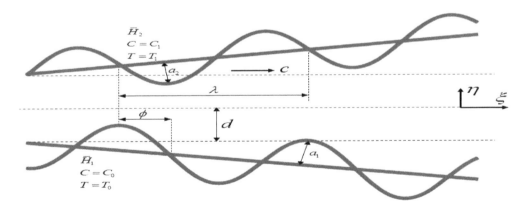

Figure 1. Geometry for peristaltic pumping of nanofluids through a tapered microchannel.

For an incompressible viscous nanofluid the balance of mass, momentum, nanoparticle temperature and volumetric concentration are presented as [56–59]:

$$\frac{\partial \bar{U}}{\partial \bar{\xi}} + \frac{\partial \bar{V}}{\partial \bar{\eta}} = 0, \tag{2}$$

$$\rho_f\left[\frac{\partial}{\partial t'} + \bar{U}\frac{\partial}{\partial \bar{\xi}} + \bar{V}\frac{\partial}{\partial \bar{\eta}}\right]\bar{U} = -\frac{\partial \bar{P}}{\partial \bar{\xi}} + \mu\left[\frac{\partial^2}{\partial \bar{\xi}^2} + \frac{\partial^2}{\partial \bar{\eta}^2}\right]\bar{U} - \frac{\mu}{k}\bar{U}, \tag{3}$$

$$\rho_f\left[\frac{\partial}{\partial t'} + \bar{U}\frac{\partial}{\partial \bar{\xi}} + \bar{V}\frac{\partial}{\partial \bar{\eta}}\right]\bar{V} = -\frac{\partial \bar{P}}{\partial \bar{\eta}} + \mu\left[\frac{\partial^2}{\partial \bar{\xi}^2} + \frac{\partial^2}{\partial \bar{\eta}^2}\right]\bar{V} - \frac{\mu}{k}\bar{V}, \tag{4}$$

$$(\rho c')_f\left[\frac{\partial}{\partial t'} + \overline{U}\frac{\partial}{\partial\overline{\xi}} + \overline{V}\frac{\partial}{\partial\overline{\eta}}\right]\overline{T} = \kappa\left[\frac{\partial^2}{\partial\overline{\xi}^2} + \frac{\partial^2}{\partial\overline{\eta}^2}\right]\overline{T} - \frac{\partial\overline{q}_r}{\partial\overline{\eta}} + \mu\left[4\left(\frac{\partial\overline{U}}{\partial\overline{\xi}}\right)^2 + \left(\frac{\partial\overline{V}}{\partial\overline{\xi}} + \frac{\partial\overline{U}}{\partial\overline{\eta}}\right)^2\right]$$
$$+ (\rho c')_p\left[D_B\left(\frac{\partial\overline{C}}{\partial\overline{\xi}}\frac{\partial\overline{T}}{\partial\overline{\xi}} + \frac{\partial\overline{C}}{\partial\overline{\eta}}\frac{\partial\overline{T}}{\partial\overline{\eta}}\right) + \frac{D_T}{T_m}\left[\left(\frac{\partial\overline{T}}{\partial\overline{\xi}}\right)^2 + \left(\frac{\partial\overline{T}}{\partial\overline{\eta}}\right)^2\right]\right], \tag{5}$$

$$\left[\frac{\partial}{\partial t'} + \overline{U}\frac{\partial}{\partial\overline{\xi}} + \overline{V}\frac{\partial}{\partial\overline{\eta}}\right]\overline{C} = D_B\left[\frac{\partial^2}{\partial\overline{\xi}^2} + \frac{\partial^2}{\partial\overline{\eta}^2}\right]\overline{C} + \frac{D_T}{T_m}\left[\frac{\partial^2}{\partial\overline{\xi}^2} + \frac{\partial^2}{\partial\overline{\eta}^2}\right]\overline{T}, \tag{6}$$

in which $\overline{U}$, $\overline{V}$ are the components of axial velocity along $\overline{\xi}$ and $\overline{\eta}$ directions correspondingly, t', d/dt', $\overline{P}$, ρ_f, ρ_p, $\overline{T}$, κ, k, D_B, D_T, T_m, $\overline{C}$ and $\tau\left(=\frac{(\rho c')_p}{(\rho c')_f}\right)$ are the dimensional time, material time derivative, dimensional pressure, density of the fluid, density of the particle, nanoparticle temperature, thermal conductivity, permeability of porous medium, Brownian diffusion coefficient, themophoretic diffusion coefficient, mean temperature, nanoparticle volumetric volume fraction and the ratio of the effective heat capacity of nanoparticle material and heat capacity of the fluid with ρ being the density. Additionally, T_0, T_1, C_0 and C_1 are the temperature and nanoparticle volume fraction at the lower and upper walls.

3. Convective Boundary Conditions

The convective boundary conditions [60,61] are utilized using Newton's cooling law as:

$$\overline{U} = \frac{\sqrt{\overline{k}}}{\overline{\alpha}}\frac{\partial\overline{U}}{\partial\overline{\eta}}, \; -\overline{k}_h\frac{\partial\overline{T}}{\partial\overline{\eta}} = \overline{h}_h\left(T_0 - \overline{T}\right) \text{ and } -\overline{k}_m\frac{\partial\overline{C}}{\partial\overline{\eta}} = \overline{h}_m\left(C_0 - \overline{C}\right) \text{ at } \overline{\eta} = \overline{H}_1, \tag{7}$$

$$\overline{U} = -\frac{\sqrt{\overline{k}}}{\overline{\alpha}}\frac{\partial\overline{U}}{\partial\overline{\eta}}, \; -\overline{k}_h\frac{\partial\overline{T}}{\partial\overline{\eta}} = \overline{h}_h\left(\overline{T} - T_1\right) \text{ and } -\overline{k}_m\frac{\partial\overline{C}}{\partial\overline{\eta}} = \overline{h}_m\left(\overline{C} - C_1\right) \text{ at } \overline{\eta} = \overline{H}_2, \tag{8}$$

where $\overline{k}$, $\overline{\alpha}$, $\overline{h}_h$, $\overline{h}_m$, $\overline{k}_h$ and $\overline{k}_m$ are the permeability of the porous walls (Darcy number), slip coefficient at the surface of the porous walls, the heat transfer coefficients, mass transfer coefficients respectively, the thermal conductivity and the mass conductivity.

4. Non-Dimensional Analysis

In order to depict the nanoliquid flow in the following non-dimensional measures are introduced in Equations (1)–(8). $\left(u = \frac{\overline{U}}{c}, \; u = \frac{\partial\psi}{\partial y}, \; v = \frac{\overline{V}}{c}, \; v = -\delta\frac{\partial\psi}{\partial y}\right)$ are the velocity components in direction of $\left(x = \frac{\overline{\xi}}{\lambda}, \; y = \frac{\overline{\eta}}{d}\right)$, ψ is the stream function, $t = \frac{ct'}{\lambda}$ is the dimensionless time, $h_1 = \frac{\overline{H}_1}{d}$ and $h_2 = \frac{\overline{H}_2}{d}$ represent the dimensionless form of the lower and upper channel, and $R = \frac{\rho_f c d}{\mu}$ is the Reynolds number, $p = \frac{d^2\overline{P}}{c\lambda\mu}$ is the dimensionless pressure, $a = \frac{a_1}{d}$ is the amplitude of lower wavy wall, $b = \frac{a_2}{d}$ is the amplitude of upper wavy wall, $m = \frac{\lambda\overline{m}}{d}$ is the dimensionless non-uniform parameter, $\delta = \frac{d}{\lambda}$ is the wave number, $Sc = \frac{v}{D_B}$ is the Schmidt number, $K = \frac{\overline{k}}{d^2}$ is the Permeability parameter, $\theta = \frac{\overline{T}-T_0}{T_1-T_0}$ is the dimensionless nanoparticle temperature, $\sigma = \frac{\overline{C}-C_0}{C_1-C_0}$ is the nanoparticle volumetric concentration or dimensionless rescaled nanoparticle volume fraction, $\Pr = \frac{\mu c_f}{\kappa}$ is the Prandtl number, $N_b = \frac{\tau D_B(C_1-C_0)}{v}$ is the Brownian motion parameter $N_t = \frac{\tau D_T(T_1-T_0)}{T_m v}$ is the thermophoresis parameter, $Ec = \frac{c^2}{c_f T_m}$ is the Eckert number, $Br = \Pr Ec$ is the Brinkman number, and $R_n = \frac{16\overline{\sigma}T_0^3}{3\overline{k}\mu c_f}$ is the thermal radiation parameter and also, applying the long wavelength and low Reynolds number approximations, we attain:

$$\frac{\partial p}{\partial x} = \frac{\partial^3\psi}{\partial y^3} - \left(\frac{1}{K}\right)\frac{\partial\psi}{\partial y}, \tag{9}$$

$$\frac{\partial p}{\partial y} = 0, \tag{10}$$

$$(1 + Rn\mathrm{Pr})\frac{\partial^2 \theta}{\partial y^2} + Br\left(\frac{\partial^2 \psi}{\partial y^2}\right)^2 + (Nb\mathrm{Pr})\left(\frac{\partial \sigma}{\partial y}\frac{\partial \theta}{\partial y}\right) + (Nt\mathrm{Pr})\left(\frac{\partial \theta}{\partial y}\right)^2 = 0, \tag{11}$$

$$\frac{\partial^2 \sigma}{\partial y^2} + \left(\frac{Nt}{Nb}\right)\frac{\partial^2 \theta}{\partial y^2} = 0. \tag{12}$$

Equation (10) shows that p is not dependent on x. Reducing the pressure gradient term from Equations (9) and (10), it yields:

$$\frac{\partial^4 \psi}{\partial y^4} - \left(\frac{1}{K}\right)\frac{\partial^2 \psi}{\partial y^2} = 0. \tag{13}$$

Additionally, it is noticed that the continuity equation is routinely fulfilled.

$$h_1 = -1 - mx - a\cos(\pi(x-t)+\phi)\sin(\pi(x-t)+\phi) \text{ and}$$
$$h_2 = 1 + mx + b\cos(\pi(x-t))\sin(\pi(x-t))$$

$$\psi = -\frac{F}{2}, \frac{\partial \psi}{\partial y} = L\frac{\partial^2 \psi}{\partial y^2}, \frac{\partial \theta}{\partial y} = B_h \theta \text{ and } \frac{\partial \sigma}{\partial y} = B_m \sigma \text{ at } y = h_1,$$
$$\psi = \frac{F}{2}, \frac{\partial \psi}{\partial y} = -L\frac{\partial^2 \psi}{\partial y^2}, \frac{\partial \theta}{\partial y} = B_h(1-\theta) \text{ and } \frac{\partial \sigma}{\partial y} = B_m(1-\sigma) \text{ at } y = h_2. \tag{14}$$

where $L = \frac{\sqrt{k}}{d\alpha}$ is the velocity slip parameter, $B_h = \frac{\bar{h}_h \bar{d}}{\bar{k}_h}$ is the heat transfer Biot number and $B_m = \frac{\bar{h}_m \bar{d}}{\bar{k}_m}$ is the mass transfer Biot number.

5. Analytical Solution

The solution of the Equation (13) subject to the conditions in Equation (14) is obtained as:

$$\psi(y) = \begin{aligned} &-(F(\cosh Ny - \sinh Ny)(2(\cosh 2Ny - \sinh 2Ny) - 2(\cosh(N(h_1+h_2)) + \sinh(N(h_1+h_2)))\\ &+(h_1+h_2-2y)N(\cosh Ny + \sinh Ny)(\cosh Nh_1 + \sinh Nh_1 + \cosh Nh_2 + \sinh Nh_2)\\ &-LN^2(\cosh Ny + \sinh Ny)(h_1+h_2-2y)(\cosh Nh_1 + \sinh Nh_1 - \cosh Nh_2 - \sinh Nh_2)))\\ &\Big/\left(2\left(\begin{array}{c}(2+LN^2h_1-LN^2h_2)(\cosh Nh_1 + \sinh Nh_1 - \cosh Nh_2 - \sinh Nh_2)\\ -(Nh_1-Nh_2)(\cosh Nh_1 + \sinh Nh_1 + \cosh Nh_2 + \sinh Nh_2)\end{array}\right)\right)\end{aligned} \tag{15}$$

The integration of Equation (12) with respect to y, we obtain

$$\frac{\partial \sigma}{\partial y} + \frac{Nt}{Nb}\frac{\partial \theta}{\partial y} = f(x). \tag{16}$$

Solving Equations (11) and (12) and substituting in Equation (16) subject to boundary conditions of Equations (14), the dimensionless nanoparticle temperature field is attained as

$$\theta(y) = \begin{aligned}&A_8 + A_9(\cosh(A_1A_5N_by) - \sinh(A_1A_5N_by)) - \frac{A_2^3A_4(\cosh(2Ny)+\sinh(2Ny))}{4N^2+(\cosh(2A_1A_5N_bN)+\sinh(2A_1A_5N_bN))}\\ &-\frac{A_3^2A_4(\cosh(2Ny)-\sinh(2Ny))}{4N^2-(\cosh(2A_1A_5N_bN)+\sinh(2A_1A_5N_bN))} - \frac{(A_6\beta+2A_2A_3A_4)y}{A_1A_5N_b},\end{aligned} \tag{17}$$

and the nanoparticle volumetric concentration is obtained as:

$$\sigma(y) = \begin{aligned}&A_{10} + A_{11}y - \frac{A_9N_t(\cosh(A_1A_5N_by)-\sinh(A_1A_5N_by))e^{-A_1A_5N_by}}{N_b}\\ &+\frac{A_4A_3^2N_t(\cosh(2Ny)-\sinh(2Ny))}{4N^2N_b-2A_1A_5NN_b^2} + \frac{A_4A_2^2N_t(\cosh(2Ny)+\sinh(2Ny))}{4N^2N_b+2A_1A_5NN_b^2}.\end{aligned} \tag{18}$$

The coefficient of nanoparticle heat transfer at the lower wall is specified by

$$Z = h_{1x}\theta_y. \tag{19}$$

The above mentioned constants are elaborated in the Appendix A.

6. Computational Results and Discussion

In general, exact solutions for temperature, nanoparticle volumetric concentrations and coefficient of nanoparticle temperature depend on the value of $f(x)$. First of all, $f(x)$ can be influenced by hiring for θ and σ from Equation (16). It ought to be noticed that observing the value of $f(x)$ analytically from Equation (16) in terms of the other parameters set is a very difficult task and it may be impossible. Nevertheless, with the help of MATHEMATICA/MATLAB software, the numerical solutions are still available. The numerical value of $f(x)$ plays an important role in plotting the graphs for variation of the streamlines, nanoparticle temperature distribution, nanoparticle volumetric concentration and coefficient of nanoparticle temperature.

To analyze the results, instantaneous volume rate $F(x, t)$ is considered as varying exponentially with the relation (Kikuchi [62])

$$F = \Theta e^{-At}, \tag{20}$$

where Θ is the mean flow rate or flow constant, A is the blood flow constant. Figures 2–20 were plotted to examine the stream function $(\psi(x, y))$, nanoparticle temperature $(\theta(y))$, nanoparticle concentration $(\sigma(y))$ and heat transfer coefficient $(Z(x))$. Additionally, it is noticed that the flow rate for the non-positive and positive flow rate $F < 0$ or $F > 0$ may be according to $\Theta < 0$ or $\Theta > 0$. It was detected through an experiment performed by Kikuchi [62] that the flow rate decreases exponentially with time however mean flow rate does not depend on the structural details of the channel.

6.1. Thermal and Concentration Profiles

Effects of permeability parameter (K), slip parameter (L), mean flow rate (Θ), non-uniform parameter (m), Brinkman number (Br), Prandtl number (Pr), heat transfer Biot number (B_h), mass transfer Biot number (B_m), thermal radiation (R_n) and thermophoresis parameter (N_t) on temperature profile are analyzed through Figures 2–11. In accordance with physical laws, the energy fluency requires to destroy cancer cells greatly depends on the number of nanoparticles temperature within the cell. Additionally, the role of this study is to improve correlations and estimation methods for calculating magnitudes of upper and lower tapered wavy wall boundaries of heat transfer in and around the individual blood vessels. The analysis did not consider vessel size and any experimental values because flow oscillations due to the heartbeat are not present in these small vessels. The aim of this study is to improve correlations and estimation methods for scheming magnitudes and upper and lower limits of heat and mass transfer in and around individual blood vessels.

The nanoparticle temperature and concentration profiles resulted in the vessel exit are shown in Figure 2 for three different permeability parameter values such as $(K \to 0, K = 0.2, K \to \infty)$. It is noticed, theoretically, the absence of a permeability parameter shows very few heat exchanges in the blood vessel, but the particles movement is raised in the blood vessel. In Figure 3, we noticed the effects of slip parameter (L) on the nanoparticle temperature and concentration profiles for fixed values of other parameters. Three different slip parameters are used in the nanoparticle temperature and concentration distribution such as $L = 0$, $L = 0.1$ and $L = 0.2$. It is important to note with the enhancement in the velocity slip parameter, the nanoparticle temperature and concentration at any point in the flow medium enhances, but the behavior of temperature profile decreases and at the same time nanoparticle concentration increases when the velocity slip parameter rises. The effects of flow constant (Θ) on nanoparticle temperature and concentration profiles are shown in Figure 4. It is noticed that presence of a flow constant increases the nanoparticle temperature and also enhances uniformly in the boundaries of the channel. However, the nanoparticle concentration shows the revised behavior in nature of temperature distribution. From Figure 5, which elucidates the effect of the non-uniform parameter (m) on the nanoparticle temperature and concentration profiles, it is exposed that when the non-uniform parameter increases, the nanoparticle temperature of blood flow consistently reduces with the flow medium.

The nanoparticle displacement increases with increasing of the non-uniform parameter. These physical changes play crucial roles in the treatment of thermotherapy. In Figure 6, the causes of Brinkman number (Br) on nanoparticle temperature and concentration are captured. It is noticed that the temperature of the fluid increases with the increase of Brinkman number. It is well known about nanofluids that when the nanoparticle temperature rises, the distance between molecules increases due to cohesive force decreases. Therefore, viscosity of nanofluids decreases when the nanoparticle temperature increases. On the flip side, absence of Brinkman number shows the maximum displacement of the particles.

The effect of the Prandtl number (Pr) on nanoparticle temperature and concentration are depicted in Figure 7. It is observed that with an increase in Pr, the temperature of the fluid increases. It indicates that nanofluids can have significantly better heat transfer characteristics than the base fluids. Additionally, it is noticed that the nanoparticle concentration decreases with increasing the Prandtl number. This indicates that enhances in Prandtl number is accompanied by an enrichment of the heat transfer rate at the tapered wavy wall of the blood vessel. The fundamental physics behind this can be depicted as follows. When the blood achieves a higher Prandtl number, its thermal conductivity is dropped down and so its heat conduction capacity is reduced. Simultaneously, the heat transfer rate at the vessel wall is enhanced. We noticed from Figure 8 that the temperature enhances with rise of heat transfer Biot number (B_h) at the upper portion of the channel, but the influence is reversed at the lower portion of the channel. Further, it can be noted that the temperature at the upper wall is maximum and it reduces slowly towards the lower wall.

The small value of heat transfer Biot number shows the conduction nature, while high values of heat transfer Biot number indicates that the convection is the main heat transfer mechanism. Any rate of nanoparticle concentration reduces with increase of the heat transfer Biot number. Figure 9 reveals that the nanoparticle temperature and concentration enhance as mass transfer Biot number increases. Figure 10 illuminates the influence of the thermal radiation on nanoparticle temperature and concentration distribution. This figure highlights that thermal radiation enhances during blood flow in the channel, thereby the nanoparticle temperature of the tapered asymmetric wavy channel is reduced by increase of thermal radiation. Additionally, the converse situation occurred in the nanoparticle concentration profile. It shows that the external radiation dilutes the temperature, and at the same time movement of the particle increases.

This concept may be very useful in the treatment of heart transfer mechanism. Figure 11 illuminates a very significant influence of the thermophoresis parameter on the nanoparticle temperature and concentration profiles. It is well known that the strength of thermophoresis rises due to temperature gradient enhancement, which increases the blood flow in the channel. At the same time, the nanoparticle concentration of the particle displacement reduces with increases of the thermophoresis parameter.

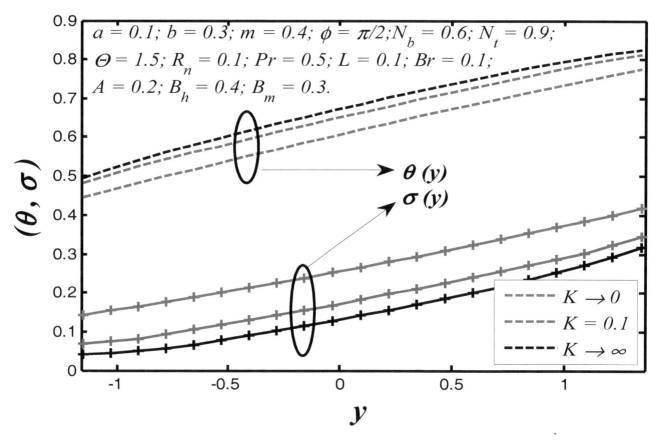

Figure 2. Nanoparticle temperature and concentration profiles $\theta(y)$ for K

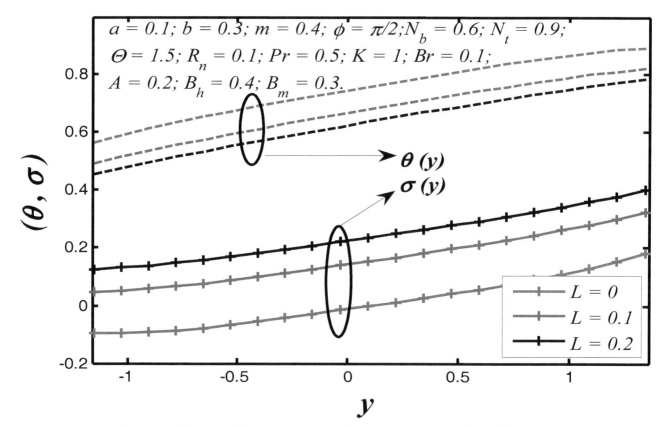

Figure 3. Nanoparticle temperature and concentration profiles $\theta(y)$ for L.

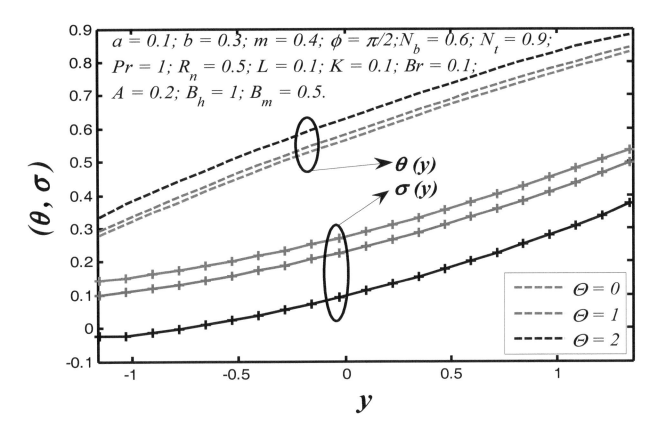

Figure 4. Nanoparticle temperature and concentration profiles $\theta(y)$ for Θ.

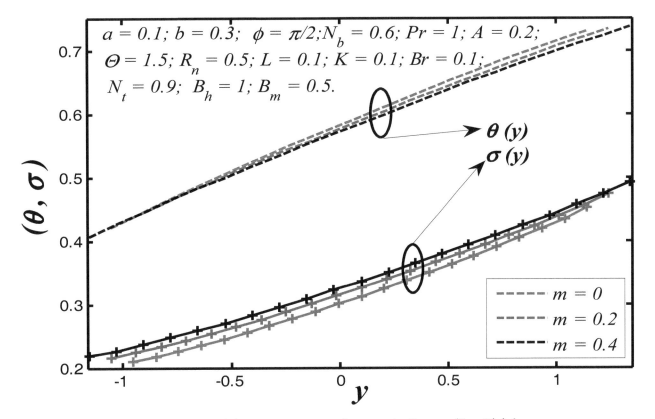

Figure 5. Nanoparticle temperature and concentration profiles $\theta(y)$ for m.

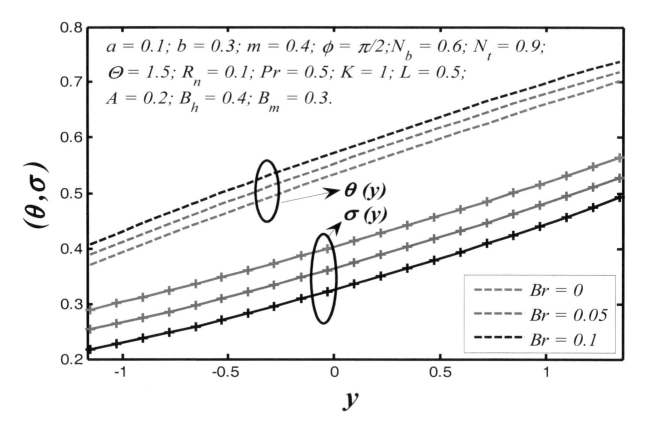

Figure 6. Nanoparticle temperature and concentration profiles $\theta(y)$ for Br.

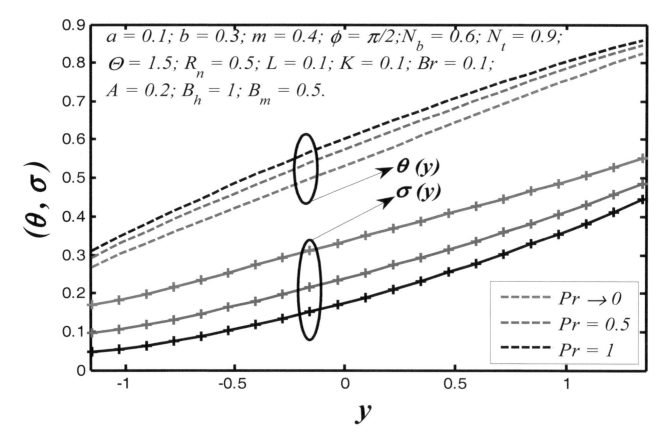

Figure 7. Nanoparticle temperature and concentration profiles $\theta(y)$ for Pr.

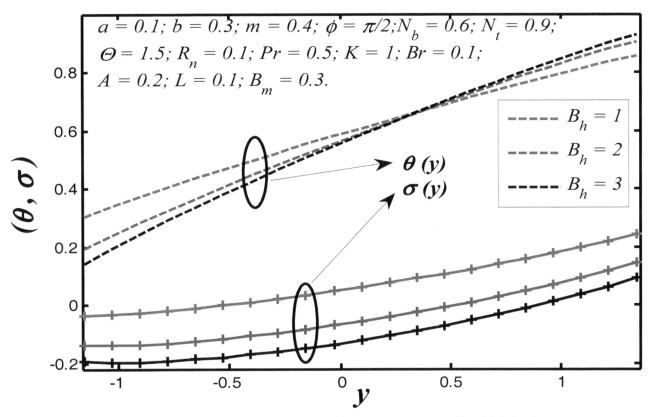

Figure 8. Nanoparticle temperature and concentration profiles $\theta(y)$ for B_h.

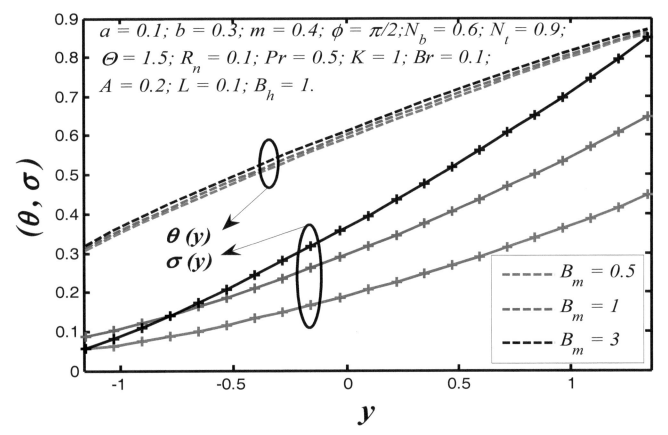

Figure 9. Nanoparticle temperature and concentration profiles $\theta(y)$ for B_m.

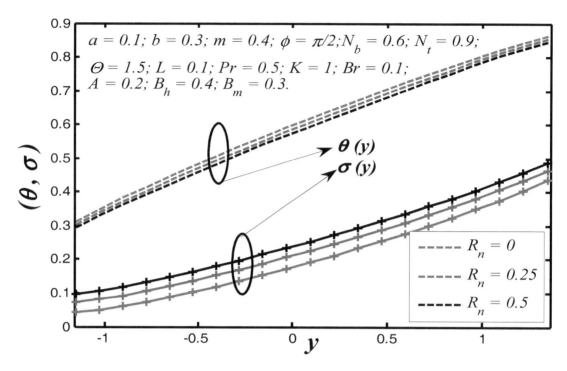

Figure 10. Nanoparticle temperature and concentration profiles $\theta(y)$ for R_n.

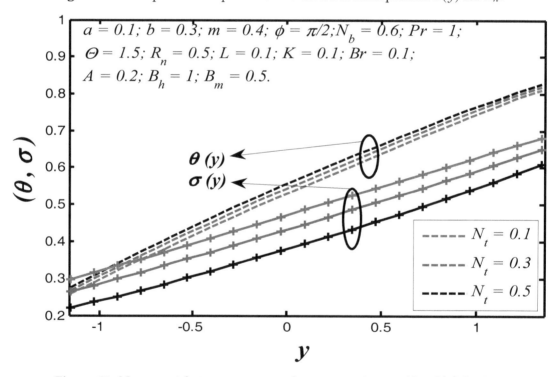

Figure 11. Nanoparticle temperature and concentration profiles $\theta(y)$ for N_t.

6.2. Nanoparticle Heat Transfer Coefficient

The effects of various parameters on the nanoparticle heat transfer coefficient at the upper wall are represented in Figures 12–18. The nanoparticle heat transfer coefficients for a viscous nanofluid in the tapered wavy channel depends on many physical quantities related to the fluid or the geometry of the system through which the fluid is flowing. It is observed that the heat transfer coefficient is in oscillatory behavior which may be due to contraction and equation of walls. The absolute value of heat transfer coefficient increases with the increase of L and Br while it decreases with increasing m, B_h, B_m, N_b and R_n.

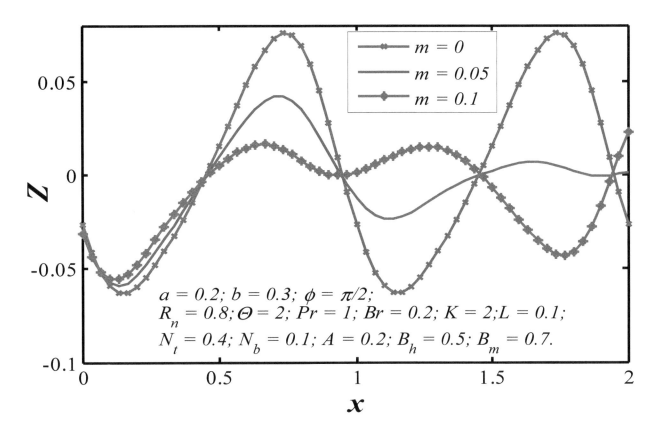

Figure 12. Nanoparticle heat transfer coefficient $Z(x)$ profiles for various values of m.

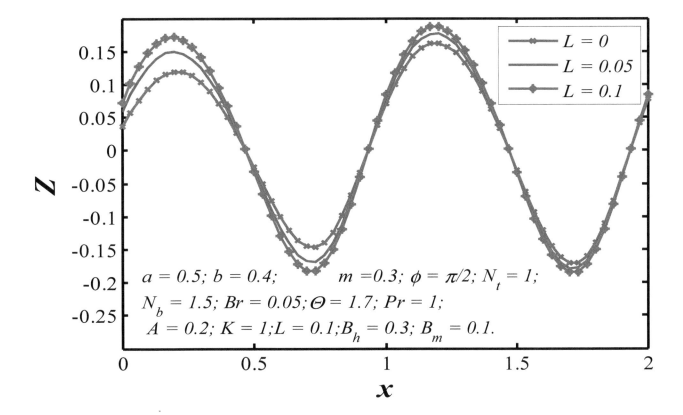

Figure 13. Nanoparticle heat transfer coefficient $Z(x)$ profiles for various values of L.

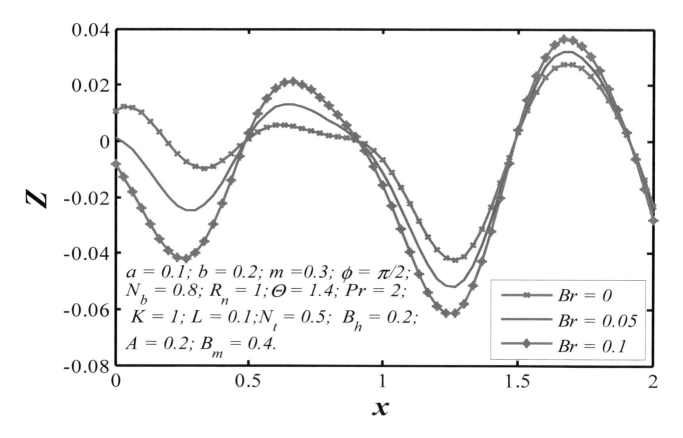

Figure 14. Nanoparticle heat transfer coefficient $Z(x)$ profiles for various values of Br.

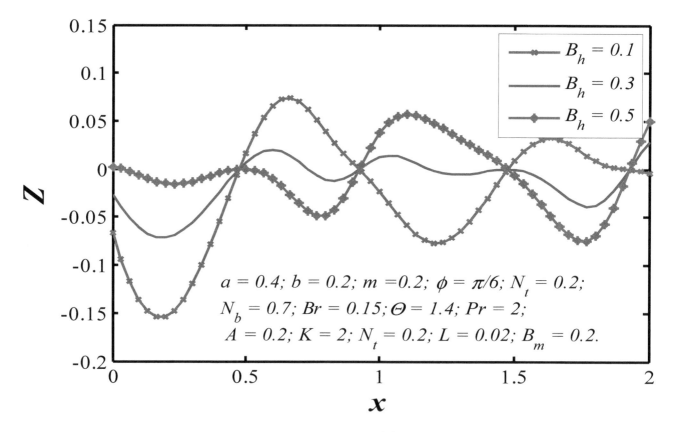

Figure 15. Nanoparticle heat transfer coefficient $Z(x)$ profiles for various values of B_h.

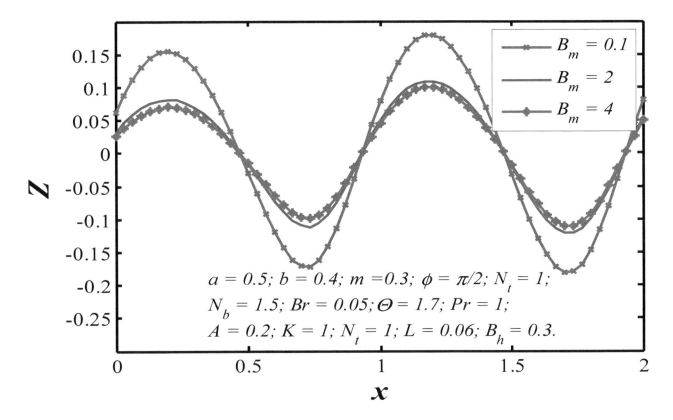

Figure 16. Nanoparticle heat transfer coefficient $Z(x)$ profiles for various values of B_m.

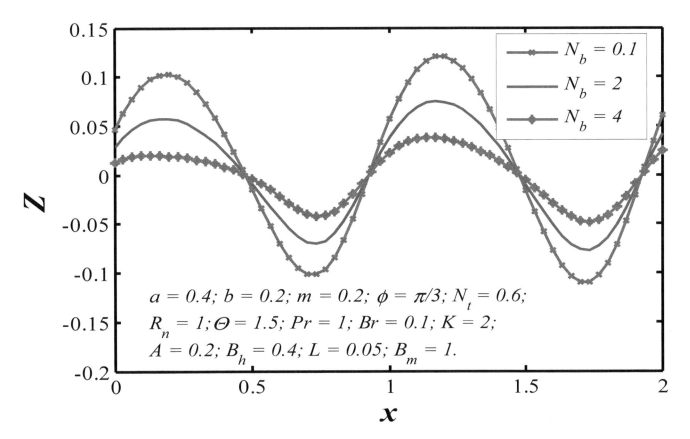

Figure 17. Nanoparticle heat transfer coefficient $Z(x)$ profiles for various values of N_b.

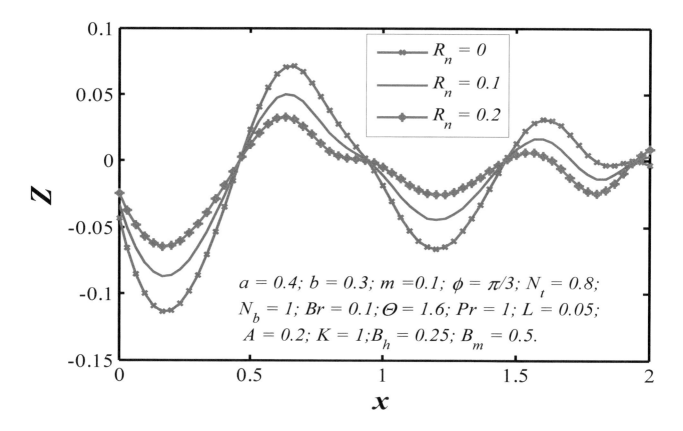

Figure 18. Nanoparticle heat transfer coefficient $Z(x)$ profiles for various values of R_n.

6.3. Trapping

In this subsection, the streamlines for the tapered asymmetry channels are shown in Figure 19. The effects of the non-uniform parameter m on trapping are presented in Figure 19a,b. One can observe that the size of the trapped bolus increases with an increase in m. The effect of slip parameter L on trapping can be seen in Figure 19b,c. It is observed that by increasing the value of velocity slip parameter L, the circulation of trapped bolus increases, at the same time size of the bolus is reduced. To see the effects of permeability parameter K on trapping, Figure 19c,d was illustrated. It is noted that an increase in the permeability parameter increases the size of channel, but the size of the trapped bolus decreases. The streamline patterns in the wave frame for viscous nanofluid for different values of Blood flow rate parameter Θ are shown in Figure 19d,e. It is observed that for small values of Θ only one trapped bolus is formed. It is also observed that the bolus near the upper and lower wavy walls increase eventually with the tapered micro channel.

6.4. Validation

The results of present mathematical model obtained by direct analytical approach were authenticated with the numerical solutions computed by MATLAB through BVP4c command. A validation was completed in Figure 20 and it is portrayed for nanoparticle temperature and concentration distribution at fixed values of pertinent parameters. Additionally, it is noticed that the analytical solution for the entire values of the tapered wavy channel width has a good correlation with the numerical solution computed by the MATLAB. The proposed mathematical formulation has very good correlation in axial velocity with Mishra and Rao [63] in the absence of $a = 0, b = 0$ and $K \to \infty$.

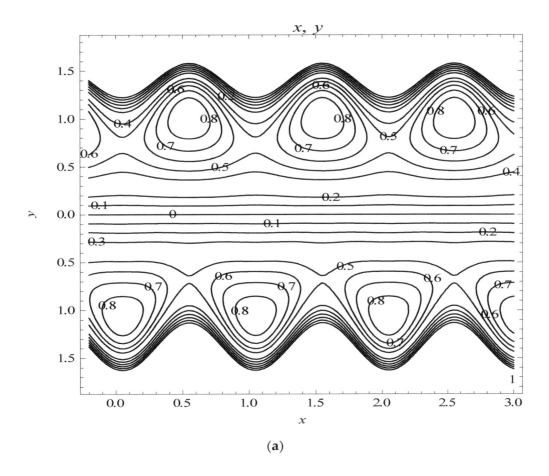

(a)

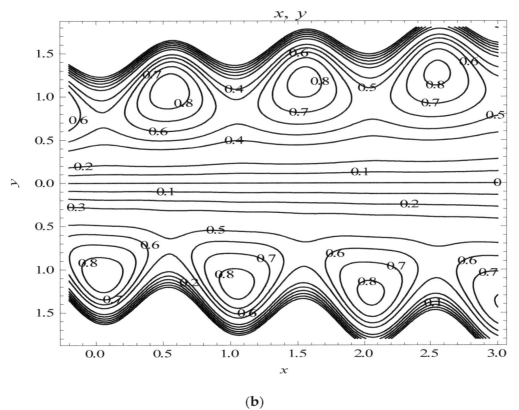

(b)

Figure 19. *Cont.*

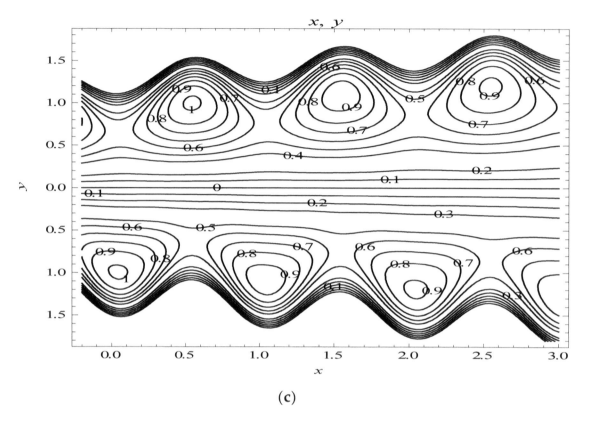

(c)

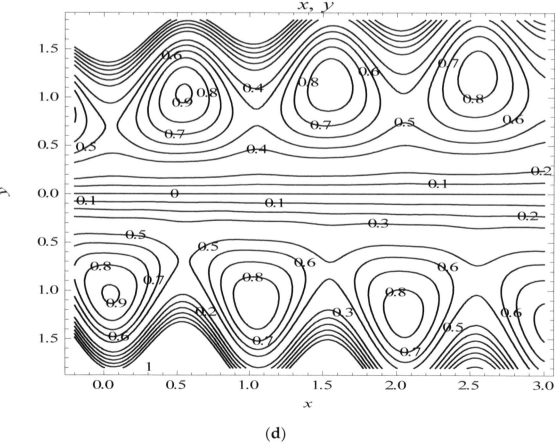

(d)

Figure 19. *Cont.*

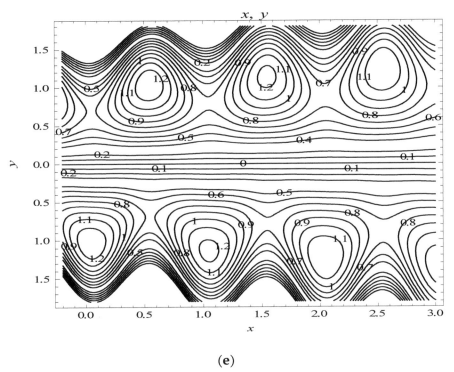

(e)

Figure 19. Streamlines when $a = 0.4, b = 0.3, \phi = \pi/2, t = 0.3, A = 0.2$, (a) $m = 0, L = 0.1, K = 0.1$, $\Theta = 1.25$, (b) $m = 0.1, L = 0.1, K = 0.1, \Theta = 1.25$, (c) $m = 0.1, L = 0.15, K = 0.1, \Theta = 1.25$, (d) $m = 0.1, L = 0.15, K \rightarrow \infty, \Theta = 1.25$, (e) $m = 0.1, L = 0.15, K \rightarrow \infty, \Theta = 1.5$.

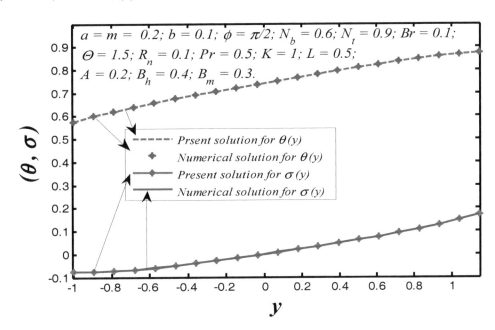

Figure 20. Comparison between numerical and present solutions for temperature and nanoparticle volume fraction profiles.

7. Conclusions

The analytical and numerical solutions of Equations (9)–(12) were estimated for stream function, nanoparticle temperature and concentration distribution. The expressions for nanoparticle temperature, volumetric fraction, heat transfer coefficient profile and stream function were discussed graphically. The following observations were noticed.

- Nanoparticle heat transfer between the tapered walls strongly depends on Brinkman number because the tissue presents the chief resistance to heat flow.

- Thermal radiation contains the potential to contribute a significant change in the nanoparticle temperature distribution.

- With increasing the radiation parameter, the nanoparticle temperature and heat transfer coefficient enhance.

- The nanoparticle temperature reduces with enhancing the Prandtl number, however, reverse behavior is noticed for nanoparticle concentration.

- Heat transfer coefficient depends on the flow, thermal and geometrical nature of flow regime.

- The trapping phenomenon also alters with changing the magnitude of slip and permeability parameters.

- The findings of the present models can be utilized to engineer smart peristaltic pumps which can be applicable for transporting drugs and delivery of nanoparticles.

Author Contributions: Conceptualization, D.T.; Investigation, J.P.; Methodology, A.K.T.; Visualization, R.E.; Writing—review and editing, S.M.S.

Acknowledgments: R. Ellahi thanks Sadiq M. Sait, the Director Office of Research Chair Professors, King Fahd University of Petroleum and Minerals, Dhahran, Saudi Arabia, to honor him with the Chair Professor at KFUPM.

Nomenclature

Symbol	description	Unit
(a_1, a_2)	Dimensional amplitude of the lower and upper walls	m
c	Wave speed	m/s
$\overline{C}$	Nanoparticle volumetric volume fraction	Kg/m^3
C_0, C_1	Nanoparticle concentration at the lower and upper walls	Kg/m^3
D_B	Brownian diffusion coefficient	m/s
D_T	Themophoretic diffusion coefficient	m^2/s
$\overline{d}$	Dimensionless half width of the channel	m
$\overline{h}_h$	Heat transfer coefficient	W/m^{2}K (or) kg/s^{3}K
$\overline{h}_m$	Mass transfer coefficient	m/s
k	Permeable of porous medium	H/m
$\overline{k}$	Permeability of the porous wavy wall	Darcy (or) m^2
$\overline{k}_h$	Thermal conductivity of wavy wall	W/mK
$\overline{k}_m$	Mass conductivity of wavy wall	W/mK
$\overline{m}$	Dimensional non-uniform parameter	m
$\overline{P}$	Dimensional pressures	Pa (or) N/m^2 (or) kg/ms^2
$\overline{q}_r$	Uni-directional thermal radiative flux	kg/s^3 (or) W/m^2
t'	Dimensional time	s
$\overline{T}$	Nanoparticle temperature	K
T_m	Mean temperature	K
(T_0, T_1)	Temperature at the lower and upper walls	K
$\overline{U}, \overline{V}$	Velocity components in the wave frame	m/s
$\overline{\xi}, \overline{\eta}$	Rectangular coordinates	m
ρ_f	Density of the fluid	Kg/m^3
ρ_p	Density of the particle	Kg/m^3
μ	Dynamic Viscosity	kg/m.s
κ	Thermal conductivity of the fluid	m^2/s
λ	Wave length	m

Dimensionless parameters:

$\bar{\alpha}$	Slip coefficient at the surface of the porous walls
A	Blood flow constant
$(a,\ b)$	Dimensionless amplitude of the lower and upper walls
B_h	Heat transfer Biot number
B_m	Mass transfer Biot number
Br	Brinkman number
Ec	Eckert number
F	Dimensionless flow rate
$\left(\overline{H}_1,\ \overline{H}_2\right)$	Lower and upper wall boundaries of the micro- asymmetric channel
$(h_1,\ h_2)$	Dimensionless lower and upper wall shapes in wave frame
L	Slip parameter
m	Dimensionless non-uniform parameter
N_t	Thermophoresis parameter
N_b	Brownian motion parameter
p	Dimensionless pressure
Pr	Prandtl number
R	Reynolds number
R_n	Thermal radiation
Sc	Schmidt number
t	Dimensionless time
$(u,\ v)$	Velocity components in the wave frame $(x,\ y)$
Θ	Constant flow rate
σ	Dimensionless rescaled nanoparticle volume fraction
θ	Dimensionless nanoparticle temperature
ψ	Stream function
ϕ	Phase difference
K	Permeability parameter
δ	Wave number

Appendix A

The following constants are utilized in the solution of the manuscript.

$$N = 1/\sqrt{K},\ A_1 = f(x),$$

$$A_2 = \frac{-F}{(LN^2h_1 + 2 - LN^2h_2)\left(\begin{array}{c}\cosh(Nh_1) + \sinh(Nh_1)\\ -\cosh(Nh_2) - \sinh(Nh_2)\end{array}\right) + (Nh_2 - Nh_1)\left(\begin{array}{c}\cosh(Nh_1) + \sinh(Nh_1)\\ +\cosh(Nh_2) + \sinh(Nh_2)\end{array}\right)},$$

$$A_3 = \frac{F(\cosh(N(h_1 + h_2)) + \sinh(N(h_1 + h_2)))}{(LN^2h_1 + 2 - LN^2h_2)\left(\begin{array}{c}\cosh(Nh_1) + \sinh(Nh_1)\\ -\cosh(Nh_2) - \sinh(Nh_2)\end{array}\right) + (Nh_2 - Nh_1)\left(\begin{array}{c}\cosh(Nh_1) + \sinh(Nh_1)\\ +\cosh(Nh_2) + \sinh(Nh_2)\end{array}\right)},$$

$$A_4 = \frac{BrN^4}{1 + R_n Pr},\ A_5 = \frac{Pr}{1 + R_n Pr},$$

$$A_6 = \frac{(B_h h_1 - 1)2A_2 A_3 A_5}{A_1 A_5 N_b} + \frac{A_2^2 A_4(\cosh(2Nh_1) + \sinh(2Nh_1))(B_h - 2N)}{4N^2 + 2A_1 A_5 N_b N} + \frac{A_3^2 A_4(\cosh(2Nh_1) - \sinh(2Nh_1))(B_h + 2N)}{4N^2 - 2A_1 A_6 N_b N},$$

$$A_7 = -B_h - \frac{(B_h h_2 + 1)(A_5 \beta + 2A_2 A_3 A_5)}{A_1 A_5 N_b} - \frac{A_2^2 A_4(B_h + 2N)(\cosh(2Nh_2) + \sinh(2Nh_2))}{4N^2 + 2A_1 A_5 N_b N} + \frac{A_3^2 A_4(\cosh(2Nh_2) - \sinh(2Nh_2))(B_h - 2N)}{4N^2 - 2A_1 A_6 N_b N},$$

$$A_8 = \frac{A_7 + A_8(\cosh(A_1 A_5 N_b h_2) - \sinh(A_1 A_5 N_b h_2))(B_h - A_1 A_5 N_b)}{B_h},$$

$$A_9 = \frac{A_6 + A_7}{(\cosh(A_1 A_5 N_b h_1) - \sinh(A_1 A_5 N_b h_1))(B_h + A_1 A_5 N_b) - (\cosh(A_1 A_5 N_b h_2) - \sinh(A_1 A_5 N_b h_2))(B_h - A_1 A_5 N_b)}$$

$$A_{10} = -\frac{A_{13}}{B_m} - \frac{(B_m h_2 + 1)(A_{12} + A_{13})}{B_m(B_m(h_1 - h_2) - 2)}, \quad A_{11} = \frac{A_{12} + A_{13}}{B_m(h_1 - h_2) - 2},$$

$$A_{12} = A_9\left(A_1 A_5 N_t + \frac{B_m N_t}{N_b}\right)(\cosh(A_1 A_5 N_b h_1) - \sinh(A_1 A_5 N_b h_1))$$
$$- \frac{A_2^2 A_4 N_t (B_m - 2N)(\cosh(2Nh_1) + \sinh(2Nh_1))}{4N^2 N_b + 2A_1 A_5 N N_b^2}$$
$$- \frac{A_3^2 A_4 N_t (\cosh(2Nh_1) - \sinh(2Nh_1))(B_m + 2N)}{4N^2 N_b - 2A_1 A_5 N N_b^2},$$

$$A_{13} = A_9(\cosh(A_1 A_5 N_b h_2) - \cosh(A_1 A_5 N_b h_2))\left(A_1 A_5 N_t - \frac{B_m N_t}{N_b}\right)$$
$$- B_m + \frac{A_2^2 A_4 N_t (B_m + 2N)(\cosh(2Nh_2) - \cosh(2Nh_2))}{4N^2 N_b + 2A_1 A_5 N N_b^2}$$
$$+ \frac{A_3^2 A_4 N_t (\cosh(2Nh_2) - \cosh(2Nh_2))(B_m - 2N)}{4N^2 N_b - 2A_1 A_5 N N_b^2},$$

References

1.   Burns, J.C.; Parkes, T. Peristaltic motion. *J. Fluid Mech.* **1967**, *29*, 731–743. [CrossRef]
2.   Zien, T.F.; Ostrach, S. A long wave approximation to peristaltic motion. *J. Biomech.* **1970**, *3*, 63–75. [CrossRef]
3.   Raju, K.K.; Devanathan, R. Peristaltic motion of a non-Newtonian fluid. *Rheol. Acta* **1972**, *11*, 170–178. [CrossRef]
4.   Ellahi, R.; Zeeshan, A.; Hussain, F.; Asadollahi, A. Peristaltic blood flow of couple stress fluid suspended with nanoparticles under the influence of chemical reaction and activation energy. *Symmetry* **2019**, *11*, 276. [CrossRef]
5.   Zeeshan, A.; Ijaz, N.; Abbas, T.; Ellahi, R. The sustainable characteristic of Bio-bi-phase flow of peristaltic transport of MHD Jeffery fluid in human body. *Sustainability* **2018**, *10*, 2671. [CrossRef]
6.   Hussain, F.; Ellahi, R.; Zeeshan, A.; Vafai, K. Modelling study on heated couple stress fluid peristaltically conveying gold nanoparticles through coaxial tubes: A remedy for gland tumors and arthritis. *J. Mol. Liq.* **2018**, *268*, 149–155. [CrossRef]
7.   Choi, S.U.S.; Eastman, J.A. Enhancing Thermal Conductivity of Fluids with Nanoparticles. In *Proceedings of Enhancing Thermal Conductivity of Fluids with Nanoparticles, San Francisco, CA, USA*; American Society of Mechanical Engineers, FED: New York, NY, USA, 1995; Volume 231, pp. 99–105. Available online: https://www.osti.gov/biblio/196525-enhancing-thermal-conductivity-fluids-nanoparticles (accessed on 21 June 2019).
8.   Masuda, H.; Ebata, A.; Teramae, K.; Hishinuma, N. Alteration of thermal conductivity and viscosity of liquids by dispersing ultra-fine particles. *Netsu Bussei.* **1993**, *7*, 227–233. [CrossRef]
9.   Buongiorno, J.; Hu, W. Nanofluid Coolants for Advanced Nuclear Power Plants. In Proceedings of the nternational Congress on Advances in Nuclear Power Plants (ICAPP'05), Seoul, Korea, 15–19 May 2005.
10. Buongiorno, J. Convective transport in nanofluids. *J. Heat Transf.* **2005**, *128*, 240–250. [CrossRef]
11. Akbar, N.S.; Nadeem, S. Endoscopic effects on the peristaltic flow of a nanofluid. *Commun. Theor. Phys.* **2011**, *56*, 761–768. [CrossRef]
12. Akbar, N.S. Peristaltic Sisko nanofluid in an asymmetric channel. *Appl. Nanosci.* **2014**, *4*, 663–673. [CrossRef]
13. Tripathi, D.; Beg, O.A. A study on peristaltic flow of nanofluids: Application in drug delivery systems. *Int. J. Heat Mass Transf.* **2014**, *70*, 61–70. [CrossRef]
14. Akbar, N.S.; Nadeem, S.; Khan, Z.H. Numerical simulation of peristaltic flow of a Carreau nanofluid in an asymmetric channel. *Alexandria Eng. J.* **2013**, *53*, 191–197. [CrossRef]
15. Bég, O.A.; Tripathi, D. Mathematica simulation of peristaltic pumping with double-diffusive convection in nanofluids: A bio-nano-engineering model. *Proc. Inst. Mech. Eng. Part N J Nanoeng. Nanosyst.* **2012**, *225*, 99–114. [CrossRef]
16. Akbar, N.S.; Tripathi, D.; Bég, A.O. Modeling nanoparticle geometry effects on peristaltic pumping of medical magnetohydrodynamic nanofluids with heat transfer. *J. Mechan. Med. Bio.* **2016**, *16*, 1650088. [CrossRef]

17. Reddy, M.G.; Makinde, O.D. Magnetohydrodynamic peristaltic transport of Jeffrey nanofluid in an asymmetric channel. *J. Mol. Liq.* **2016**, *223*, 1242–1248. [CrossRef]

18. Akbar, N.S.; Huda, A.B.; Tripathi, D. Thermally developing MHD peristaltic transport of nanofluids with velocity and thermal slip effects. *Eur. Phys. J. Plus.* **2016**, *131*, 332. [CrossRef]

19. Nadeem, S.; Riaz, A.; Ellahi, R.; Akbar, N.S.; Zeeshan, A. Heat and mass transfer analysis of peristaltic flow of nanofluid in a vertical rectangular duct by using the optimized series solution and genetic algorithm. *J. Comput. Theor. Nanosci.* **2014**, *11*, 1133–1149. [CrossRef]

20. Ellahi, R.; Riaz, A.; Nadeem, S. A theoretical study of Prandtl nanofluid in a rectangular duct through peristaltic transport. *Appl. Nanosci.* **2014**, *4*, 753–760. [CrossRef]

21. Ellahi, R.; Bhatti, M.M.; Riaz, A.; Sheikholeslami, M. Effects of magnetohydrodynamics on peristaltic flow of Jeffrey fluid in a rectangular duct through a porous medium. *J. Por. Med.* **2014**, *17*, 143–157. [CrossRef]

22. Kothandapani, M.; Prakash, J. Influence of heat source, thermal radiation and inclined magnetic field on peristaltic flow of a hyperbolic tangent nanofluid in a tapered asymmetric channel. *IEEE Trans. NanoBiosci.* **2015**, *14*, 385–392. [CrossRef]

23. Nadeem, S.; Riaz, A.; Ellahi, R.; Akbar, N.S. Effects of heat and mass transfer on peristaltic flow of a nanofluid between eccentric cylinders. *Appl. Nanosci.* **2014**, *4*, 393–404. [CrossRef]

24. Prakash, J.; Sharma, A.; Tripathi, D. Thermal radiation effects on electroosmosis modulated peristaltic transport of ionic nanoliquids in biomicrofluidics channel. *J. Mol. Liq.* **2018**, *249*, 843–855. [CrossRef]

25. Tripathi, D.; Shashi, B.; Bég, O.A.; Akbar, N.S. Transient peristaltic diffusion of nanofluids: A model of micropumps in medical engineering. *J. Hydrodyn.* **2018**, *30*, 1001–1011. [CrossRef]

26. Tripathi, D.; Sharma, A.; Bég, O.A. Joule heating and buoyancy effects in electro-osmotic peristaltic transport of aqueous nanofluids through a microchannel with complex wave propagation. *Adv. Powder Technol.* **2018**, *29*, 639–653. [CrossRef]

27. Prakash, J.; Siva, E.P.; Tripathi, D.; Kuharat, S.; Bég, O.A. Peristaltic pumping of magnetic nanofluids with thermal radiation and temperature-dependent viscosity effects: Modelling a solar magneto-biomimetic nanopump. *Renew. Energ.* **2019**, *133*, 1308–1326. [CrossRef]

28. Prakash, J.; Tripathi, D. Electroosmotic flow of Williamson ionic nanoliquids in a tapered microfluidic channel in presence of thermal radiation and peristalsis. *J. Mol. Liq.* **2018**, *256*, 352–371. [CrossRef]

29. Prakash, J.; Jhorar, R.; Tripathi, D.; Azese, M.N. Electroosmotic flow of pseudoplastic nanoliquids via peristaltic pumping. *J. Braz. Soc. Mech. Sci. Eng.* **2019**, *41*, 61. [CrossRef]

30. Mosayebidorcheh, S.; Hatami, M. Analytical investigation of peristaltic nanofluid flow and heat transfer in an asymmetric wavy wall channel (Part I: Straight channel). *Int. J. Heat Mass Transf.* **2018**, *126*, 790–799. [CrossRef]

31. Abbasi, F.M.; Gul, M.; Shehzad, S.A. Hall effects on peristalsis of boron nitride-ethylene glycol nanofluid with temperature dependent thermal conductivity. *Physica E Low Dimens. Syst Nanostruct.* **2018**, *99*, 275–284. [CrossRef]

32. Ranjit, N.K.; Shit, G.C.; Tripathi, D. Joule heating and zeta potential effects on peristaltic blood flow through porous micro vessels altered by electrohydrodynamic. *Microvasc. Res.* **2018**, *117*, 74–89. [CrossRef]

33. Sadiq, M.A. MHD stagnation point flow of nanofluid on a plate with anisotropic slip. *Symmetry* **2019**, *11*, 132. [CrossRef]

34. Ellahi, R. The effects of MHD and temperature dependent viscosity on the flow of non-Newtonian nanofluid in a pipe: Analytical solutions. *Appl. Math. Model.* **2013**, *37*, 1451–1457. [CrossRef]

35. Zeeshan, A.; Shehzad, N.; Abbas, A.; Ellahi, R. Effects of radiative electro-magnetohydrodynamics diminishing internal energy of pressure-driven flow of titanium dioxide-water nanofluid due to entropy generation. *Entropy* **2019**, *21*, 236. [CrossRef]

36. Hussain, F.; Ellahi, R.; Zeeshan, A. Mathematical models of electro magnetohydrodynamic multiphase flows synthesis with nanosized hafnium particles. *Appl. Sci.* **2018**, *8*, 275. [CrossRef]

37. Ellahi, R.; Zeeshan, A.; Hussain, F.; Abbas, T. Study of shiny film coating on multi-fluid flows of a rotating disk suspended with nano-sized silver and gold particles: A comparative analysis. *Coatings* **2018**, *8*, 422. [CrossRef]

38. Harvey, R.W.; Metge, D.W.; Kinner, N.; Mayberry, N. Physiological considerations in applying laboratory-determined buoyant densities to predictions of bacterial and protozoan transport in groundwater, Results of in-situ and laboratory tests. *Enviorn. Sci. Technol.* **1997**, *31*, 289–295. [CrossRef]

39. Mishra, M.; Rao, A.R. Peristaltic transport in a channel with a porous peripheral layer: Model of a flow in gastrointestinal tract. *J. Biomech.* **2005**, *38*, 779–789. [CrossRef]

40. Mekheimer, K.S. Nonlinear peristaltic transport through a porous medium in an inclined planar channel. *J. Por. Med.* **2003**, *6*, 13. [CrossRef]

41. Siddiqui, A.M.; Ansari, A.R. A note on the swimming problem of a singly flagellated microorganism in a fluid flowing through a porous medium. *J. Porous Med.* **2005**, *8*, 551–556. [CrossRef]

42. Wernert, V.; Schäf, O.; Ghobarkar, H.; Denoyel, R. Adsorption properties of zeolites for artificial kidney applications. *Microporous Mesoporous Mat.* **2005**, *83*, 101–113. [CrossRef]

43. Jafari, A.; Zamankhan, P.; Mousavi, S.M.; Kolari, P. Numerical investigation of blood flow part II: In capillaries. *Commun. Nonlinear Sci. Numeri. Simulat.* **2009**, *14*, 1396–1402. [CrossRef]

44. Goerke, A.R.; Leung, J.; Wickramasinghe, S.R. Mass and momentum transfer in blood oxygenators. *Che. Eng. Sci.* **2002**, *57*, 2035–2046. [CrossRef]

45. Mneina, S.S.; Martens, G.O. Linear phase matched filter design with causal real symmetric impulse response. *AEU Int. J. Electron. Commun.* **2009**, *63*, 83–91. [CrossRef]

46. Andoh, Y.H.; Lips, B. Prediction of porous walls thermal protection by effusion or transpiration cooling. An analytical approach. *Appl. Thermal. Eng.* **2003**, *23*, 1947–1958. [CrossRef]

47. Runstedtler, A. On themodified Stefan–Maxwell equation for isothermal multi component gaseous diffusion. *Chemical Eng. Sci.* **2006**, *61*, 5021–5029. [CrossRef]

48. Uddin, M.J.; Khan, W.A.; Ismail, A.I.M. Free convection boundary layer flow from a heated upward facing horizontal flat plate embedded in a porous medium filled by a nanofluid with convective boundary condition. *Transp. Porous Med.* **2012**, *92*, 867–881. [CrossRef]

49. Chamkha, A.J.; Abbasbandy, S.; Rashad, A.M.; Vajravelu, K. Radiation effects on mixed convection over a wedge embedded in a porous medium filled with a nanofluid. *Transp. Porous Med.* **2011**, *91*, 261–279. [CrossRef]

50. Kuznetsov, A.V.; Nield, D.A. Effect of local thermal non-equilibrium on the onset of convection in a porous medium layer saturated by a nanofluid. *Transp. Porous Med.* **2010**, *83*, 425–436. [CrossRef]

51. Akbar, N.S. Double-diffusive natural convective peristaltic flow of a Jeffrey nanofluid in a porous channel. *Heat Trans. Res.* **2014**, *45*, 293–307. [CrossRef]

52. Nadeem, S.; Riaz, A.; Ellahi, R.; Akbar, N.S. Mathematical model for the peristaltic flow of nanofluid through eccentric tubes comprising porous medium. *Appl. Nanosci.* **2014**, *4*, 733–743. [CrossRef]

53. Bhatti, M.M.; Zeeshan, A.; Ellahi, R.; Shit, G.C. Mathematical modeling of heat and mass transfer effects on MHD peristaltic propulsion of two-phase flow through a Darcy-Brinkman-Forchheimer porous medium. *Adv. Powder Technol.* **2018**, *29*, 1189–1197. [CrossRef]

54. Alamri, S.Z.; Ellahi, R.; Shehzad, N.; Zeeshan, A. Convective radiative plane Poiseuille flow of nanofluid through porous medium with slip: An application of Stefan blowing. *J. Mol. Liq.* **2019**, *273*, 292–304. [CrossRef]

55. Shehzad, N.; Zeeshan, A.; Ellahi, R.; Rashidid, S. Modelling study on internal energy loss due to entropy generation for non-Darcy Poiseuille flow of silver-water nanofluid: An application of purification. *Entropy* **2018**, *20*, 851. [CrossRef]

56. Kothandapani, M.; Prakash, J. The peristaltic transport of Carreau nanofluids under effect of a magnetic field in a tapered asymmetric channel: Application of the cancer therapy. *J. Mech. Med. Bio.* **2015**, *15*, 1550030. [CrossRef]

57. Hayat, T.; Abbasi, F.M.; Al-Yami, M.; Monaquel, S. Slip and Joule heating effects in mixed convection peristaltic transport of nanofluid with Soret and Dufour effects. *J. Mol. Liq.* **2014**, *194*, 93–99. [CrossRef]

58. Kothandapani, M.; Prakash, J. Effects of thermal radiation parameter and magnetic field on the peristaltic motion of Williamson nanofluids in a tapered asymmetric channel. *Int. J. Heat Mass Transf.* **2015**, *51*, 234–245. [CrossRef]

59. Hayat, T.; Yasmin, H.; Ahmad, B.; Chen, B. Simultaneous effects of convective conditions and nanoparticles on peristaltic motion. *J. Mol. Liq.* **2014**, *193*, 74–82. [CrossRef]

60. Makinde, O.D. Thermal stability of a reactive viscous flow through a porous-saturated channel with convective boundary conditions. *Appl. Therm. Eng.* **2009**, *29*, 1773–1777. [CrossRef]

61. Parti, M. Mass transfer Biot numbers. *Periodica Polytechnica Mech. Eng.* **1994**, *38*, 109–122.

62. Kikuchi, Y. Effect of leukocytes and platelets on blood flow through a parallel array of microchannels: Micro-and Macroflow relation and rheological measures of leukocytes and platelate acivities. *Microvasc. Res.* **1995**, *50*, 288–300. [CrossRef]
63. Mishra, M.; Rao, A.R. Peristaltic transport of a Newtonian fluid in an asymmetric channel. *Z. Angew. Math. Phys.* **2003**, *54*, 532–550. [CrossRef]

8

Impact of Nonlinear Thermal Radiation and the Viscous Dissipation Effect on the Unsteady Three-Dimensional Rotating Flow of Single-Wall Carbon Nanotubes with Aqueous Suspensions

Muhammad Jawad [1], Zahir Shah [1], Saeed Islam [1], Jihen Majdoubi [2], I. Tlili [3], Waris Khan [4] and Ilyas Khan [5,*]

[1] Department of Mathematics, Abdul Wali Khan University, Mardan, Khyber Pakhtunkhwa 23200, Pakistan; muhammadjawad175@yahoo.com (M.J.); zahir1987@yahoo.com (Z.S.); saeedislam@awkum.edu.pk (S.I.)

[2] Department of Computer Science, College of Science and Humanities at Alghat Majmaah University, Al-Majmaah 11952, Saudi Arabia; j.majdoubi@mu.edu.sa

[3] Department of Mechanical and Industrial Engineering, College of Engineering, Majmaah University, Al-Majmaah 11952, Saudi Arabia; l.tlili@mu.edu.sa

[4] Department of Mathematics, Kohat University of Science and technology, Kohat, KP 26000, Pakistan; Wariskhan758@yahoo.com

[5] Faculty of Mathematics and Statistics, Ton Duc Thang University, Ho Chi Minh City 72915, Vietnam

* Correspondence: ilyaskhan@tdt.edu.vn

Abstract: The aim of this article is to study time dependent rotating single-wall electrically conducting carbon nanotubes with aqueous suspensions under the influence of nonlinear thermal radiation in a permeable medium. The impact of viscous dissipation is taken into account. The basic governing equations, which are in the form of partial differential equations (PDEs), are transformed to a set of ordinary differential equations (ODEs) suitable for transformations. The homotopy analysis method (HAM) is applied for the solution. The effect of numerous parameters on the temperature and velocity fields is explanation by graphs. Furthermore, the action of significant parameters on the mass transportation and the rates of fiction factor are determined and discussed by plots in detail. The boundary layer thickness was reduced by a greater rotation rate parameter in our established simulations. Moreover, velocity and temperature profiles decreased with increases of the unsteadiness parameter. The action of radiation phenomena acts as a source of energy to the fluid system. For a greater rotation parameter value, the thickness of the thermal boundary layer decreases. The unsteadiness parameter rises with velocity and the temperature profile decreases. Higher value of ϕ augments the strength of frictional force within a liquid motion. For greater R and θw; the heat transfer rate rises. Temperature profile reduces by rising values of Pr.

Keywords: unsteady rotating flow; porous medium; aqueous suspensions of CNT's; nonlinear thermal radiation; viscous dissipation effect; HAM

1. Introduction

The recent period of technology and science has been totally affected by nanofluids, and they possess an important role in various engineering and machinery uses, like biomedical applications, detergency, transferences, industrialized cooling, microchip technology and nuclear reactions. By adding nanoparticles, mathematicians and physicists developed to a way to increase the capacity of heat transfer in the base fluid. The investigators applied advanced techniques, like in the base fluids, by adding ultra-fine solid particles. Heat transfer and single-phase higher thermal conductivity coefficients are greater in nanofluids as compared to bottom liquids. Khan [1] has examined Buongiorno's model with heat transfer and mass for nanofluid flow. Mahdy et al. [2] have applied Buongiorno's model for nanofluid flow with heat transfer through an unsteady contracting cylinder. Malvandi et al. [3] have described flow through a vertical annular pipe. In [4–6], researchers have examined the flow of nanofluids through a stretching sheet. Ellahi [7] has discussed nanofluid flow through a pipe with the MHD effect. Nanofluid flow through a cone has been deliberated by Nadeem et al. [8]. Abolbashari et al. [9] have debated entropy generation for the analytical modelling of Casson nanofluid flow. A mixture of suspended metallic nanoscale particles with a base fluid is known as a mixture of nanoparticles, the word nanofluid was invented by Choi [10]. Nanofluids were formulated for heat transport, momentum and mass transfer by Buongiorno, in order to obtain four equations of two components for non-homogeneous equilibrium models [11]. Most of the research available on the nanofluid problem is cited in works [12–14]. Presently, in these studies, few significant attempts have been presented for nonlinear thermal radiation [15]. Kumar et al. [15] have reported the rotating nanofluid flow problem with deliberation of dissimilar types of nanoparticles, including entropy generation to use the second law of thermodynamics. Nadeem et al. [16] have studied the inclusion of nanoparticles of titanium and copper oxide, which related to the rotating fluid flow problem. Mabood et al. [17] have studied the flow of rotating nanofluid and the impact of magnetic and heat transfer. Shah et al. [18,19] have deliberated a rotating system nanofluid flow with hall current and thermal radiation. Gireesha et al. [20] have examined a single-wall nanotube in an unsteady rotating flow with heat transfer and nonlinear thermal radiation. Currently, Ishaq et al. [21] have discussed the thermal radiation effect with respect to the entropy generation of unsteady nanofluid thin film flow on a porous stretching sheet. In the field of nanoscience, the study of the flow of fluid with nanoparticles (nanofluids) holds a great amount of attention. Nanofluids scattered with 10^9 nm sized materials are arranged in fluids such as nanofibers, droplets, nanoparticles, nanotubes etc. Solid phase and liquid phase are the two period systems. To augment the thermal conductivity of fluids, nanofluids can be applied, and they thrive in stable fluids, exhibiting good writing and dispersion properties on hard materials [22,23]. Sheikholeslami et al. [24–26] have described the significance of nanofluids in in nanotechnology. Yadav et al. [27–29] have deliberated nanofluids with the MHD effect by using dissimilar phenomena, further study of heat transfer enhancement, numerical simulation, stability and instability with linear and non-linear flows of nanofluid.

The Darcy–Forchheimer flow of radiative carbon nanotubes (CNTs) in nanofluid in a rotating frame has been investigated by Shah et al. [30–33]. Dawar et al. [33] have studied CNTs, Casson MHD and nanofluid with radiative heat transfer in rotating channels. Khan et al. [34] have studied three-dimensional Williamson nanofluid flow over a linear stretching surface. Shah et al. [35] have described the analysis of a micropolar nanofluid flow with radiative heat and mass transfer. Khan et al. [36,37] have discussed the Darcy–Forchheimer flow of micropolar nanofluid between two plates. Shah et al. [38] have described MHD thin film flow of Williamson fluid over an unsteady permeable stretching surface. Jawad et al. [39] have investigated the Darcy–Forchheimer flow of MHD nanofluid thin film flow with Joule dissipation and Navier's partial slip. Khan et al. [40] have investigated the slip flow of Eyring–Powell nanoliquid. Hammed et al. [41] have described a combined magnetohydrodynamic and electric field effect on an unsteady Maxwell nanofluid flow. Dawar et al. [42] have described unsteady squeezing flow of MHD CNT nanofluid in rotating channels. Khan et al. [43] have investigated the Darcy–Forchheimer flow of MHD CNT nanofluid with radiative thermal aspects. Sheikholeslami et al. [44] have investigated the uniform magnetic force impact on water based nanofluid with thermal aspects in a porous enclosure. Feroz et al. [45] have examined the entropy generation of carbon nanotube flow in a rotating channel. Alharbi et al. [46] have described entropy generation in MHD Eyring–Powell fluid flow over an unsteady oscillatory porous stretching surface with thermal radiation. Liao [47] learned in 1992 that this method was a fast way to find the approximate solution and that for the solution of nonlinear problems it was a better fit.

The main objective of this article is to investigate the augmentation of heat transfer in time dependent rotating single-wall carbon nanotubes with aqueous suspensions under the influence of nonlinear thermal radiation in a permeable medium. The impact of viscous dissipation is taken into account. For the solution, the homotopy analysis method (HAM) [48–50] is applied. The effect of numerous parameters on the temperature and velocity fields are explained by graphs.

2. Problem Formulation

We considered a three-dimensional time dependent electrically conducting incompressible laminar rotating flow of nanofluid. Aqueous suspensions of single-wall CNTs based on water were assumed as a nanoscale materials with a porous medium. The x, y and z variables are the Cartesian coordinates where $\Omega(t)$ is angular velocity of the rotating fluid, which is measured about the z-axis. The surface velocities in the x and y axes are taken as $u_w(x,t) = \frac{bx}{(1-\delta t)}$ and $v_w(x,t)$ respectively, and $w_w(x,t)$ is the velocity in the z-axis, known as the mass flux wall velocity. The governing equations under these assumptions are taken in the form as [4,14,18]

$$\frac{\partial u}{\partial x} + \frac{\partial v}{\partial y} + \frac{\partial w}{\partial z} = 0 \tag{1}$$

$$\frac{\partial u}{\partial t} + u\frac{\partial u}{\partial x} + v\frac{\partial u}{\partial y} + w\frac{\partial u}{\partial z} + \frac{2\Omega v}{1-\delta t} = -\frac{1}{\rho}\frac{\partial P}{\partial x} + \frac{\mu_{nf}}{\rho_{nf}}\frac{\partial^2 u}{\partial z^2} - \frac{\sigma\beta_0^2}{1-\delta t}u - \frac{\mu_{nf}}{k^*}\frac{u}{1-\delta t} \tag{2}$$

$$\frac{\partial v}{\partial t} + u\frac{\partial v}{\partial x} + v\frac{\partial v}{\partial y} + w\frac{\partial v}{\partial z} + \frac{2\Omega u}{1-\delta t} = -\frac{1}{\rho}\frac{\partial P}{\partial y} + \frac{\mu_{nf}}{\rho_{nf}}\frac{\partial^2 v}{\partial z^2} - \frac{\sigma\beta_0^2}{1-\delta t}v - \frac{\mu_{nf}}{k^*}\frac{v}{1-\delta t} \tag{3}$$

$$\frac{\partial w}{\partial t} + u\frac{\partial w}{\partial x} + v\frac{\partial w}{\partial y} + w\frac{\partial w}{\partial z} = -\frac{1}{\rho}\frac{\partial P}{\partial z} + \frac{\mu_{nf}}{\rho_{nf}}\frac{\partial^2 w}{\partial z^2} \tag{4}$$

$$\frac{\partial T}{\partial t} + u\frac{\partial T}{\partial x} + v\frac{\partial T}{\partial y} + w\frac{\partial T}{\partial z} = \alpha_{nf}\frac{\partial^2 T}{\partial z^2} + \frac{1}{(\rho c_p)_{nf}}\frac{\partial q_r}{\partial z} + \frac{\mu_{nf}}{(\rho c_p)_{nf}}\left[\left(\frac{\partial u}{\partial z}\right)^2 + \left(\frac{\partial v}{\partial z}\right)^2\right] \tag{5}$$

The corresponding boundary conditions are given as

$$\begin{aligned} u &= u_w(x,t), v = 0, w = 0, T = T_w \quad at \quad z = 0 \\ u &\to 0, v \to 0, w \to 0, T \to T_\infty \quad at \quad z \to \infty. \end{aligned} \tag{6}$$

where x, y and z, are the directions of the velocity components. The angular velocity of the nanofluid is denoted as Ω, nanofluid dynamic viscosity is denoted as μ_{nf}, the nanofluid density is denoted as ρ_{nf}, the nanofluid thermal diffusivity is denoted as α_{nf}, temperature of the nanofluid is denoted as T, T_w and T_∞ denotes the wall and outside surface temperature, respectively.

The expression of radiative heat flux in Equation (5) is written as:

$$q_r = -\frac{4\sigma^*}{3(\rho c_p)_{nf}k^*}\frac{\partial T^4}{\partial z} = -\frac{16\sigma^*}{3k^*}T^3\frac{\partial T}{\partial z} \tag{7}$$

where the mean absorption coefficient is denoted by k^* and the Stefan–Boltzman constant is denoted by σ^*. By using Equation (7) in Equation (5), this produces the following equation:

$$\begin{aligned} &\frac{\partial T}{\partial t} + u\frac{\partial T}{\partial x} + v\frac{\partial T}{\partial y} + w\frac{\partial T}{\partial z} \\ &= \alpha_{nf}\frac{\partial^2 T}{\partial z^2} + \frac{16\sigma^*}{(\rho c_p)_{nf}k^*}\left[T^3\frac{\partial^2 T}{\partial z^2} + 3T^2\left(\frac{\partial T}{\partial z}\right)^2\right] + \frac{\mu_{nf}}{(\rho c_p)_{nf}}\left[\left(\frac{\partial u}{\partial z}\right)^2 + \left(\frac{\partial v}{\partial z}\right)^2\right] \end{aligned} \tag{8}$$

α_{nf}, μ_{nf} and ρ_{nf} are interrelated with ϕ, which is denoted as:

$$
\begin{aligned}
\rho_{nf} &= \left(1 - \phi + \phi\left(\frac{(\rho_s)_{CNT}}{\rho_f}\right)\right), \mu_{nf} = \frac{\mu_f}{(1-\phi)^{2.5}}, \alpha_{nf} \\
&= (1-\phi)\left(\rho_f\right)_f + \phi(\rho_s)_{CNT}, \\
(\rho C_p)_{nf} &= (1-\phi)(\rho C_p)_f + \phi(\rho C_p)_{CNT}, \frac{k_{nf}}{k_f} \\
&= \frac{1-\phi+2\phi\left(\frac{k_{CNT}}{k_{CNT}-k_f}\right)\ln\left(\frac{k_{CNT}+k_f}{2k_f}\right)}{1-\phi+2\phi\left(\frac{k_f}{k_{CNT}-k_f}\right)\ln\left(\frac{k_{CNT}+k_f}{2k_f}\right)}
\end{aligned}
\tag{9}
$$

The volumetric heat capacity of the CNT and base fluid is denoted by $(\rho C_p)_{CNT}$ and $(\rho C_p)_f$ respectively. Nanofluid thermal conductivity is denoted as k_{nf}, base fluid thermal conductivity is denoted by k_f, CNT thermal conductivity is denoted by k_{CNT}, nanoparticle volume fraction is denoted by ϕ, the base fluid density viscosity is denoted by ρ_f and CNT density is denoted by ρ_{CNT}.

Similarity transformations are now introduce as:

$$
\begin{aligned}
u &= \frac{bx}{(1-\delta t)}f'(\eta), \ v = \frac{bx}{(1-\delta t)}g(\eta), w = -\sqrt{\frac{bv}{(1-\alpha t)}}f(\eta), \ \eta = \sqrt{\frac{b}{v(1-\alpha t)}}z \\
T &= T_\infty(1 + (1-\delta t)\theta(\eta))
\end{aligned}
\tag{10}
$$

Inserting Equation (10) in Equations (2)–(6), we have:

$$
\begin{aligned}
&\frac{1}{(1-\phi)^{2.5}\left((1-\phi)+\phi\frac{(\rho_s)_{CNT}}{\rho_f}\right)}f^{iv} - \frac{\beta_0^2}{\lambda}f'' - \frac{\mu_{nf}}{(1-\phi)^{2.5}k^*b}f'' \\
&- \left[\lambda\left(f' + \frac{\eta}{2}f''\right) + f'f'' - ff'' - 2\frac{\Omega}{b}g'\right] = 0
\end{aligned}
\tag{11}
$$

$$
\begin{aligned}
&\frac{1}{(1-\phi)^{2.5}\left((1-\phi)+\phi\frac{(\rho_s)_{CNT}}{\rho_f}\right)}g''' - \frac{\beta_0^2}{\lambda}g' + \frac{\mu_{nf}}{(1-\phi)^{2.5}k^*b}g' \\
&- \left[\lambda\left(g' + \frac{\eta}{2}g''\right) + gf'' - fg'' - 2\frac{\Omega}{b}f''\right] = 0
\end{aligned}
\tag{12}
$$

$$
\begin{aligned}
&\theta'' + R\left[(1+(\theta_w-1)\theta)^3\theta'' + 3(1+(\theta_w-1)\theta)^2(\theta_w-1)\theta'^2\right] + \frac{Ec}{Pr}\left(f''^2 + g'^2\right) \\
&- \frac{1}{Pr}\left[\lambda\frac{\eta}{2}\theta' - f\theta'\right] = 0
\end{aligned}
\tag{13}
$$

The boundary conditions are written as:

$$
\begin{aligned}
f(0) &= 0, f'(0) = 1, g(0) = 0, \theta(0) = 1, \quad at \quad \eta = 0 \\
f'(\eta) &\to 0, g(\eta) \to 0, f(\eta) \to 0, \theta(\eta) \to 0. \quad at \quad \eta \to \infty.
\end{aligned}
\tag{14}
$$

Here, Ω denotes the rotation parameter and is defined as $\Omega = \frac{\omega}{b}$, and λ represents the unsteadiness parameter, defined as $\lambda = \frac{\delta}{b}$. R is radiation parameter, defined as $R = \frac{16\sigma^* T_\infty^3}{3k_{nf}k^*}$. Pr is Prandtl number, defined as $Pr = \frac{\alpha_{nf}}{v_{nf}}$. Ec represents the Eckert number, defined as $Ec = \frac{u_w^2}{c_p(T-T_\infty)}$. The temperature ratio parameter is denoted by θ_w and $\theta_w = \frac{T_w}{T_\alpha}$.

Physical Quantities of Interest

In the given problem, the physical quantities of interest are C_{fx}, C_{fy} and Nu_x, which is denoted as:

$$
C_{fx}\frac{\tau_{wx}}{\rho_f u_w^2(x,t)}, C_{fy}\frac{\tau_{wy}}{\rho_f u_w^2(x,t)}, Nu_x = \frac{xq_w}{(T_w - T_\infty)}
\tag{15}
$$

The surface heat flux is denoted by q_w, and τ_{wx} and τ_{wy} are surface shear stress, which are written as:

$$\tau_{wx} = \mu_{nf}\left(\frac{\partial u}{\partial z}\right)_{z=0}, \tau_{wy} = \mu_{nf}\left(\frac{\partial v}{\partial z}\right)_{z=0} \ and \ q_w = -k_{nf}\left(\frac{\partial T}{\partial z}\right) + (q_r)_{z=0}. \tag{16}$$

By the use of Equations (15) and (16), then we have:

$$\sqrt{Re_x}C_{fx} = \frac{1}{(1-\phi)^{2.5}}f''(0), \sqrt{Re_x}C_{fy} = \frac{1}{(1-\phi)^{2.5}}g'(0), \frac{Nu_x}{\sqrt{Re_x}} = \frac{k_{nf}}{k_f}\left(-\left[1+Rd\theta_w^3\right]\theta'(0)\right) \tag{17}$$

$Re_x = u_w x/v$ is the Reynolds number.

3. HAM Solution

Liao [46,47] proposed the homotopy analysis technique in 1992. In order to attain this method, he used the ideas of a topology called homotopy. For the derivation he used two homotopic functions, where one of them can be continuously distorted into another. He used Ψ_1, Ψ_2 for binary incessant purposes and X and Y are dual topological plane, additionally, if Ψ_1 and Ψ_2 map from X to Y, then Ψ_1 is supposed to be homotopic to Ψ_2. If a continuous function of ψ is produced:

$$\Psi : X \times [0, 1] \to Y \tag{18}$$

So that $x \in X$

$$\Psi[x, 0] = \Psi_1(x) \ and \ \Psi[x, 1] = \Psi_2(x) \tag{19}$$

where Ψ is homotopic. Here, we use the HAM to solve Equations (11)–(13), consistent with the boundary restrain (14). The preliminary guesses are selected as follows:
The linear operators are denoted as $L_{\hat{f}}$ and $L_{\hat{\theta}}$, and are represented as

$$L_{\hat{f}}(\hat{f}) = \hat{f}''', L_{\hat{g}}(\hat{g}) = g'', L_{\hat{\theta}}(\hat{\theta}) = \hat{\theta}'' \tag{20}$$

Which has the following applicability:

$$L_{\hat{f}}(e_1 + e_2\eta + e_3\eta^2 + e_4\eta^3) = 0, L_{\hat{g}}(e_5 + e_6\eta + e_7\eta^2) = 0$$
$$L_{\hat{\theta}}(e_8 + e_9\eta) = 0 \tag{21}$$

where the representation of coefficients is included in the general solution by $e_i(i = 1- > 7)$.
The corresponding non-linear operators are sensibly selected as $N_{\hat{f}}, N_{\hat{g}}$ and $N_{\hat{\theta}}$, and identify in the form:

$$N_{\hat{f}}\left[\hat{f}(\eta;\zeta), \hat{g}(\eta;\zeta)\right] = \frac{1}{(1-\phi)^{2.5}\left((1-\phi)+\phi\frac{(\rho_s)_{CNT}}{\rho_f}\right)}\hat{f}_{\eta\eta\eta\eta} - \frac{\beta_0^2}{\lambda}\hat{f}_{\eta\eta} - \frac{\mu_{nf}}{(1-\phi)^{2.5}k^*b}\hat{f}_{\eta\eta}$$
$$-\left[\lambda\left(\hat{f}_{\eta\eta} + \frac{\eta}{2}\hat{f}_{\eta\eta\eta}\right) + \hat{f}_\eta\hat{f}_{\eta\eta} - \hat{f}\hat{f}_{\eta\eta} - 2\frac{\Omega}{b}\hat{g}_\eta\right] \tag{22}$$

$$N_{\hat{g}}\left[\hat{f}(\eta;\zeta), \hat{g}(\eta;\zeta)\right] = \frac{1}{(1-\phi)^{2.5}\left((1-\phi)+\phi\frac{(\rho_s)_{CNT}}{\rho_f}\right)}\hat{g}_{\eta\eta\eta} - \frac{\beta_0^2}{\lambda}\hat{g}_\eta + \frac{\mu_{nf}}{(1-\phi)^{2.5}k^*b}\hat{g}_\eta$$
$$-\left[\lambda(\hat{g}_\eta + \frac{\eta}{2}\hat{g}_{\eta\eta}) + \hat{g}\hat{f}_{\eta\eta} - \hat{f}\hat{g}_{\eta\eta} - 2\frac{\Omega}{b}\hat{f}_{\eta\eta}\right] \tag{23}$$

$$N_{\hat{\theta}}\left[\hat{f}(\eta;\zeta), \hat{g}(\eta;\zeta), \hat{\theta}(\eta;\zeta)\right] = \hat{\theta}_{\eta\eta} + R\left[(1 + (\theta_w - 1)\hat{\theta})^3\hat{\theta}_{\eta\eta} + 3(1 + (\theta_w - 1)\hat{\theta})^2(\theta_w - 1)\hat{\theta}_\eta^2\right]$$
$$+\frac{Ec}{Pr}\left(\hat{f}_{\eta\eta}^2 + \hat{g}_\eta^2\right) - \frac{1}{Pr}[\lambda\frac{\eta}{2}\hat{\theta}_\eta - \hat{f}\hat{\theta}_\eta] \tag{24}$$

For Equations (8–10), the 0th-order scheme takes the form

$$(1 - \zeta)L_{\hat{f}}\left[\hat{f}(\eta;\zeta) - \hat{f}_0(\eta)\right] = p\hbar_{\hat{f}}N_{\hat{f}}\left[\hat{f}(\eta;\zeta), \hat{g}(\eta;\zeta)\right] \tag{25}$$

$$(1 - \zeta)L_{\hat{g}}[\hat{g}(\eta; \zeta) - \hat{g}_0(\eta)] = p\hbar_{\hat{g}}N_{\hat{g}}\left[\hat{f}(\eta; \zeta), \hat{g}(\eta; \zeta)\right] \tag{26}$$

$$(1 - \zeta)\, L_{\hat{\theta}}\left[\hat{\theta}(\eta; \zeta) - \hat{\theta}_0(\eta)\right] = p\hbar_{\hat{\theta}}N_{\hat{\theta}}\left[\hat{f}(\eta; \zeta), \hat{g}(\eta; \zeta), \hat{\theta}(\eta; \zeta)\right] \tag{27}$$

where the boundary constrains are:

$$
\begin{aligned}
&\hat{f}(\eta; \zeta)\Big|_{\eta=0} = 0, \ \frac{\partial \hat{f}(\eta; \zeta)}{\partial \eta}\Big|_{\eta=0} = 1, \ \hat{g}(\eta; \zeta)|_{\eta=0} = 0 \\
&\hat{\theta}(\eta; \zeta)\Big|_{\eta=0} = 1, \ \frac{\partial \hat{f}(\eta; \zeta)}{\partial \eta}\Big|_{\eta\to\infty} \to 0, \ \hat{g}(\eta; \zeta)|_{\eta\to\infty} \to 0 \\
&\hat{f}(\eta; \zeta)\Big|_{\eta\to\infty} \to 0, \ \hat{\theta}(\eta; \zeta)|_{\eta\to\infty} \to 0.
\end{aligned}
\tag{28}
$$

where the embedding restriction is $\zeta \in [0, 1]$, to normalize for the solution convergence $\hbar_{\hat{f}}, \hbar_{\hat{g}}$ and $\hbar_{\hat{\theta}}$ are used. When $\zeta = 0$ and $\zeta = 1$, we get:

$$\hat{f}(\eta; 1) = \hat{f}(\eta), \hat{g}(\eta; 1) = \hat{g}(\eta), \hat{\theta}(\eta; 1) = \hat{\theta}(\eta) \tag{29}$$

Expanding $\hat{f}(\eta; \zeta), \hat{g}(\eta; \zeta)$ and $\hat{\theta}(\eta; \zeta)$ through Taylor's series for $\zeta = 0$,

$$
\begin{aligned}
\hat{f}(\eta; \zeta) &= \hat{f}_0(\eta) + \sum_{n=1}^{\infty} \hat{f}_n(\eta)\zeta^n \\
\hat{g}(\eta; \zeta) &= \hat{g}_0(\eta) + \sum_{n=1}^{\infty} \hat{g}_n(\eta)\zeta^n \\
\hat{\theta}(\eta; \zeta) &= \hat{\theta}_0(\eta) + \sum_{n=1}^{\infty} \hat{\theta}_n(\eta)\zeta^n.
\end{aligned}
\tag{30}
$$

$$\hat{f}_n(\eta) = \frac{1}{n!}\frac{\partial \hat{f}(\eta; \zeta)}{\partial \eta}\Big|_{p=0}, \hat{g}_n(\eta) = \frac{1}{n!}\frac{\partial \hat{g}(\eta; \zeta)}{\partial \eta}\Big|_{p=0}, \hat{\theta}_n(\eta) = \frac{1}{n!}\frac{\partial \hat{\theta}(\eta; \zeta)}{\partial \eta}\Big|_{p=0}. \tag{31}$$

The boundary constrains are:

$$\hat{f}_w(0) = \hat{f}_w'(0) = \hat{g}(0) = 0, \hat{\theta}_w(0) = 0, \text{at} \eta = 0 \tag{32}$$

$$\hat{f}_w'(\infty) = \hat{g}_w(\infty) = \hat{\theta}_w(\gamma) = \hat{f}_w(\infty) \to 0 \text{at} \eta \to \infty. \tag{33}$$

Resulting in:

$$
\begin{aligned}
\mathfrak{R}_n^{\hat{f}}(\eta) &= \frac{1}{(1-\phi)^{2.5}\left((1-\phi)+\phi\frac{(\rho_s)_{CNT}}{\rho_f}\right)}\hat{f}_{n-1}^{iv} - \frac{\beta_0^2}{\lambda}\hat{f}_{n-1}'' - \frac{\mu_{nf}}{(1-\phi)^{2.5}k^*b}\hat{f}_{n-1}'' \\
&\quad - \left[\lambda\left(\hat{f}_{n-1}'' + \frac{\eta}{2}\hat{f}_{n-1}'''\right) + \sum_{j=0}^{w-1}\hat{f}_{w-1-j}'\hat{f}_j'' - \sum_{j=0}^{w-1}\hat{f}_{w-1-j}\hat{f}_j'' - 2\frac{\Omega}{b}\hat{g}_{n-1}'\right]
\end{aligned}
\tag{34}
$$

$$
\begin{aligned}
\mathfrak{R}_n^{\hat{g}}(\eta) &= \frac{1}{(1-\phi)^{2.5}\left((1-\phi)+\phi\frac{(\rho_s)_{CNT}}{\rho_f}\right)}\hat{g}_{n-1}''' - \frac{\beta_0^2}{\lambda}\hat{g}_{n-1}' + \frac{\mu_{nf}}{(1-\phi)^{2.5}k^*b}\hat{g}_{n-1}' \\
&\quad - \left[\lambda\left(\hat{g}_{n-1}' + \frac{\eta}{2}\hat{g}_{n-1}''\right) + \sum_{j=0}^{w-1}\hat{g}_{w-1-j}\hat{f}_j' - \sum_{j=0}^{w-1}\hat{f}_{w-1-j}\hat{g}_j'' - 2\frac{\Omega}{b}\hat{f}_{n-1}''\right]
\end{aligned}
\tag{35}
$$

$$
\begin{aligned}
\mathfrak{R}_n^{\hat{\theta}}(\eta) &= (\hat{\theta}''_{n-1}) + R\left[\left(1 + (\theta_w - 1)\hat{\theta}_{n-1}\right)^3\hat{\theta}''_{n-1} + 3(\theta_w - 1)\left(1 + (\theta_w - 1)\hat{\theta}_{n-1}\right)^2\hat{\theta}'_{n-1}\right] + \frac{Ec}{Pr}\left(\hat{f}''2 + \hat{g}'^2\right) \\
&\quad - \frac{1}{Pr}\left[\lambda\frac{\eta}{2}\hat{\theta}'_{n-1} - \hat{f}\hat{\theta}'_{n-1}\right]
\end{aligned}
\tag{36}
$$

where

$$\chi_n = \begin{cases} 0, & \text{if } \zeta \leq 1 \\ 1, & \text{if } \zeta > 1. \end{cases} \tag{37}$$

4. Results and Discussion

In this section, we described the physical resources of the modeled problems and their impact on $f'(\eta)$, $g(\eta)$, and $\theta(\eta)$, which are identified in Figures 1–13. In Figure 1, a graphical representation of the problem is shown.

4.1. Velocity Profile $f'(\eta)$ and $g(\eta)$

The influence of Ω, β, ϕ and λ on the velocity profile is presented in Figures 1–8. The impact of Ω on $f'(\eta)$ and $g(\eta)$ is presented in Figures 1 and 2. For greater values of Ω, $f'(\eta)$ is increased. Actually, increasing the rotation parameter increases kinetic energy, which in result augmented the velocity profile. Indeed, $g(\eta)$ is decreased due to a greater rotation parameter rate, as compared to the stretching rate, which had a greater rotation rate for a larger value of Ω. Hence, the velocity field increases for larger rotation effects. Figures 3 and 4 represent the influence of ϕ on the $f'(\eta)$ and $g(\eta)$ profile. With an increase in ϕ, the velocity profile decreases. It is examined that $f'(\eta)$ and $g(\eta)$ decrease uniformly in nanofluids with a rise in ϕ. This is due to the detail that an increase in the ϕ increases the density of the nanofluid and this results in equally slowing the fluid $f'(\eta)$ and $g(\eta)$ profiles. Figures 5 and 6 describe the effect of λ on $f'(\eta)$ and $g(\eta)$. It is defined in the figures that with increases in λ, the velocity profiles decrease consistently. It is also indicated from the figures that the velocity profiles intensify for rising λ, while we can examine an opposing influence of λ on $f'(\eta)$ and $g(\eta)$ inside the nanofluid and in the thickness of the layer. Figures 7 and 8 represent the influence of β on $f'(\eta)$ and $g(\eta)$. With increases in β, the velocity profiles of fluid film decrease consistently. It was also detected that a rise in β resulted in a decrease in the fluid profiles $f'(\eta)$ and $g(\eta)$ of the nanofluid, as well as for the layer thickness. The purpose behind such an influence of β is for the stimulation of a delaying body force, stated as Lorentz force, due to the presence of β in an electrically conducting nanofluid layer. Since β suggests the ratio of hydromagnetic body force and viscous force, a larger value of β specifies a stronger hydromagnetic body force, which has a trend to slow the fluid flow.

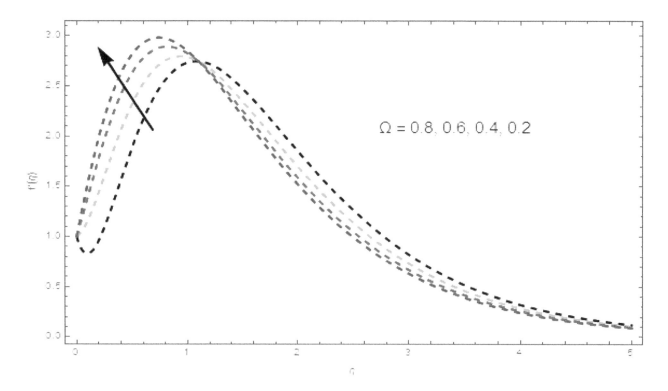

Figure 1. The influence of the rotation parameter Ω on $f'(\eta)$ when $\beta = 0.2, \phi = 0.1, \lambda = 0.3$.

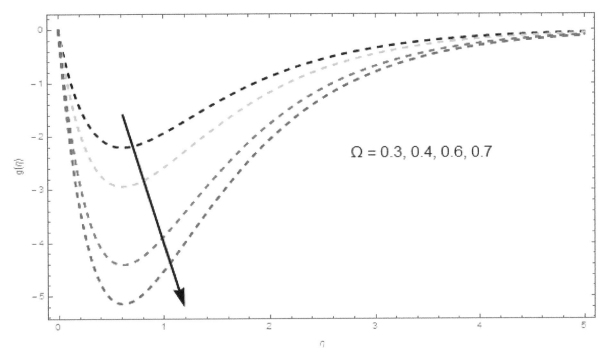

Figure 2. The effect of the rotation rate parameter Ω on $g(\eta)$ when $\beta = 0.2, \phi = 0.1, \lambda = 0.7$.

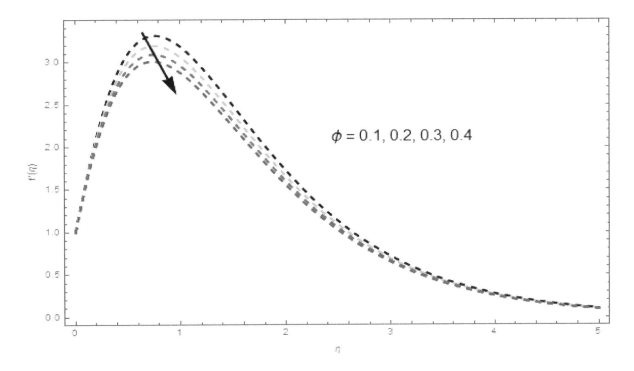

Figure 3. The effect of nanoparticle volume friction ϕ on $f'(\eta)$ when $\beta = 0.9, \lambda = 0.5, \Omega = 0.1$.

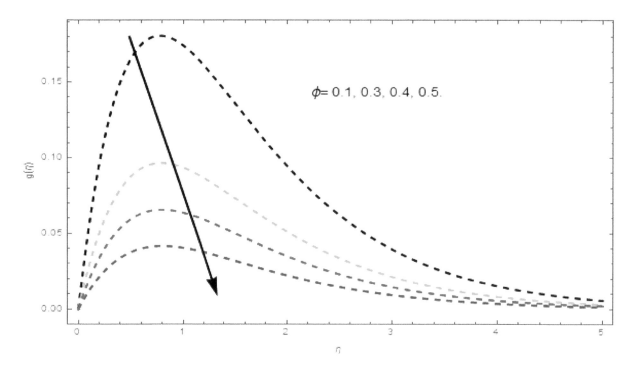

Figure 4. The impact of nanoparticle volume friction ϕ on $g(\eta)$ when $\beta = 0.1, \lambda = 0.2, \Omega = 0.7$.

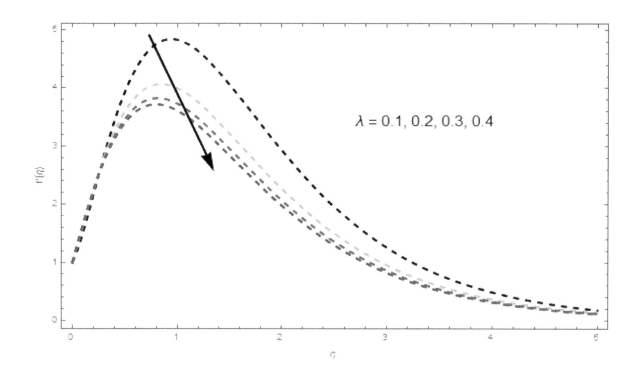

Figure 5. The influence of the unsteadiness parameter λ on $f'(\eta)$ when $\beta = 0.9, \phi = 0.1, \Omega = 0.1$.

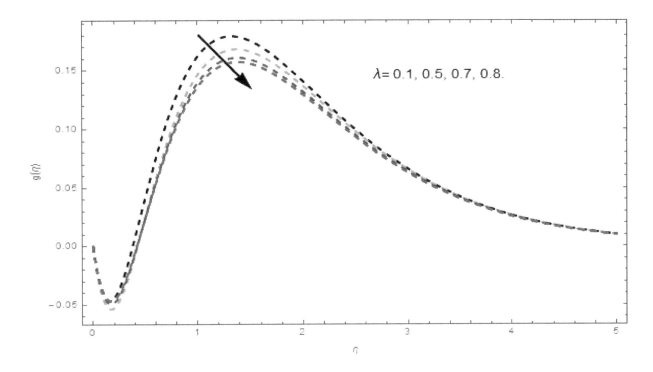

Figure 6. The effect of the unsteadiness parameter λ on $g(\eta)$ when $\beta = 0.2, \phi = 0.1, \Omega = 0.2$.

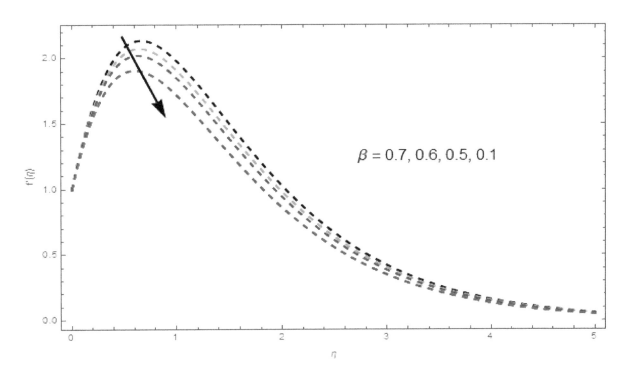

Figure 7. The effect of magnetic field β on $f'(\eta)$ when $\lambda = 0.2, \phi = 0.1, \Omega = 0.1$.

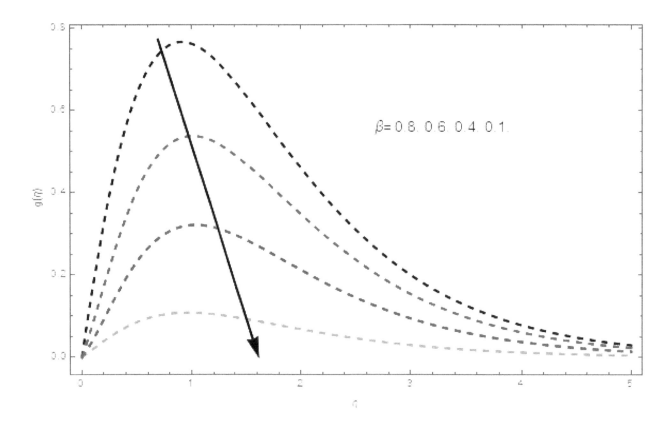

Figure 8. The effect of magnetic field β on $g(\eta)$ when $\lambda = 0.7, \phi = 0.1, \Omega = 0.1$.

4.2. Temperature Profile $\theta(\eta)$

The impact of $\theta(\eta)$ on physical parameters λ, Ec, θ_w, Rd and Pr is defined in Figures 9–13. Figure 9 represent the impact of λ on the $\theta(\eta)$ profile. Figure 9 represents that the increase in the λ momentum boundary layer thickness decreases. Figures 10 and 11 disclosed the responses of $\theta(\eta)$ for θ_w and Rd. The influence of Rd on $\theta(\eta)$ is presented in Figure 11. By increasing Rd, the temperature of the nanofluid boundary layer area is increased. It is observed in Figure 11 that $\theta(\eta)$ is augmented with a rise in the Rd. An enhanced Rd parameter leads to a release of heat energy in the flow direction, therefore the fluid $\theta(\eta)$ is increased. Graphical representation identifies that $\theta(\eta)$ is increased when we increase the ratio strength and thermal radiation temperature. The impact of $\theta(\eta)$ on Pr is given in Figure 12, where $\theta(\eta)$ decreases with large value of Pr and for smaller value increases. The variation of $\theta(\eta)$ for the variation of Pr is illustrated such that Pr specifies the ratio of momentum diffusivity to thermal diffusivity. It is realized that $\theta(\eta)$ is reduced with a rising Pr. Moreover, by suddenly rising Pr, the thermal boundary layer thickness decreases. Figure 13 identifies that for increasing Ec, the $\theta(\eta)$ attainment is enlarged, which supports the physics. By increasing Ec, the heat stored in the liquid is dissipated, due to the enhanced temperature, whereas $\theta(\eta)$ increases with increasing values of the Eckert number and the thermal boundary layer thickness of the nanofluid becomes larger.

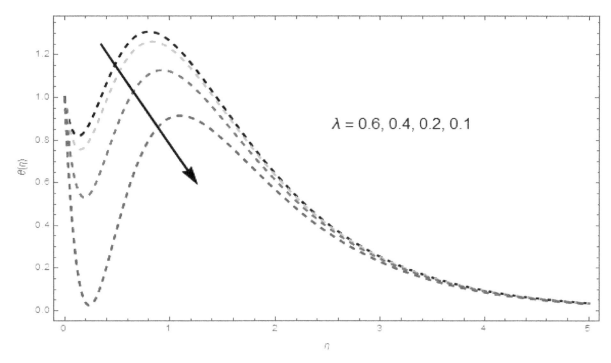

Figure 9. The effect of the unsteadiness parameter (λ) on $\theta(\eta)$ when $Rd = 0.1, Ec = 0.2, Pr = 7, \theta_w = 1.2$.

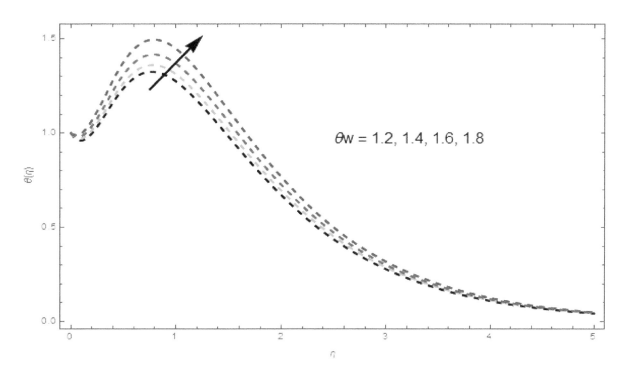

Figure 10. The influence of the temperature ratio parameter on $\theta(\eta)$ when $Rd = 0.5, \lambda = 0.6, Pr = 6.6, Ec = 0.2$.

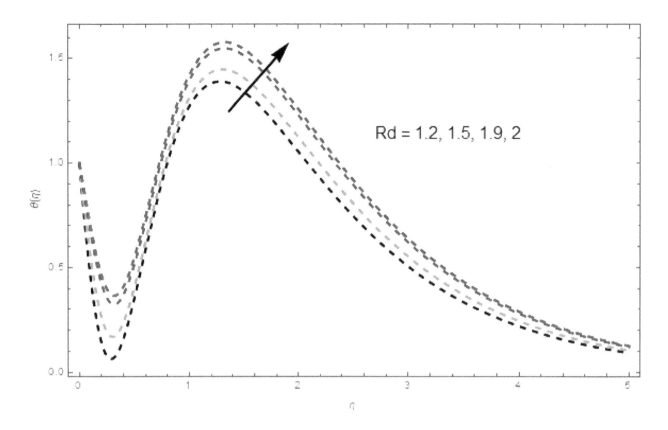

Figure 11. The influence of the radiation parameter (Rd) on $\theta(\eta)$ when $\theta_w = 0.5, \lambda = 0.6, \mathrm{Pr} = 6.6, Ec = 0.6$.

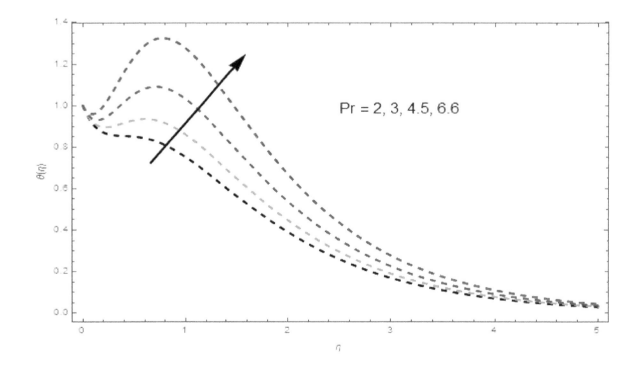

Figure 12. The influence of Prandtl number (Pr) on $\theta(\eta)$ when $Rd = 0.5, \lambda = 0.6, \theta_w = 6.6, Ec = 0.2$.

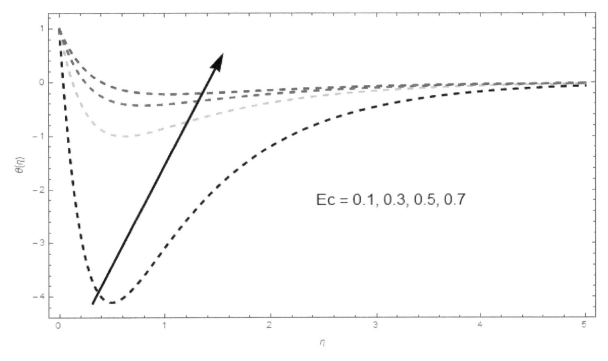

Figure 13. The impact of the Eckert number (Ec) on $\theta(\eta)$ when $Rd = 0.4, \lambda = 0.6, \theta_w = 0.6, \mathrm{Pr} = 1.8$.

4.3. Table Discussion

The effect of skin friction in the x and y directions is shown in Table 1, due to change of parameters $\phi, \beta, \lambda, \rho_s$ and ρ_f. We can see in Table 1 that the skin friction coefficient rises with rising values of λ, β, ρ_s and ϕ. The skin friction coefficient decreases with increasing values of ρ_f, as shown in Table 1. The impact of $Rd, \theta_w, \mathrm{Pr}$ and Ec on $-\frac{k_{nf}}{k_f}\left[1 + Rd\theta_w^3\right]\theta'(0)$ are calculated numerically. It was detected that a higher value of Rd and θ_w increases the heat flux, while a higher value of Pr and Ec decreases it, as shown in Table 2. Some physical properties of CNTs are shown in Tables 3 and 4.

Table 1. Skin friction $f''(0)$ versus various values of embedded parameters.

ϕ	β	λ	ρ_s	ρ_f	C_{fx}	C_{fy}
0.1	0.3	0.5	0.4	0.4	1.80946	1.80946
0.3					2.36646	2.36646
0.5					3.23183	3.23183
0.1	0.1				1.80946	1.80946
	0.3				1.83280	1.83280
	0.5				1.86280	1.86280
	0.1	0.1			1.80946	1.80946
		0.3			1.81835	1.81835
		0.5			1.82867	1.82867
		0.1	0.1		1.80946	1.80946
			0.3		1.82759	1.82759
			0.5		1.84571	1.84571
			0.1	0.1	1.80946	1.80946
				0.3	1.79496	1.79496
				0.5	1.78530	1.78530

Table 2. Nusslet number versus various values of embedded parameters.

R	θ_w	Pr	Ec	$-\frac{k_{nf}}{k_f}[1+Rd\theta_w^3]\theta'(0)$
0.1	0.5	0.7	0.6	0.408170
0.3				0.452415
0.5				0.497617
0.1	0.1			0.408170
	0.3			0.418514
	0.5			0.429786
	0.1	0.1		0.408170
		0.3		0.401541
		0.5		0.393572
		0.1	0.1	0.408170
			0.3	0.420824
			0.5	0.430319

Table 3. Thermophysical properties of different base fluids and carbon nanotubes (CNTs).

Physical Properties	Base Fluid	Nanoparticles	
	Water/Ethylene Glycol)	SWCNT	MWCNT
$\rho(kg/m^3)$	=	2,600	1,600
$c_p(J/kgK)$	=	425	796
$k(W/mK)$	=	6,600	3,000

Table 4. Variation of thermal conductivities of CNT nanofluids with the diameter of CNTs.

Diameter of CNTs	Thermal Conductivity of Nanofluids
50 mm	1.077
200 μm	1.078
20 μm	1.083
5 μm	1.085
50 nm	1.085
500 pm	1.087

5. Conclusions

Exploration on nanoparticle presentation has established more deliberation in mechanical and industrial engineering, due to the probable uses of nanoparticles in cooling devices to produce an increase in continuous phase fluid thermal performance. The radiation phenomena acts as a source of energy to the fluid system. In this research, the time dependent rotating single-wall electrically conducting carbon nanotubes with aqueous suspensions under the influence of nonlinear thermal radiation in a permeable medium was investigated. The basic governing equations, in the form of partial differential equations (PDEs), were transformed into to a set of ordinary differential equations (ODEs) suitable for transformations. The optimal approach was used for the solution. The main concluding points are given below:

- The thermal boundary layer thickness is reduced by a greater rotation rate parameter.
- Velocity and temperature profile decrease due to increases in the unsteadiness parameter.
- A greater ϕ increases the asset of frictional force within a fluid motion.
- The heat transfer rate rises for greater Rd and θ_w values.
- The skin friction coefficient increases with increasing values of ϕ and β.
- A greater value of Rd and θ_w increases the heat flux, while a greater value of Pr and Ec decreases it.
- By enhancing Pr, $\theta(\eta)$ is reduced.

Author Contributions: M.J. and Z.S. modeled the problem and wrote the manuscript. S.I., I.T. and I.K. thoroughly checked the mathematical modeling and English corrections. W.K. and Z.S. solved the problem using Mathematica software. M.J., J.M. and I.T. contributed to the results and discussions. All authors finalized the manuscript after its internal evaluation.

Acknowledgments: The fourth author would like to thank Deanship of Scientific Research, Majmaah University for supporting this work under the Project Number No 1440-43.

References

1. Khan, W.A. Buongiorno model for nanofluid Blasius flow with surface heat and mass fluxes. *J. Thermophys. Heat Transf.* **2013**, *27*, 134–141. [CrossRef]

2. Mahdy, A.; Chamkha, A. Heat transfer and fluid flow of a non-Newtonian nano fluid over an unsteady contracting cylinder employing Buongiorno'smodel. *Int. J. Numer. Method Heat Fluid Flow* **2015**, *25*, 703–723. [CrossRef]

3. Malvandi, A.; Moshizi, S.A.; Soltani, E.G.; Ganji, D.D. Modified Buongiorno's model for fully developed mixed convection flow of nanofluids in a vertical annular pipe. *Comput. Fluids* **2014**, *89*, 124–132. [CrossRef]

4. Hayat, T.; Ashraf, M.B.; Shehzad, S.A.; Abouelmaged, E.I. Three dimensional flow of Erying powell nanofluid over an exponentially stretching sheet. *Int. J. Numer. Method Heat Fluid Flow* **2015**, *25*, 333–357. [CrossRef]

5. Nadeem, S.; Haq, R.U.; Akbar, N.S.; Lee, C.; Khan, Z.H. Numerical study of boundary layer flow and heat transfer of Oldroyed-B nanofluid towards a stretching sheet. *PLoS ONE* **2013**, *8*, e69811. [CrossRef] [PubMed]

6. Rosmila, A.B.; Kandasamy, R.; Muhaimin, I. Lie symmetry groups transformation for MHD natural convection flow of nanofluid over linearly porous stretching sheet in presence of thermal stratification. *Appl. Math. Mech. Engl. Ed.* **2012**, *33*, 593–604. [CrossRef]

7. Ellahi, R. The effects of MHD an temperature dependent viscosity on the flow of non-Newtonian nanofluid in a pipe analytical solutions. *Appl. Math. Model.* **2013**, *37*, 1451–1467. [CrossRef]

8. Nadeem, S.; Saleem, S. Series solution of unsteady Erying Powell nanofluid flow on a rotating cone. *Indian J. Pure Appl. Phys.* **2014**, *52*, 725–737.

9. Abolbashari, M.H.; Freidoonimehr, N.; Rashidi, M.M. Analytical modeling of entropy generation for Casson nano-fluid flow induced by a stretching surface. *Adv. Powder Technol.* **2015**, *6*, 542–552. [CrossRef]

10. Choi, S.U.S.; Siginer, D.A.; Wang, H.P. Enhancing thermal conductivity of fluids with nanoparticle developments and applications of non-Newtonian flows. *ASME N. Y.* **1995**, *66*, 99–105.

11. Buongiorno, J. Convective transport in nanofluids. *ASME J Heat Transf.* **2005**, *128*, 240–250. [CrossRef]

12. Kumar, K.G.; Rudraswamy, N.G.; Gireesha, B.J.; Krishnamurthy, M.R. Influence of nonlinear thermal radiation and viscous dissipation on three-dimensional flow of Jeffrey nanofluid over a stretching sheet in the presence of Joule heating. *Nonlinear Eng.* **2017**, *6*, 207–219.

13. Rudraswamy, N.G.; Shehzad, S.A.; Kumar, K.G.; Gireesha, B.J. Numerical analysis of MHD three-dimensional Carreau nanoliquid flow over bidirectionally moving surface. *J. Braz. Soc. Mech. Sci. Eng.* **2017**, *23*, 5037–5047. [CrossRef]

14. Gireesha, B.J.; Kumar, K.G.; Ramesh, G.K.; Prasannakumara, B.C. Nonlinear convective heat and mass transfer of Oldroyd-B nanofluid over a stretching sheet in the presence of uniform heat source/sink. *Results Phys.* **2018**, *9*, 1555–1563. [CrossRef]

15. Kumar, K.G.; Gireesha, B.J.; Manjunatha, S.; Rudraswamy, N.G. Effect of nonlinear thermal radiation on double-diffusive mixed convection boundary layer flow of viscoelastic nanofluid over a stretching sheet. *IJMME* **2017**, *12*, 18.

16. Nadeem, S.; Rehman, A.U.; Mehmood, R. Boundary layer flow of rotating two phase nanofluid over a stretching surface. *Heat Transf. Asian Res.* **2016**, *45*, 285–298. [CrossRef]

17. Mabood, F.; Ibrahim, S.M.; Khan, W.A. Framing the features of Brownian motion and thermophoresis on radiative nanofluid flow past a rotating stretching sheet with magnetohydrodynamics. *Results Phys.* **2016**, *6*, 1015–1023. [CrossRef]

18. Shah, Z.; Islam, S.; Gul, T.; Bonyah, E.; Khan, M.A. The electrical MHD and hall current impact on micropolar nanofluid flow between rotating parallel plates. *Results Phys.* **2018**, *9*, 1201–1214. [CrossRef]

19. Shah, Z.; Gul, T.; Khan, A.M.; Ali, I.; Islam, S. Effects of hall current on steady three dimensional non-newtonian nanofluid in a rotating frame with brownian motion and thermophoresis effects. *J. Eng. Technol.* **2017**, *6*, 280–296.

20. Gireesha, B.J.; Ganesh, K.; Krishanamurthy, M.R.; Rudraswamy, N.G. Enhancement of heat transfer in an unsteady rotating flow for the aqueous suspensions of single wall nanotubes under nonlinear thermal radiation. *Numer. Study* **2018**. [CrossRef]

21. Ishaq, M.; Ali, G.; Shah, Z.; Islam, S.; Muhammad, S. Entropy Generation on Nanofluid Thin Film Flow of Eyring–Powell Fluid with Thermal Radiation and MHD Effect on an Unsteady Porous Stretching Sheet. *Entropy* **2018**, *20*, 412. [CrossRef]

22. Sarit, K.D.; Stephen, U.S.; Choi Wenhua, Y.U.; Pradeep, T. *Nanofluids Science and Technology*; Wiley-Interscience: New Your, NY, USA, 2007; Volume 397.

23. Wong, K.F.V.; Leon, O.D. Applications of nanofluids: Current and future. *Adv. Mech. Eng.* **2010**, *2*, 519659. [CrossRef]

24. Sheikholeslami, M.; Haq, R.L.; Shafee, A.; Zhixiong, L. Heat transfer behavior of Nanoparticle enhanced PCM solidification through an enclosure with V shaped fins. *Int. J. Heat Mass Transf.* **2019**, *130*, 1322–1342. [CrossRef]

25. Sheikholeslami, M.; Gerdroodbary, M.B.; Moradi, R.; Shafee, A.; Zhixiong, L. Application of Neural Network for estimation of heat transfer treatment of Al2O3-H2O nanofluid through a channel. *Comput. Methods Appl. Mech. Eng.* **2019**, *344*, 1–12. [CrossRef]

26. Sheikholeslami, M.; Mehryan, S.A.M.; Shafee, A.; Sheremet, M.A. Variable magnetic forces impact on Magnetizable hybrid nanofluid heat transfer through a circular cavity. *J. Mol. Liquids* **2019**, *277*, 388–396. [CrossRef]

27. Yadav, D.; Lee, D.; Cho, H.H.; Lee, J. The onset of double-diffusive nanofluid convection in a rotating porous medium layer with thermal conductivity and viscosity variation: A revised model. *J. Porous Media* **2016**, *19*, 31–46. [CrossRef]

28. Yadav, D.; Nam, D.; Lee, J. The onset of transient Soret-driven MHD convection confined within a Hele-Shaw cell with nanoparticles suspension. *J. Taiwan Inst. Chem. Eng.* **2016**, *58*, 235–244. [CrossRef]

29. Yadav, D.; Lee, J. The onset of MHD nanofluid convection with Hall current effect. *Eur. Phys. J. Plus* **2015**, *130*, 162–184. [CrossRef]

30. Shah, Z.; Dawar, A.; Islam, S.; Khan, I.; Ching, D.L.C. Darcy-Forchheimer Flow of Radiative Carbon Nanotubes with Microstructure and Inertial Characteristics in the Rotating Frame. *Case Stud. Therm. Eng.* **2018**, *12*, 823–832. [CrossRef]

31. Shah, Z.; Bonyah, E.; Islam, S.; Gul, T. Impact of thermal radiation on electrical mhd rotating flow of carbon nanotubes over a stretching sheet. *AIP Adv.* **2019**, *9*, 015115. [CrossRef]

32. Shah, Z.; Dawar, A.; Islam, S.; Khan, I.; Ching, D.L.C.; Khan, Z.A. Cattaneo-Christov model for Electrical MagnetiteMicropoler Casson Ferrofluid over a stretching/shrinking sheet using effective thermal conductivity model. *Case Stud. Therm. Eng.* **2018**. [CrossRef]

33. Dawar, A.; Shah, Z.; Islam, S.; Idress, M.; Khan, W. Magnetohydrodynamic CNTs Casson Nanofl uid and Radiative heat transfer in a Rotating Channels. *J. Phys. Res. Appl.* **2018**, *1*, 017–032.

34. Khan, A.S.; Nie, Y.; Shah, Z.; Dawar, A.; Khan, W.; Islam, S. Three-Dimensional Nanofluid Flow with Heat and Mass Transfer Analysis over a Linear Stretching Surface with Convective Boundary Conditions. *Appl. Sci.* **2018**, *8*, 2244. [CrossRef]

35. Shah, Z.; Islam, S.; Ayaz, H.; Khan, S. Radiative Heat and Mass Transfer Analysis of Micropolar Nanofluid Flow of Casson Fluid between Two Rotating Parallel Plates with Effects of Hall Current. *ASME J. Heat Transf.* **2019**, *141*, 022401. [CrossRef]

36. Khan, A.; Shah, Z.; Islam, S.; Khan, S.; Khan, W.; Khan, Z.A. Darcy–Forchheimer flow of micropolar nanofluid between two plates in the rotating frame with non-uniform heat generation/absorption. *Adv. Mech. Eng.* **2018**, *10*, 1687814018808850. [CrossRef]

37. Shah, Z.; Bonyah, E.; Islam, S.; Khan, W.; Ishaq, M. Radiative MHD thin film flow of Williamson fluid over an unsteady permeable stretching. *Heliyon* **2018**, *4*, e00825. [CrossRef] [PubMed]

38. Jawad, M.; Shah, Z.; Islam, S.; Islam, S.; Bonyah, E.; Khan, Z.A. Darcy-Forchheimer flow of MHD nanofluid thin film flow with Joule dissipation and Navier's partial slip. *J. Phys. Commun.* **2018**. [CrossRef]

39. Khan, N.; Zuhra, S.; Shah, Z.; Bonyah, E.; Khan, W.; Islam, S. Slip flow of Eyring-Powell nanoliquid film containing graphene nanoparticles. *AIP Adv.* **2018**, *8*, 115302. [CrossRef]

40. Hammed, K.; Haneef, M.; Shah, Z.; Islam, I.; Khan, W.; Asif, S.M. The Combined Magneto hydrodynamic and electric field effect on an unsteady Maxwell nanofluid Flow over a Stretching Surface under the Influence of Variable Heat and Thermal Radiation. *Appl. Sci.* **2018**, *8*, 160. [CrossRef]

41. Dawar, A.; Shah, Z.; Khan, W.; Idrees, M.; Islam, S. Unsteady squeezing flow of MHD CNTS nanofluid in rotating channels with Entropy generation and viscous Dissipation. *Adv. Mech. Eng.* **2019**, *10*, 1–18. [CrossRef]

42. Khan, A.; Shah, Z.; Islam, S.; Dawar, A.; Bonyah, E.; Ullah, H.; Khan, Z.A. Darcy-Forchheimer flow of MHD CNTs nanofluid radiative thermal behaviour andconvective non uniform heat source/sink in the rotating frame with microstructureand inertial characteristics. *AIP Adv.* **2018**, *8*, 125024. [CrossRef]

43. Sheikholeslami, M.; Shah, Z.; Shafi, A.; Khan, I.; Itili, I. Uniform magnetic force impact on water based nanofluid thermal behavior in a porous enclosure with ellipse shaped obstacle. *Sci. Rep.* 2019. [CrossRef]

44. Feroz, N.; Shah, Z.; Islam, S.; Alzahrani, E.O.; Khan, W. Entropy Generation of Carbon Nanotubes Flow in a Rotating Channel with Hall and Ion-Slip Effect Using Effective Thermal Conductivity Model. *Entropy* **2019**, *21*, 52. [CrossRef]

45. Alharbi, S.O.; Dawar, A.; Shah, Z.; Khan, W.; Idrees, M.; Islam, S.; Khan, I. Entropy Generation in MHD Eyring–Powell Fluid Flow over an Unsteady Oscillatory Porous Stretching Surface under the Impact of Thermal Radiation and Heat Source/Sink. *Appl. Sci.* **2018**, *8*, 2588. [CrossRef]

46. Liao, S.J. On Homotopy Analysis Method for Nonlinear Problems. *Appl. Math. Comput.* **2004**, *147*, 499–513. [CrossRef]

47. Nasir, N.; Shah, Z.; Islam, S.; Bonyah, E.; Gul, T. Darcy Forchheimer nanofluid thin film flow of SWCNTs and heat transfer analysis over an unsteady stretching sheet. *AIP Adv.* **2019**, *9*, 015223. [CrossRef]

48. Tlili, I.; Khan, W.A.; Khan, I. Multiple slips effects on MHD SA-Al2O3 and SA-Cu non-Newtonian nanofluids flow over a stretching cylinder in porous medium with radiation and chemical reaction. *Results Phys.* **2018**, *8*, 213–221. [CrossRef]

49. Khan, N.S.; Shah, Z.; Islam, S.; Khan, I.; Alkanhal, T.A.; Tlili, I. Entropy Generation in MHD Mixed Convection Non-Newtonian Second-Grade Nanoliquid Thin Film Flow through a Porous Medium with Chemical Reaction and Stratification. *Entropy* **2019**, *21*, 139. [CrossRef]

50. Fiza, M.; Islam, S.; Ullah, H.; Shah, Z.; Chohan, F. An Asymptotic Method with Applications to Nonlinear Coupled Partial Differential Equations. *Punjab Univ. J. Math.* **2018**, *50*, 139–151.

9

Two-Phase Couette Flow of Couple Stress Fluid with Temperature Dependent Viscosity Thermally Affected by Magnetized Moving Surface

Rahmat Ellahi [1,2,*], **Ahmed Zeeshan** [2], **Farooq Hussain** [2,3] and **Tehseen Abbas** [4]

[1] Center for Modeling & Computer Simulation, Research Institute, King Fahd University of Petroleum & Minerals, Dhahran 31261, Saudi Arabia

[2] Department of Mathematics & Statistics, Faculty of Basic and Applied Sciences (FBAS), International Islamic University (IIUI), Islamabad 44000, Pakistan; ahmad.zeeshan@iiu.edu.pk (A.Z.); farooq.hussain@buitms.edu.pk (F.H.)

[3] Department of Mathematics, Faculty of Arts and Basic Sciences (FABS), Balochistan University of Information Technology, Engineering, and Management Sciences (BUITEMS), Quetta 87300, Pakistan

[4] Department of Mathematics, University of Education Lahore, Faisalabad Campus, Faisalabad 38000, Pakistan; tehseen.abbas@ue.edu.pk

* Correspondence: rellahi@alumni.ucr.edu

Abstract: The Couette–Poiseuille flow of couple stress fluid with magnetic field between two parallel plates was investigated. The flow was driven due to axial pressure gradient and uniform motion of the upper plate. The influence of heating at the wall in the presence of spherical and homogeneous Hafnium particles was taken into account. The temperature dependent viscosity model, namely, Reynolds' model was utilized. The Runge–Kutta scheme with shooting was used to tackle a non-linear system of equations. It was observed that the velocity decreased by increasing the values of the Hartman number, as heating of the wall reduced the effects of viscous forces, therefore, resistance of magnetic force reduced the velocity of fluid. However, due to shear thinning effects, the velocity was increased by increasing the values of the viscosity parameter, and as a result the temperature profile also declined. The suspension of inertial particles in an incompressible turbulent flow with Newtonian and non-Newtonian base fluids can be used to analyze the biphase flows through diverse geometries that could possibly be future perspectives of proposed model.

Keywords: couple stress fluid; Hafnium particles; Couette–Poiseuille flow; shooting method; magnetic field

1. Introduction

Diverse forms of flow paths appear when fluid flow is diverted by debris blocking streams. Such multiphase flows take place naturally due to the various factors on plateaus. The physical occurrence of multiphase flows includes chemical processes, pharmaceutical, wastewater management, and power generation. Consequently, the multiphase flows have attracted the attention of scientists and engineers due to the frequently arising issues in industrial and mechanical problems. For instance, couple stress fluid flow under the influence of heat between two parallel walls was examined by Farooq et al. [1]. Mahabaleshwar et al. [2] have investigated the magnetohydrodynamics (MHD) couple stress fluid over the flat sheet affected by the radiation. Exact solutions for the velocity were derived using a power series method for two different models. The First case described the surface temperature while the second case dealt with heat flux. Saad and Ashmway [3] have studied the flow of an unsteady couple stress fluid between two plates. The fluid flows with constant motion of the upper plate which was initially at rest. Influence of lubrication on walls was pondered in such a way that the couple stresses on the boundaries had no impact at all. A suitable transform helps to obtain the velocity of fluid numerically. Akhtar and Shah [4] have presented the exact results for three different types of fundamental flows by taking couple stress fluid as a base fluid. Khan et al. [5] reported an incompressible flow of MHD couple stress in which thermally charged fluid was disturbed by transversely applied magnetic fields. The unsteady Couette flow of non-uniform magnetic field has been investigated by Asghar and Ahamd [6]. Shaowei and Mingyu [7] have devoted their efforts for the study of the Couette flow of Maxwell fluid. Integral and Weber transforms have been used to analyze the physical phenomenon. The Couette flow through a symmetric channel was numerically tackled by Eegunjobi et al. [8]. Few core investigations on Couette flow [9–12] and couple stress fluid [13–15] are listed for those working in the same regimes.

Moreover, Poply et al. [16] have examined the temperature-dependent fluid properties of MHD flow with heat transfer. Ellahi et al. [17] have considered two different viscosity models for their investigations of heated flow. They chose third-grade nanofluid flow through coaxial cylinders. Homotopy analysis method is used to produce a closed form solution. In Reference [18] authors have discussed a temperature dependent thick flow between two opposite walls of uneven configurations. The viscosity of two-dimensional flow was assumed to be decreasing exponentially subject to temperature rise. The study contained the simultaneous effects of radiation and a porous medium. A steady-state flow of fourth-grade fluid in a cylinder was analyzed by Nadeem and Ali [19] and offered a comparative analysis in it. Ellahi et al. [20] studied the thermally charged couple stress fluid suspended with spherically homogenous metallic Hafnium particles for bi-phase flow along slippery walls. The rough surfaces of the walls is tackled with the lubrication effects. Variation in the viscosity of viscoelastic fluid by the Runge–Kutta technique with the shooting technique can be seen in Reference [21]. Makinde [22] focused on the impact of viscosity on the steady fluid flow with gravitational effects. The overhead surface was assumed to be at a constant temperature while the adjunct surface of the plate was heated with some external source. A few core investigations for viscous dissipation can be found in References [23,24].

Furthermore, to enhance the thermal performance, different types of nanoparticles having sizes from 1–100 nm have been utilized in bi-phase fluids. For example, Karimipour et al. [25] have studied the role of miscellaneous nanoparticles for heat transfer flow with MHD. Hosseini et al. [26] repotted a unique model on thermal conductivity of nanofluids. Nasiri et al. [27] have proposed a particle hydrodynamics approach for nano-fluid flows. Safaei et al. [28] have examined nanoplatelets–silver/water nanofluids in fully developed turbulent flows of graphene. All said investigations including References [29–33] end up stating that the presence of nanoparticles always sped up the heat transfer rate.

In the current article, we aim to study the magnetized multiphase Couette-Poiseuille flow of non-Newtonian couple stress fluid suspended by metallic particles of Hafnium with temperature dependent viscosity. The viscosity of the base fluid is exponentially decreasing due to the heating effects at the lower wall of the channel which is at rest. However, the motion of the upper wall causes the multiphase (i.e., solid–liquid) transport. The contribution of the pressure gradient simultaneously distinguishes the investigation further. The humble effort will not only speak about the mechanical and industrial multi-phase flows but would also fill the gap yet not available in the existing literature on the topic under consideration.

2. Mathematical Analysis

Consider a plane Couette flow between two opposite flat plates at $\eta = \pm h$, as shown in Figure 1. Flow is investigated in (ξ, η) plane in such a way that ξ-axis lies in the middle and along the plates. It is a well-established fact [34] that when the flow is generated by the constantly moving upper plate, then only can the unidirectional disturbance in the ξ-direction occur. The axial velocity [u, 0, 0] was along the ξ-direction, whereas lateral velocity was in the η-direction is zero. When the metallic particles of

Hafnium were suspended in couple stress fluid under the influence of higher temperature of the lower wall, then the governing equations in component form [35] can be expressed as:

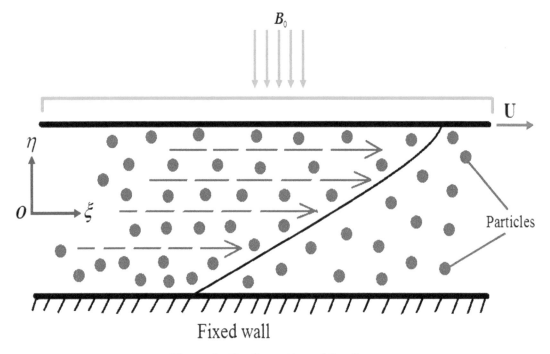

Figure 1. Configuration of the flow.

(i). For fluid phase

$$\frac{\partial u_f}{\partial \xi} = 0,$$
(1)

$$\frac{\partial}{\partial \eta}\left(\mu_s \frac{\partial u_f}{\partial \eta}\right) - \eta_1\left(\frac{\partial^4 u_f}{\partial \eta^4}\right) + \frac{CS}{(1-C)}\left(u_p - u_f\right) - \frac{\sigma B_0^2}{(1-C)}u_f = \frac{\partial p}{\partial \xi}.$$
(2)

(ii). For particle phase

$$\frac{\partial u_p}{\partial \xi} = 0,$$
(3)

$$u_f = u_p + \frac{1}{S}\left(\frac{\partial p}{\partial \xi}\right).$$
(4)

(iii). Energy equation

$$\frac{\partial^2 \Theta}{\partial \eta^2} + \frac{\mu_s}{k}\left(\frac{\partial u_f}{\partial \eta}\right)^2 = \frac{\eta_1}{k}\left(\frac{\partial u_f}{\partial \eta}\right)\left(\frac{\partial^3 u_f}{\partial \eta^3}\right).$$
(5)

where, C denotes concentration of the particles, μ_s is the viscosity of solid-liquid, μ_0 viscosity of the base liquid, η_1 is a material constant associated with couple stress fluid, σ is the electric conductivity of the fluid, B_0^2 is the magnetic strength, Θ is temperature, and k is the thermal conductivity of the fluid whereas, ξ and η are, respectively, axial and lateral coordinates. Moreover, S denotes the drag coefficient of interaction for the force exerted by particle on the fluid, and is given by Tam [36]:

$$S = \frac{4.5\,\mu_0}{r}\lambda(C),$$
(6)

$$\lambda(C) = \frac{4 + 3\sqrt{8C - 3C^2} + 3C}{(2 - 3C)^2}.$$
(7)

where, in Equation (6) the radius of the Hafnium particles is denoted by r.

Boundary Conditions

The flow interaction at the surfaces of the parallel plates are denoted by the following:

$$\left.\begin{array}{l} (i).\ u_f(\eta) = 0, \\ (ii).\ \dfrac{\partial^2 u_f}{\partial \eta^2} = 0, \\ (iii).\Theta(\eta) = \Theta_0. \end{array}\right\} ; \text{ When } \eta = -h, \tag{8}$$

$$\left.\begin{array}{l} (iv).\ u_f(\eta) = U, \\ (v).\ \dfrac{\partial^2 u_f}{\partial \eta^2} = 0, \\ (vi).\Theta(\eta) = \Theta_w. \end{array}\right\} ; \text{ When } \eta = h. \tag{9}$$

By using the following appropriate quantities:

$$\frac{u_f}{U} = u_f^*;\ \frac{u_p}{U} = u_p^*;\ \frac{\eta}{h} = \eta^*\ ;\ \frac{\xi}{h} = \xi^*;\ \frac{\mu_s}{\mu_0} = \mu^*\ ;\ \frac{hp}{\mu_0 U} = p^*\ ;\ B_r = \frac{U^2 \mu_0}{k(\Theta_w - \Theta_0)};$$
$$\gamma = \sqrt{\frac{\mu_0}{\eta_1}}h;\ M = \sqrt{\frac{\sigma}{\mu_0}}hB_0;\ m = \frac{\mu_0}{h^2 S};\ \Theta^* = \frac{\Theta - \Theta_0}{(\Theta_w - \Theta_0)}. \tag{10}$$

Equations (1)–(5), in non-dimensional form after neglecting asterisk can be written as

$$\frac{dp}{d\xi} = \frac{d}{d\eta}\left(\mu \frac{du_f}{d\eta}\right) - \frac{1}{\gamma^2}\left(\frac{d^4 u_f}{d\eta^4}\right) + \frac{C}{m}\frac{(u_p - u_f)}{(1 - C)} - \frac{M^2}{(1 - C)}u_f, \tag{11}$$

$$u_p = u_f - m\frac{dp}{d\xi}, \tag{12}$$

$$\frac{d^2\Theta}{d\eta^2} + \mu B_r\left(\frac{du_f}{d\eta}\right)^2 = \frac{B_r}{\gamma^2}\left(\frac{du_f}{d\eta}\right)\left(\frac{d^3 u_f}{d\eta^3}\right). \tag{13}$$

In which, M is the Hartmann number, γ is the couple stress parameter, m is the drag constant and B_r is he Brinkman number.

3. Results and Discussion

3.1. Variable Viscosity

The Reynolds' model for temperature dependent viscosity [37] can be defined as

$$\mu_s(\Theta) = \mu_0 e^{-\alpha(\Theta - \Theta_0)}. \tag{14}$$

In view of expression given in (10), the non-dimensional form of Equation (14), after dropping asterisk is obtained as

$$\mu(\Theta) = e^{-\alpha(\Theta_w - \Theta_0)\Theta} = e^{-\beta\Theta}, \tag{15}$$

where $\beta = \alpha(\Theta_w - \Theta_0)$.

Obviously, for the convergence of Equation (15), $\beta \in [0\ 1]$.

By Walter' lemma, the Maclaurin' series of Equation (15) can be linearized as

$$\mu(\Theta) = 1 - \beta\Theta. \tag{16}$$

In view of Equations (12) and (16), Equations (11) and (13) provide the set of nonlinear coupled differential equations involving the viscosity of the fluid deeply affected by the presence of heat applied at the wall along with a constant pressure gradient at each point of the channel (i.e., $\frac{dp}{d\xi} = P$) as follows

$$\frac{d^4 u_f}{d\eta^4} + \gamma^2 \beta \left(\frac{d\Theta}{d\eta}\right)\left(\frac{du_f}{d\eta}\right) + \gamma^2 (\beta\Theta - 1)\frac{d^2 u_f}{d\eta^2} + \frac{M^2 \gamma^2}{(1-C)} u_f + \frac{\gamma^2 P}{(1-C)} = 0, \tag{17}$$

$$\frac{d^2 \Theta}{d\eta^2} + B_r (1 - \beta\Theta)\left(\frac{du_f}{d\eta}\right)^2 = \frac{B_r}{\gamma^2}\left(\frac{du_f}{d\eta}\right)\left(\frac{d^3 u_f}{d\eta^3}\right). \tag{18}$$

On the same contrast, Equations (8) and (9), in view of (10), are acquired as

$$\left.\begin{array}{l} (i).\ u_f(\eta) = 0, \\ (ii).\ \frac{\partial^2 u_f}{\partial\eta^2} = 0, \\ (iii).\Theta(\eta) = 0. \end{array}\right\};\ When\ \eta = -1, \tag{19}$$

$$\left.\begin{array}{l} (iv).\ u_f(\eta) = 1, \\ (v).\ \frac{\partial^2 u_f}{\partial\eta^2} = 0, \\ (vi).\Theta(\eta) = 1. \end{array}\right\};\ When\ \eta = 1. \tag{20}$$

3.2. Numerical Procedure

The set of non-linear differential Equations (17) and (18) with the boundary conditions (19) and (20) are solved by employing the most efficient numerical procedure consist of Runge–Kutta method and the shooting scheme [38] using MATLAB software. It is an iterative scheme, in which each step possible error can be successively reduced by changing higher order derivatives.

Let:

$$u_f = f_1 \tag{21}$$

be the velocity of the fluid phase, then the derivatives of u_f, in terms of system of first ordinary differential equations (ODEs) can be expressed as:

$$f_2 = \frac{du_f}{d\eta} = f_1', \tag{22}$$

$$f_3 = \frac{d^2 u_f}{d\eta^2} = f_2', \tag{23}$$

$$f_4 = \frac{d^3 u_f}{d\eta^3} = f_3', \tag{24}$$

$$\Theta = f_5, \tag{25}$$

$$f_6 = \frac{d\Theta}{d\eta} = f_5', \tag{26}$$

here the sign of prime (') at the top indicates the derivative with respect to "η". In view of Equations (22)–(26) the fluid phase differential equation is transformed as:

$$f_4' = \gamma^2 (1 - \beta(f_5)) f_3 - \gamma^2 \beta(f_2)(f_6) - \left(\frac{\gamma^2 M^2}{1-C}\right) f_1 - \left(\frac{\gamma^2}{1-C}\right) P, \tag{27}$$

$$f_6' = \frac{B_r}{\gamma^2}\ (f_2)(f_4) + B_r\ (\beta(f_5) - 1)(f_2)^2. \tag{28}$$

The transformed set of conditions are given as:

$$
\left.
\begin{array}{l}
(i). \ f_1 = 0, \\
(ii). \ f_2 = k_1, \\
(iii). \ f_3 = 0, \\
(iv). \ f_4 = k_2, \\
(v). \ f_5 = 0, \\
(vi). \ f_6 = k_3.
\end{array}
\right\} ; \ When \ \eta = -1, \tag{29}
$$

$$
\left.
\begin{array}{l}
(i). \ f_1 = 1 \\
(ii). \ f_2 = k_4, \\
(iii). \ f_3 = 0, \\
(iv). \ f_4 = k_5, \\
(v). \ f_5 = 1, \\
(vi). \ f_6 = k_6.
\end{array}
\right\} ; \ When \ \eta = 1. \tag{30}
$$

where $k_1, k_2, k_3, k_4, k_5,$ and k_6 can be easily determined during the routine numerical procedure.

3.3. Graphical Illustration

To see the effects of physical parameters for Reynolds model on velocity and temperature, Figures 2–6 have been displayed. The range of all physical parameters available in the existing literatures are as follows: the range of Hartmann number is $0 < M < 1$ [39], the Brinkman number B_r varies from 0.5 to 2.0 [40], the range of couple stress parameter γ is 0.5 to 2.0 [41], the range of concentration of the metallic particles' C is 0 to 0.2 [42], and the range of viscous parameter β lies between 0 to 1.

The role of transversely applied magnetic fields can be sighted in Figure 2. It is found that the velocity of fluid decreases by increasing the values of Hartmann number. It is in accordance with the physical expectation, as increased in the Hartmann number, means to strengthen the magnetic field lines which result to impede the flow. Therefore, the obtained results validate the expected outcomes. In Figure 3, addition of some extra metallic particles to the system that expedites the flow is observed. It is found that velocity of fluid escalates for higher values of C. It is very much obvious as the constant movement of the upper wall does not allow the particle to exert an extra drag force to attenuate the base fluid motion. Thus, particle-to-particle interaction and fluid–particle interaction gets meager, which causes the frisky movement of the Hafnium particles in the base fluid. In Figure 4, we show that an increase in the couple stress parameter weakens the rotational field of couple stress fluid particles. It was revealed that the velocity profile increased by increasing the values of the couple stress parameter. It is because of friction force that fails to gain enough strength which can cause enough resistance to slow down the celerity of the flow. Similarly, the application of heat on the lower wall contributes in shear thinning effects which aids the fluid particles to get extra momentum. Hence, increase in the velocity of the fluid flow is vivid in display.

Figure 5 shows the impact of decreasing viscous parameter β on the flow dynamics. In Equation (14), it can be inferred that as the temperature difference mounts, the shear thinning effects on the viscosity of the base fluid aggravates. This attenuation of physical property results in the increase of the celerity of the fluid and particles movements. Figure 6 describes the role of Brinkman number B_r on the temperature. It is seen that higher values of Brinkman heats up the fluid by surging the temperature. However, the quite opposite behavior was observed for the case of viscosity parameter as shown in Figure 7. It was revealed that the temperature of the fluid declines for the higher values of β. This temperature decline was in fact due to the rapid movement of the couple stress fluid.

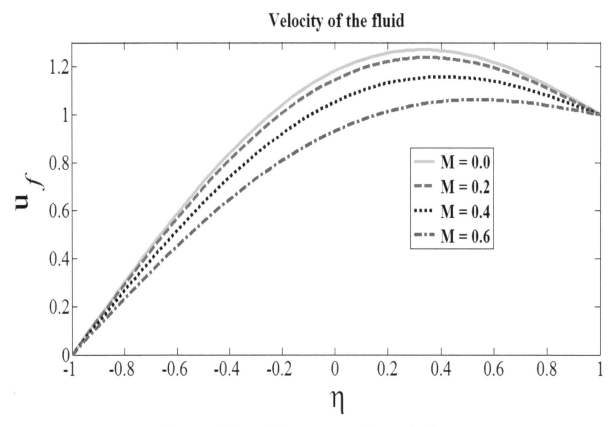

Figure 2. Effects of Hartmann number on the flow.

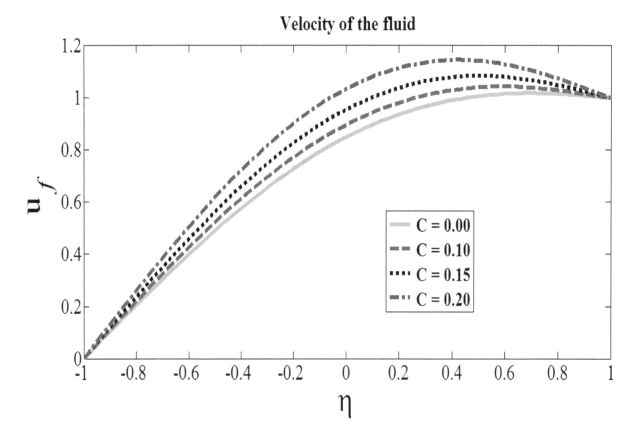

Figure 3. Effects of metallic particle concentration on the flow.

Velocity of the fluid

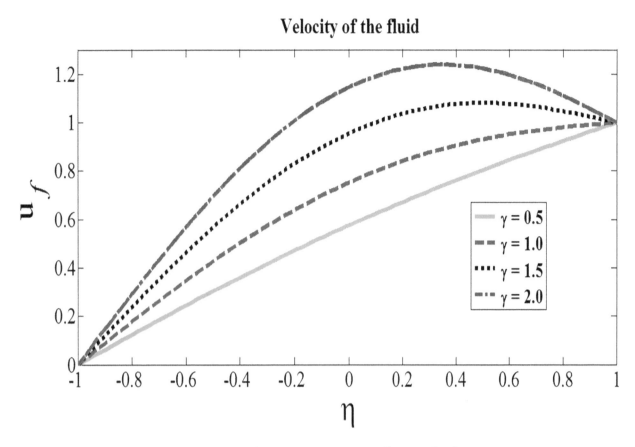

Figure 4. Couples stress parameter affecting the flow.

Velocity of the fluid

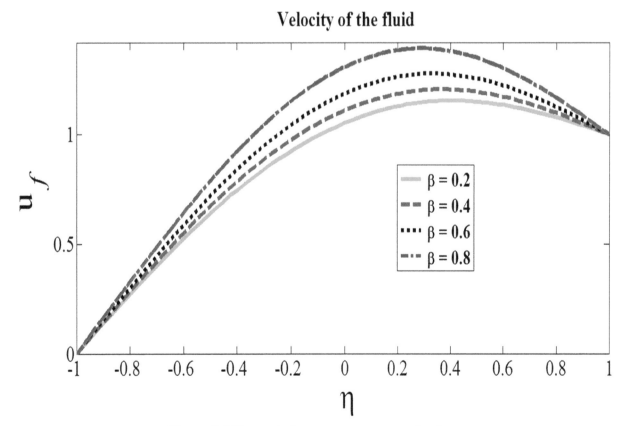

Figure 5. Effects of viscous parameter on the flow.

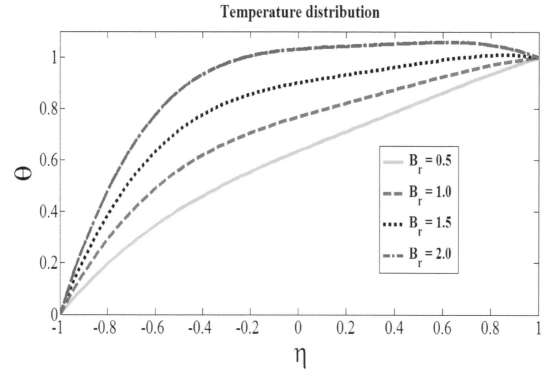

Figure 6. Role of Brinkman number on the temperature.

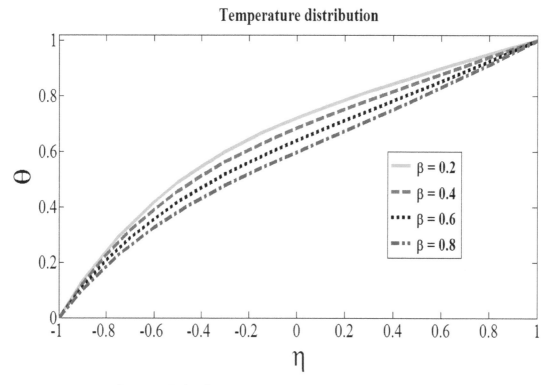

Figure 7. Role of viscous parameter on the temperature.

3.4. Validation

The numerical results are being presented in Tables 1–3. The variation in the velocities of both phases against couple stress parameter when $M = 1.0$, $C = 0.4$, and $B_r = 2.0$ are kept fixed are given in Table 1 whereas the variation in the velocities for single- and two-phase flows at different points of the domain when $M = 1.0$, $\gamma = 2.0$, and $B_r = 2.0$ are specified in Table 2. The thermal variation at the different points of given domain when $M = 1.0$ can be seen in Table 3. In all three table, one can conclude that temperature and velocities for both fluid and nanoparticles were an increasing function of metallic particles concentration C, couple stress parameter γ, and the Brinkman number B_r. The results extracted by numerical computation were found to be in excellent agreement with graphical illustrations and also satisfied all the subjected conditions. This provides a useful check that the presented solutions are correct.

Table 1. Variation in the velocities of both phases for Newtonian case and couple stress fluid.

y	u_p Newtonian Fluid $(\gamma = 0.0)$	u_p Couple Stress Fluid $(\gamma = 2.0)$	u_f Newtonian Fluid $(\gamma = 0.0)$	u_f Couple Stress Fluid $(\gamma = 2.0)$
−1.0	1.0000	1.0000	0.0000	0.0000
−0.6	1.2000	1.3221	0.2000	0.3221
−0.2	1.4000	1.5826	0.4000	0.5826
0.2	1.6000	1.7698	0.6000	0.7698
0.6	1.8000	1.8998	0.8000	0.8998
1.0	2.0000	2.0000	1.0000	1.0000

Table 2. Variation in the velocities for single- and two-phase flows.

y	u_f Single Phase $(C = 0.0)$	u_p Solid–Liquid Phase $(C = 0.4)$	u_f Solid–Liquid Phase $(C = 0.4)$
−1.0	0.0000	1.0000	0.0000
−0.6	0.2741	1.3221	0.3221
−0.2	0.5117	1.5826	0.5826
0.2	0.7047	1.7698	0.7698
0.6	0.8618	1.8998	0.8998
1.0	1.0000	2.0000	1.0000

Table 3. Thermal variation at the different points.

y	Θ $B_r = 0.0$	Θ $B_r = 2.0$	Θ $\gamma = 0.0$	Θ $C = 0.0$
−1.0	0.0000	0.0000	0.0000	0.0000
−0.6	0.2000	0.3916	0.3512	0.3578
−0.2	0.4000	0.6066	0.5629	0.5870
0.2	0.6000	0.7528	0.7095	0.7504
0.6	0.8000	0.8785	0.8446	0.8830
1.0	1.0000	1.0000	1.0000	1.0000

4. Conclusions

The Couette–Poiseuille flow of couple stress fluid in the presence of Hafnium particles was studied. The viscous dissipation effects were also reported. Exponentially decreasing viscosity of base fluid was presented by the Reynolds model. Transversely acting magnetic fields contributed by hindering the bi-phase flow. The key findings are described as:

➢ The flow of couple stress fluid resists for increasing values of Hartmann number.
➢ The temperature effectively variates the viscosity of the fluid to cause the shear thinning effects.
➢ The temperature of the flow mounts in response of higher values of Brinkman number.
➢ Attenuation of the viscosity results to expedite the flows.
➢ Viscosity parameter brings celerity in the velocity of bi-phase fluid due to high temperature difference.

➢ Molecules additives of base fluid reduce the force of friction and hence in both phases the velocity is galvanized.

➢ Due to the immense applications of multiphase flows in industrial and pharmaceutical, the proposed theoretical model is now available to vet relevant experimental investigations.

Author Contributions: Conceptualization, R.E.; Investigation, F.H.; Methodology, T.A.; Visualization, A.Z.

Acknowledgments: R. Ellahi thanks to Sadiq M. Sait, the Director Office of Research Chair Professors, King Fahd University of Petroleum and Minerals, Dhahran, Saudi Arabia, to honor him with the Chair Professor at KFUPM. F. Hussain is also acknowledged Higher Education Commission Pakistan to provide him indigenous scholar for the pursuance of his Ph.D. studies.

References

1. Farooq, M.; Rahim, M.T.; Islam, S.; Siddiqui, A.M. Steady Poiseuille flow and heat transfer of couple stress fluids between two parallel inclined plates with variable viscosity. *J. Assoc. Arab Univ. Basic Appl. Sci.* **2013**, *14*, 9–18. [CrossRef]

2. Mahabaleshwar, U.S.; Sarris, I.E.; Hill, A.; Lorenzini, G.; Pop, I. An MHD couple stress fluid due to a perforated sheet undergoing linear stretching with heat transfer. *Int. J. Heat Mass. Transf.* **2017**, *105*, 157–167. [CrossRef]

3. Saad, H.; Ashmawy, E.A. Unsteady plane Couette flow of an incompressible couple stress fluid with slip boundary conditions. *Int. J. Med. Health Sci.* **2016**, *3*, 85–92. [CrossRef]

4. Akhtar, S.; Shah, N.A. Exact solutions for some unsteady flows of a couple stress fluid between parallel plates. *Ain Shams Eng. J.* **2018**, *9*, 985–992. [CrossRef]

5. Khan, N.A.; Khan, H.; Ali, S.A. Exact solutions for MHD flow of couple stress fluid with heat transfer. *J. Egypt. Math. Soc.* **2016**, *24*, 125–129. [CrossRef]

6. Asghar, S.; Ahmad, A. Unsteady Couette flow of viscous fluid under a non-uniform magnetic field. *Appl. Math. Lett.* **2012**, *25*, 1953–1958. [CrossRef]

7. Shaowei, W.; Mingyu, X. Exact solution on unsteady Couette flow of generalized Maxwell fluid with fractional derivative. *Acta Mech.* **2006**, *187*, 103–112. [CrossRef]

8. Eegunjobi, A.S.; Makinde, O.D.; Tshehla, M.S.; Franks, O. Irreversibility analysis of unsteady Couette flow with variable viscosity. *J. Hydrodyn. B* **2015**, *27*, 304–310. [CrossRef]

9. Ellahi, R.; Wang, X.; Hameed, M. Effects of heat transfer and nonlinear slip on the steady flow of Couette fluid by means of Chebyshev Spectral Method. *Z. Naturforsch A.* **2014**, *69*, 1–8. [CrossRef]

10. Ellahi, R.; Shivanian, E.; Abbasbandy, S.; Hayat, T. Numerical study of magnetohydrodynamics generalized Couette flow of Eyring-Powell fluid with heat transfer and slip condition. *Int. J. Numer. Method Heat Fluid Flow* **2016**, *26*, 1433–1445. [CrossRef]

11. Zeeshan, A.; Shehzad, N.; Ellahi, R. Analysis of activation energy in Couette-Poiseuille flow of nanofluid in the presence of chemical reaction and convective boundary conditions. *Results Phys.* **2018**, *8*, 502–512. [CrossRef]

12. Shehzad, N.; Zeeshan, A.; Ellahi, R. Electroosmotic flow of MHD Power law Al2O3-PVC nanofluid in a horizontal channel: Couette-Poiseuille flow model. *Commun. Theor. Phys.* **2018**, *69*, 655–666. [CrossRef]

13. Hussain, F.; Ellahi, R.; Zeeshan, A.; Vafai, K. Modelling study on heated couple stress fluid peristaltically conveying gold nanoparticles through coaxial tubes: A remedy for gland tumors and arthritis. *J. Mol. Liq.* **2018**, *268*, 149–155. [CrossRef]

14. Ellahi, R.; Zeeshan, A.; Hussain, F.; Asadollahi, A. Peristaltic blood flow of couple stress fluid suspended with nanoparticles under the influence of chemical reaction and activation energy. *Symmetry* **2019**, *11*, 276. [CrossRef]

15. Ellahi, R.; Bhatti, M.M.; Fetecau, C.; Vafai, K. Peristaltic flow of couple stress fluid in a non-uniform rectangular duct having compliant walls. *Commun. Theor. Phys.* **2016**, *65*, 66–72. [CrossRef]

16. Poply, V.; Singh, P.; Yadav, A.K. A study of Temperature-dependent fluid properties on MHD free stream flow and heat transfer over a non-linearly stretching sheet. *Procedia Eng.* **2015**, *127*, 391–397. [CrossRef]

17. Ellahi, R.; Raza, M.; Vafai, K. Series solutions of non-Newtonian nanofluids with Reynolds model and Vogel's model by means of the homotopy analysis method. *Math. Comput. Model.* **2012**, *55*, 1876–1891. [CrossRef]

18. Disu, A.B.; Dada, M.S. Reynolds model viscosity on radiative MHD flow in a porous medium between two vertical wavy walls. *J. Taibah Univ. Sci.* **2017**, *11*, 548–565. [CrossRef]

19. Nadeem, S.; Ali, M. Analytical solutions for pipe flow of a fourth-grade fluid with Reynolds and Vogel's models of viscosities. *Commun. Nonlin. Sci. Numer. Simulat.* **2009**, *14*, 2073–2090. [CrossRef]

20. Ellahi, R.; Zeeshan, A.; Hussain, F.; Abbas, T. Thermally charged MHD bi-phase flow coatings with non-Newtonian nanofluid and Hafnium particles through slippery walls. *Coatings* **2019**, *9*, 300. [CrossRef]

21. Mahmoud, M.A. Chemical reaction and variable viscosity effects on flow and mass transfer of a non-Newtonian visco-elastic fluid past a stretching surface embedded in a porous medium. *Meccanica* **2010**, *45*, 835–846. [CrossRef]

22. Makinde, O.D. Laminar falling liquid film with variable viscosity along an inclined heated plate. *Appl. Math. Comput.* **2006**, *175*, 80–88. [CrossRef]

23. Jawad, M.; Shah, Z.; Islam, S.; Majdoubi, J.; Tlili, I.; Khan, W.; Khan, I. Impact of nonlinear thermal radiation and the viscous dissipation effect on the unsteady three-dimensional rotating flow of single-wall carbon nanotubes with aqueous suspensions. *Symmetry* **2019**, *11*, 207. [CrossRef]

24. Ellahi, R. A study on the convergence of series solution of non-Newtonian third grade fluid with variable viscosity: By means of homotopy analysis method. *Adv. Math. Phys.* **2012**, *2012*, 634925. [CrossRef]

25. Karimipour, A.; Orazio, A.D.; Shadloo, M.S. The effects of different nano particles of Al_2O_3 and Ag on the MHD nano fluid flow and heat transfer in a microchannel including slip velocity and temperature jump. *Phys. E* **2017**, *86*, 146–153. [CrossRef]

26. Hosseini, S.M.; Safaei, M.R.; Goodarzi, M.; Alrashed, A.A.A.A.; Nguyen, T.K. New temperature, interfacial shell dependent dimensionless model for thermal conductivity of nanofluids. *Int. J. Heat Mass Transf.* **2017**, *114*, 207–210. [CrossRef]

27. Nasiri, H.; Jamalabadi, M.Y.A.; Sadeghi, R.; Safaei, M.R.; Nguyen, T.K.; Shadloo, M.S. A smoothed particle hydrodynamics approach for numerical simulation of nano-fluid flows. *J. Therm. Anal. Calorim.* **2018**, 1–9. [CrossRef]

28. Safaei, M.R.; Ahmadi, G.; Goodarzi, M.S.; Shadloo, M.S.; Goshayeshi, H.R.; Dahari, M. Heat transfer and pressure drop in fully developed turbulent flows of graphene nanoplatelets–silver/water nanofluids. *Fluids* **2016**, *1*, 20. [CrossRef]

29. Sadiq, M.A. MHD stagnation point flow of nanofluid on a plate with anisotropic slip. *Symmetry* **2019**, *11*, 132. [CrossRef]

30. Rashidi, S.; Esfahani, J.A.; Ellahi, R. Convective heat transfer and particle motion in an obstructed duct with two side by side obstacles by means of DPM model. *Appl. Sci.* **2017**, *7*, 431. [CrossRef]

31. Shehzad, N.; Zeeshan, A.; Ellahi, R.; Rashidid, S. Modelling study on internal energy loss due to entropy generation for non-Darcy Poiseuille flow of silver-water nanofluid: An application of purification. *Entropy* **2018**, *20*, 851. [CrossRef]

32. Hassan, M.; Ellahi, R.; Bhatti, M.M.; Zeeshan, A. A comparative study of magnetic and non-magnetic particles in nanofluid propagating over a wedge. *Can. J. Phys.* **2019**, *97*, 277–285. [CrossRef]

33. Zeeshan, A.; Shehzad, N.; Abbas, A.; Ellahi, R. Effects of radiative electro-magnetohydrodynamics diminishing internal energy of pressure-driven flow of titanium dioxide-water nanofluid due to entropy generation. *Entropy* **2019**, *21*, 236. [CrossRef]

34. Ashrafi, N.; Khayat, R.E. A low-dimensional approach to nonlinear plane-Couette flow of viscoelastic fluids. *Phys. Fluids.* **2000**, *12*, 345–365. [CrossRef]

35. Srivastava, L.M.; Srivastava, V.P. Peristaltic transport of a particle-fluid suspension. *J. Biomech. Eng.* **1989**, *111*, 157–165. [CrossRef]

36. Tam, C.K.W. The drag on a cloud of spherical particles in a low Reynolds number flow. *J. Fluid Mech.* **1969**, *38*, 537–546. [CrossRef]

37. Ellahi, R. The effects of MHD and temperature dependent viscosity on the flow of non-Newtonian nanofluid in a pipe: Analytical solutions. *Appl. Math. Model.* **2013**, *37*, 1451–1457. [CrossRef]

38. Hossain, M.A.; Subba, R.; Gorla, R. Natural convection flow of non-Newtonian power-law fluid from a slotted vertical isothermal surface. *Int. J Numer. Methods Heat Fluid Flow* **2009**, *19*, 835–846. [CrossRef]

39. Makinde, O.D.; Onyejekwe, O.O. A numerical study of MHD generalized Couette flow and heat transfer with variable viscosity and electrical conductivity. *J. Magn. Magn. Mater.* **2011**, *323*, 2757–2763. [CrossRef]

40. Coelho, M.P.; Faria, J.S. On the generalized Brinkman number definition and its importance for Bingham fluids. *J. Heat Transf.* **2011**, *133*, 545051–545055. [CrossRef]

41. Swarnalathamma, B.V.; Krishna, M.V. Peristaltic hemodynamic flow of couple stress fluid through a porous medium under the influence of magnetic field with slip effect. *AIP Conf. Proc.* **2016**, *1728*, 0206031–0206039. [CrossRef]

42. Charm, S.E.; Kurland, G.S. *Blood Flow and Microcirculation*; Wiley: New York, NY, USA, 1974.

10

A Particle Method based on a Generalized Finite Difference Scheme to Solve Weakly Compressible Viscous Flow Problems

Yongou Zhang [1,2] **and Aokui Xiong** [1,2,*]

[1] Key Laboratory of High Performance Ship Technology (Wuhan University of Technology), Ministry of Education, Wuhan 430074, China

[2] School of Transportation, Wuhan University of Technology, Wuhan 430074, China

* Correspondence: xiong_ak@163.com

Abstract: The Lagrangian meshfree particle-based method has advantages in solving fluid dynamics problems with complex or time-evolving boundaries for a single phase or multiple phases. A pure Lagrangian meshfree particle method based on a generalized finite difference (GFD) scheme is proposed to simulate time-dependent weakly compressible viscous flow. The flow is described with Lagrangian particles, and the partial differential terms in the Navier-Stokes equations are represented as the solution of a symmetric system of linear equations through a GFD scheme. In solving the particle-based symmetric equations, the numerical method only needs the kernel function itself instead of using its gradient, i.e., the approach is a kernel gradient free (KGF) method, which avoids using artificial parameters in solving for the viscous term and reduces the limitations of using the kernel function. Moreover, the order of Taylor series expansion can be easily improved in the meshless algorithm. In this paper, the particle method is validated with several test cases, and the convergence, accuracy, and different kernel functions are evaluated.

Keywords: compressible viscous flow; symmetric linear equations; generalized finite difference scheme; kernel gradient free; Lagrangian approach

1. Introduction

Problems of weakly compressible flows have attracted much attention in aerospace and oceanic applications, such as wind engineering problems, turbine flow, blood flow, and water wave motion. Accurate predictions of such flows are important in computational fluid dynamics. For fluids at low Mach numbers, the ratio between the speed of flow and the speed of sound is extremely small, and therefore, density fluctuations are not obvious. As a result, such a situation can be called weakly compressible flow. Generally, there are three numerical ways to model weakly compressible flow, namely, the Eulerian approach, the Lagrangian approach, and the hybrid approach. The Eulerian approach solves for quantities at fixed locations in space, and the Lagrangian approach uses individual particles that move through both space and time and have their own physical properties, such as density, velocity, and pressure, to represent the dynamically evolving fluid flow. The flow is described by recording the time history of each fluid particle. In the present work, we propose a pure Lagrangian meshfree particle-based method based on a meshless finite difference scheme to solve weakly compressible flow problems.

The Lagrangian meshfree method is rapidly advancing and has been widely used in recent years because it can be easily adapted to modeling problems with complex or time-evolving boundaries for single or multiple phases, such as numerical simulations of dam break flow [1], hydraulic jumps [2], rising bubbles [3] and coalescing [4]. The ability of this method to model non-Newtonian fluid and large scale diffuse fluids has been demonstrated in some recent works [5,6] by introducing different symmetric models. Moreover, because the computations are based on the support domain, which is much smaller than the complete computational region, the ill conditioned system problem is rarely encountered. Among all Lagrangian meshfree methods, the smoothed particle hydrodynamics (SPH) method was one of the earliest methods developed and has been widely applied in different fields. The SPH method was first pioneered independently by Lucy [7] and Gingold and Monaghan [8] to solve astrophysical problems in 1977. Details of the SPH method as a computational fluid dynamics method can be found in recent reviews [9–12] and Liu and Liu's book [13]. Some successful applications of this method include coastal engineering, nuclear engineering, ocean engineering, and bioengineering. However, the accuracy of the conventional SPH method is unsatisfactory, and it is not easy to achieve an accurate high-order SPH approach.

As a meshfree Lagrangian method, the particle distribution generally tends to be irregular in the computations, which leads to inconsistency and low accuracy [14,15]. For that reason, in some cases, only the first-order term of the fluid dynamics equations, the Navier-Stokes equations, is solved, and the viscous term, which contains the second-order differential, is obtained through the artificial viscosity with artificial parameters in the SPH method. This issue can also occur in the incompressible SPH (ISPH) method [16]. To improve the consistency and accuracy of these methods, different modifications have been developed. After using Taylor series expansion to normalize the kernel function, the corrective smoothed particle method (CSPM) [17,18] and the modified smoothed particle method (MSPH) [19] were proposed. Both methods have better accuracy than the conventional SPH method. Nevertheless, it should be noted that both methods improve accuracy by improving the particle approximation of the kernel gradient term, which leads to more strict requirements on the kernel function. These requirements are related to the compact condition, normalization condition, and delta function behavior [20] and limit the selection of the kernel function, especially when the second-order gradient of the kernel function is required.

To avoid the solution of the gradient of the kernel function, a method with kernel gradient free (KGF) features can be developed, as discussed in detail in [21–23]; notably, a KGF-SPH method was proposed in 2015. When a particle method only involves the kernel function itself in kernel and particle approximation, the kernel gradient is not necessary in the computation, and this approach is thus referred to as a KGF method. The KGF-SPH method is used to solve for the viscous term directly without using the artificial viscosity, and the results are good for 2D models. Another KGF method is the consistent particle method (CPM) [24,25] for incompressible flow simulation. In the CPM, Poisson's equation is used in the same way as the moving-particle semi-implicit (MPS) method based on the particle number density and the difference algorithm.

The purpose of our work is to combine a finite difference scheme and the particle method for solving weakly compressible viscous flow problems. In the method, the flow is described with Lagrangian particles, and the partial differential terms in the Navier-Stokes equations are represented as the solution of a symmetric system of linear equations through a generalized finite difference (GFD) scheme. It should be noted that this method is not a completely new method, but we will simply refer to it as the finite difference particle method (FDPM) to simplify the description in the subsequent sections.

Meshless finite difference approximation was first discussed for fully arbitrary meshes by Jensen [26] in 1972. Perrone and Kao [27] also contributed to the development of this method at that time. Subsequently, a variation using the moving least squares method was proposed by Lizska and Orkisz [28], and some recent works have been published [29–31]. The meshless finite difference scheme or GFD approximation we used came from Benito, Urena and Gavete [32], and they provided a discussion of the influence of several factors in the GFD scheme. A comparison between the GFD method and the element-free Galerkin method (EFGM) in solving the Laplace equation was presented in [33,34]. The GFD method was shown to be more accurate than the EFGM and the GFD scheme was used as a Eulerian meshfree method. In the present work, the GFD scheme is utilized to build a Lagrangian meshfree particle-based method, namely, the FDPM.

The FDPM has several advantages compared with the conventional SPH. First, the method is KGF. Only the kernel function itself and the positions of each particle are used to compute the spatial differential through a set of symmetric linear equations. Second, the method can be easily extended to high orders because it is based on Taylor series expansion. We show a fourth-order scheme for the FDPM. Additionally, only a few lines of code need to change to obtain a high-order FDPM, which is simple for users, especially when the users want to start with a low-order but fast computation. Third, the second-order differential term can be obtained without additional limitations on the kernel function. Thus, the viscous term in the Navier-Stokes equations can be computed directly without introducing any artificial parameters. Fourth, the FDPM is characterized by good compatibility. Most boundary conditions in the existing Lagrangian particle-based methods, such as the SPH and MPS methods, can be used directly. In the present work, we focus on the evaluation of the convergence, efficiency, and effects of the kernel function. The method is tested by modeling flow in a pipeline, Poisseuille flow, Couette flow and flow in porous media. These classical flows are used in different ways to solve fluid dynamics problems [35–37].

The present paper is organized as follows. In Section 2, the FDPM is given to solve the Navier-Stokes equations. In Section 3, applications of the particle method are shown. Section 4 summarizes the results of this work.

2. Finite Difference Particle Method for Weakly Compressible Flow

2.1. Lagrangian Form of the Governing Equations for Weakly Compressible Viscous Flow

The Lagrangian form of the Navier-Stokes equations, i.e., the continuity equation and the momentum equation, including viscous and external forces, are defined by Equations (1) and (2), respectively. The Lagrangian form of governing equations is as follows:

$$\frac{D\rho}{Dt} = -\rho \nabla \cdot \boldsymbol{u}, \tag{1}$$

$$\frac{D\boldsymbol{u}}{Dt} = -\frac{1}{\rho} \nabla p + v_k \nabla^2 \boldsymbol{u} + \boldsymbol{F}, \tag{2}$$

where ρ is the density, t is the time, $\boldsymbol{u}$ is the particle velocity, p is the pressure, v_k is the kinematic viscosity, and $\boldsymbol{F}$ is an external body force, such as gravity. All these variables are related to the physical properties of fluid particles that can move in both space and time, rather than remain at a fixed position.

The material derivative is written as follows

$$\frac{D}{Dt} = \frac{\partial}{\partial t} + \boldsymbol{u} \cdot \nabla. \tag{3}$$

The equation of state for weakly compressible fluid flow is

$$\frac{Dp}{Dt} = c^2 \frac{D\rho}{Dt}, \tag{4}$$

where c is the speed of sound.

2.2. Generalized Finite Difference Scheme

In the FDPM, flow is described with Lagrangian particles, and the GFD approximation [32] is utilized to solve for the spatial differential terms in the governing equations.

Consider a particle i surrounded by particles $j = 1, 2, \ldots, N$, with all $N + 1$ of the particles in a compact support domain, as shown in Figure 1. For a circular support domain, r_s represents the radius of the support domain, which is called the smoothing length in the SPH method. Particles j are white circles around particle i, which is the orange circle, and Ω represents the computational domain. The closest nodes to particle i are selected as j particles, and these particles should be in the support domain at the same time.

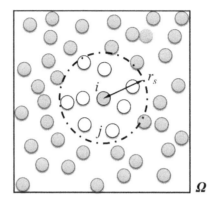

Figure 1. Computational domain Ω, support domain of particle i (point circle line), radius of the support domain r_s, and fluid particles (circles).

The values of an infinitely differentiable function F at the positions of particles i and j are defined as F_i and F_j, respectively. This F can be the pressure, velocity or density of particles in the computation. Let us expand this term as the Taylor series of F for particle i:

$$
\begin{aligned}
F_j = F_i + h_j\frac{\partial F_i}{\partial x} + k_j\frac{\partial F_i}{\partial y} + \frac{h_j^2}{2}\frac{\partial^2 F_i}{\partial x^2} + \frac{k_j^2}{2}\frac{\partial^2 F_i}{\partial y^2} + h_jk_j\frac{\partial^2 F_i}{\partial x\partial y} + \frac{h_j^3}{6}\frac{\partial^3 F_i}{\partial x^3} + \frac{k_j^3}{6}\frac{\partial^3 F_i}{\partial y^3} + \frac{h_j^2k_j}{2}\frac{\partial^3 F_i}{\partial x^2\partial y} + \frac{h_jk_j^2}{2}\frac{\partial^3 F_i}{\partial x\partial y^2} \\
+ \frac{h_j^4}{24}\frac{\partial^4 F_i}{\partial x^4} + \frac{k_j^4}{24}\frac{\partial^4 F_i}{\partial y^4} + \frac{h_j^3k_j}{6}\frac{\partial^4 F_i}{\partial x^3\partial y} + \frac{h_j^2k_j^2}{4}\frac{\partial^4 F_i}{\partial x^2\partial y^2} + \frac{h_jk_j^3}{6}\frac{\partial^4 F_i}{\partial x\partial y^3} + \cdots ,
\end{aligned}
\tag{5}
$$

where Equation (5) is for two dimensions. x and y are the spatial coordinates of the particles, and $h_j = x_j - x_i$, $k_j = y_j - y_i$.

For the fourth-order FDPM, ignoring the high-order terms, the approximation of F is denoted by f:

$$
\begin{aligned}
f_j = f_i + h_j\frac{\partial f_i}{\partial x} + k_j\frac{\partial f_i}{\partial y} + \frac{h_j^2}{2}\frac{\partial^2 f_i}{\partial x^2} + \frac{k_j^2}{2}\frac{\partial^2 f_i}{\partial y^2} + h_jk_j\frac{\partial^2 f_i}{\partial x\partial y} + \frac{h_j^3}{6}\frac{\partial^3 f_i}{\partial x^3} + \frac{k_j^3}{6}\frac{\partial^3 f_i}{\partial y^3} + \frac{h_j^2k_j}{2}\frac{\partial^3 f_i}{\partial x^2\partial y} + \frac{h_jk_j^2}{2}\frac{\partial^3 f_i}{\partial x\partial y^2} \\
+ \frac{h_j^4}{24}\frac{\partial^4 f_i}{\partial x^4} + \frac{k_j^4}{24}\frac{\partial^4 f_i}{\partial y^4} + \frac{h_j^3k_j}{6}\frac{\partial^4 f_i}{\partial x^3\partial y} + \frac{h_j^2k_j^2}{4}\frac{\partial^4 f_i}{\partial x^2\partial y^2} + \frac{h_jk_j^3}{6}\frac{\partial^4 f_i}{\partial x\partial y^3}.
\end{aligned}
\tag{6}
$$

After rearranging these equations and multiplying by a kernel function W on both sides of the equation, the sum of these expressions for all particles j is obtained:

$$
\sum_{j=1}^{N}\left(
\begin{aligned}
& f_i - f_j + h_j\frac{\partial f_i}{\partial x} + k_j\frac{\partial f_i}{\partial y} + \frac{h_j^2}{2}\frac{\partial^2 f_i}{\partial x^2} + \frac{k_j^2}{2}\frac{\partial^2 f_i}{\partial y^2} + h_jk_j\frac{\partial^2 f_i}{\partial x\partial y} \\
& + \frac{h_j^3}{6}\frac{\partial^3 f_i}{\partial x^3} + \frac{k_j^3}{6}\frac{\partial^3 f_i}{\partial y^3} + \frac{h_j^2k_j}{2}\frac{\partial^3 f_i}{\partial x^2\partial y} + \frac{h_jk_j^2}{2}\frac{\partial^3 f_i}{\partial x\partial y^2} \\
& + \frac{h_j^4}{24}\frac{\partial^4 f_i}{\partial x^4} + \frac{k_j^4}{24}\frac{\partial^4 f_i}{\partial y^4} + \frac{h_j^3k_j}{6}\frac{\partial^4 f_i}{\partial x^3\partial y} + \frac{h_j^2k_j^2}{4}\frac{\partial^4 f_i}{\partial x^2\partial y^2} + \frac{h_jk_j^3}{6}\frac{\partial^4 f_i}{\partial x\partial y^3}
\end{aligned}
\right)W(h_j, k_j, r_s) = 0,
\tag{7}
$$

where $W(h_j, k_j, r_s)$ is a kernel function in 2D and r_s represents the size of the support domain. For W, different kernel functions, including Gaussian, cubic spline, and quintic spline functions, can be found in [13]. In the following equations, we use W for the kernel function.

Function G can be defined in 2D as

$$
G(f) = \sum_{j=1}^{N}\left[\left(
\begin{aligned}
& f_i - f_j + h_j\frac{\partial f_i}{\partial x} + k_j\frac{\partial f_i}{\partial y} + \frac{h_j^2}{2}\frac{\partial^2 f_i}{\partial x^2} + \frac{k_j^2}{2}\frac{\partial^2 f_i}{\partial y^2} + h_jk_j\frac{\partial^2 f_i}{\partial x\partial y} \\
& + \frac{h_j^3}{6}\frac{\partial^3 f_i}{\partial x^3} + \frac{k_j^3}{6}\frac{\partial^3 f_i}{\partial y^3} + \frac{h_j^2k_j}{2}\frac{\partial^3 f_i}{\partial x^2\partial y} + \frac{h_jk_j^2}{2}\frac{\partial^3 f_i}{\partial x\partial y^2} \\
& + \frac{h_j^4}{24}\frac{\partial^4 f_i}{\partial x^4} + \frac{k_j^4}{24}\frac{\partial^4 f_i}{\partial y^4} + \frac{h_j^3k_j}{6}\frac{\partial^4 f_i}{\partial x^3\partial y} + \frac{h_j^2k_j^2}{4}\frac{\partial^4 f_i}{\partial x^2\partial y^2} + \frac{h_jk_j^3}{6}\frac{\partial^4 f_i}{\partial x\partial y^3}
\end{aligned}
\right)W\right]^2 = 0.
\tag{8}
$$

According to Equation (7), the norm of G equals 0, so we obtain:

$$
\frac{\partial G(f)}{\partial\left(\frac{\partial f_i}{\partial x}\right)} = 2\sum_{j=1}^{N}\Phi h_j W^2 = 0, \quad \frac{\partial G(f)}{\partial\left(\frac{\partial f_i}{\partial y}\right)} = 2\sum_{j=1}^{N}\Phi k_j W^2 = 0, \quad \frac{\partial G(f)}{\partial\left(\frac{\partial^2 f_i}{\partial x^2}\right)} = \sum_{j=1}^{N}\Phi h_j^2 W^2 = 0,
$$

$$
\frac{\partial G(f)}{\partial\left(\frac{\partial^2 f_i}{\partial y^2}\right)} = \sum_{j=1}^{N}\Phi k_j^2 W^2 = 0, \quad \frac{\partial G(f)}{\partial\left(\frac{\partial^2 f_i}{\partial x\partial y}\right)} = 2\sum_{j=1}^{N}\Phi h_j k_j W^2 = 0, \quad \frac{\partial G(f)}{\partial\left(\frac{\partial^3 f_i}{\partial x^3}\right)} = \frac{1}{3}\sum_{j=1}^{N}\Phi k_j^3 W^2 = 0,
$$

$$
\frac{\partial G(f)}{\partial\left(\frac{\partial^3 f_i}{\partial y^3}\right)} = \frac{1}{3}\sum_{j=1}^{N}\Phi k_j^3 W^2 = 0, \quad \frac{\partial G(f)}{\partial\left(\frac{\partial^3 f_i}{\partial x^2\partial y}\right)} = \sum_{j=1}^{N}\Phi h_j^2 k_j W^2 = 0, \quad \frac{\partial G(f)}{\partial\left(\frac{\partial^3 f_i}{\partial x\partial y^2}\right)} = \sum_{j=1}^{N}\Phi h_j k_j^2 W^2 = 0, \tag{9}
$$

$$
\frac{\partial G(f)}{\partial\left(\frac{\partial^4 f_i}{\partial x^4}\right)} = \frac{1}{12}\sum_{j=1}^{N}\Phi h_j^4 W^2 = 0, \quad \frac{\partial G(f)}{\partial\left(\frac{\partial^4 f_i}{\partial y^4}\right)} = \frac{1}{12}\sum_{j=1}^{N}\Phi k_j^4 W^2 = 0, \quad \frac{\partial G(f)}{\partial\left(\frac{\partial^4 f_i}{\partial x^3\partial y}\right)} = \frac{1}{3}\sum_{j=1}^{N}\Phi h_j^3 k_j W^2 = 0,
$$

$$
\frac{\partial G(f)}{\partial\left(\frac{\partial^4 f_i}{\partial x^2\partial y^2}\right)} = \frac{1}{2}\sum_{j=1}^{N}\Phi h_j^2 k_j^2 W^2 = 0, \quad \frac{\partial G(f)}{\partial\left(\frac{\partial^4 f_i}{\partial x\partial y^3}\right)} = \frac{1}{3}\sum_{j=1}^{N}\Phi h_j k_j^3 W^2 = 0,
$$

where

$$
\Phi = f_i - f_j + h_j\frac{\partial f_i}{\partial x} + k_j\frac{\partial f_i}{\partial y} + \frac{h_j^2}{2}\frac{\partial^2 f_i}{\partial x^2} + \frac{k_j^2}{2}\frac{\partial^2 f_i}{\partial y^2} + h_j k_j\frac{\partial^2 f_i}{\partial x\partial y} + \frac{h_j^3}{6}\frac{\partial^3 f_i}{\partial x^3} + \frac{k_j^3}{6}\frac{\partial^3 f_i}{\partial y^3} + \frac{h_j^2 k_j}{2}\frac{\partial^3 f_i}{\partial x^2\partial y} + \frac{h_j k_j^2}{2}\frac{\partial^3 f_i}{\partial x\partial y^2}
$$
$$
+ \frac{h_j^4}{24}\frac{\partial^4 f_i}{\partial x^4} + \frac{k_j^4}{24}\frac{\partial^4 f_i}{\partial y^4} + \frac{h_j^3 k_j}{6}\frac{\partial^4 f_i}{\partial x^3\partial y} + \frac{h_j^2 k_j^2}{4}\frac{\partial^4 f_i}{\partial x^2\partial y^2} + \frac{h_j k_j^3}{6}\frac{\partial^4 f_i}{\partial x\partial y^3}.
$$

$$
\tag{10}
$$

Equation (9) gives us the following equation:

$$
\mathbf{AD} = \mathbf{B}, \tag{11}
$$

where

$$
\mathbf{A} = \begin{bmatrix}
\sum_{j=1}^{N} h_j^2 W^2 & \sum_{j=1}^{N} h_j k_j W^2 & \sum_{j=1}^{N}\frac{1}{2}h_j^3 W^2 & \sum_{j=1}^{N}\frac{1}{2}h_j k_j^2 W^2 & \cdots & \sum_{j=1}^{N}\frac{1}{6}h_j^2 k_j^3 W^2 \\
 & \sum_{j=1}^{N} k_j^2 W^2 & \sum_{j=1}^{N}\frac{1}{2}h_j^2 k_j W^2 & \sum_{j=1}^{N}\frac{1}{2}k_j^3 W^2 & \cdots & \sum_{j=1}^{N}\frac{1}{6}h_j k_j^4 W^2 \\
 & & \sum_{j=1}^{N}\frac{1}{4}h_j^4 W^2 & \sum_{j=1}^{N}\frac{1}{4}h_j^2 k_j^2 W^2 & \cdots & \sum_{j=1}^{N}\frac{1}{12}h_j^3 k_j^3 W^2 \\
 & & & \sum_{j=1}^{N}\frac{1}{4}k_j^4 W^2 & \cdots & \sum_{j=1}^{N}\frac{1}{12}h_j k_j^5 W^2 \\
 & & & & \ddots & \vdots \\
\text{symmetric} & & & & & \sum_{j=1}^{N}\frac{1}{36}h_j^2 k_j^6 W^2
\end{bmatrix}, \tag{12}
$$

$$
\mathbf{D} = \left\{ \begin{array}{cccccc} \frac{\partial f_i}{\partial x} & \frac{\partial f_i}{\partial y} & \frac{\partial^2 f_i}{\partial x^2} & \frac{\partial^2 f_i}{\partial y^2} & \cdots & \frac{\partial^4 f_i}{\partial x\partial y^3} \end{array} \right\}^{\mathrm{T}}, \tag{13}
$$

$$
\mathbf{B} = \left\{ \begin{array}{c} \sum_{j=1}^{N}(f_j - f_i)h_j W^2 \\ \sum_{j=1}^{N}(f_j - f_i)k_j W^2 \\ \sum_{j=1}^{N}\frac{1}{2}(f_j - f_i)h_j^2 W^2 \\ \sum_{j=1}^{N}\frac{1}{2}(f_j - f_i)k_j^2 W^2 \\ \vdots \\ \sum_{j=1}^{N}\frac{1}{6}(f_j - f_i)h_j^2 k_j^3 W^2 \end{array} \right\}. \tag{14}
$$

Since matrix A is symmetrical, Equation (11) can be solved, and the solution gives the values of the spatial derivatives in matrix $\mathbf{D}$. Thus, the spatial derivatives in Equations (1) and (2) can be obtained by solving a set of symmetric linear equations, and the material derivatives in the equations can be integrated using a time integration scheme.

2.3. Particle Representation for Governing Equations

Taking particle i as an example, this section gives the particle representation of the governing equations and the solution to Equation (11). The solution includes the values of the spatial derivatives needed in the governing equations.

The coefficients of **D** and **B** are denoted by $D_m(f_i)$ and $B_m(f_i)$, respectively, with $m = 1, 2, \ldots, 5$. For example, $D_2(f_i) = \frac{\partial f_i}{\partial y}$ (the second coefficient in Equation (13)), and $B_2(f_i) = \sum_{j=1}^{N}(f_j - f_i)k_j W^2$ (the second coefficient in Equation (14)). In addition, the symmetric matrix **A** can be decomposed into the upper and lower triangular matrices $\mathbf{A} = \mathbf{LL}^{\mathsf{T}}$. The coefficients of the matrix **L** are denoted by $L(m, n)$, with m and $n = 1, 2, 3, 4, 5$.

By using the GFD scheme and Cholesky factorization to solve Equation (11), we obtain the solutions for the Lagrangian derivative terms in Equations (1) and (2) in two-dimensional form

$$\frac{\mathrm{D}\rho_i}{\mathrm{D}t} = -\rho_i D_1(u_i) - \rho_i D_2(v_i), \tag{15}$$

$$\frac{\mathrm{D}u_i}{\mathrm{D}t} = -\frac{1}{\rho_i} D_1(p_i) + v_k D_3(u_i), \tag{16}$$

$$\frac{\mathrm{D}v_i}{\mathrm{D}t} = -\frac{1}{\rho_i} D_2(p_i) + v_k D_4(v_i) + g, \tag{17}$$

where u_i and v_i are the velocity of particle i in two directions and

$$D_m(f) = \begin{cases} \frac{1}{L(m,m)}\left[Y_m(f) - \sum_{n=m+1}^{N} L(n,m)D_n(f)\right] & m = 1,2,3,4 \\ \frac{Y(f)}{L(m,m)} & m = 5 \end{cases}, \tag{18}$$

where

$$Y_m(f) = \begin{cases} \frac{b_m(f)}{L(m,m)} & m = 1 \\ \frac{1}{L(m,m)}\left[b_m(f) - \sum_{n=1}^{m-1} L(m,n)Y_n(f)\right] & m = 2,3,4,5 \end{cases}. \tag{19}$$

This method considers changes in density and is able to simulate flow at low Mach numbers, so it is used to solve weakly compressible viscous flow problems. Calculations of particle motion and time integration are performed based on second-order leapfrog integration. The equations for updating the position and velocity of particles are

$$v_i\left(t + \frac{1}{2}\Delta t\right) = v_i\left(t - \frac{1}{2}\Delta t\right) + \Delta t \frac{\mathrm{D}v_i(t)}{\mathrm{D}t}, \tag{20}$$

$$r_i(t + \Delta t) = r_i(t) + \Delta t v_i\left(t + \frac{1}{2}\Delta t\right), \tag{21}$$

where $v_i(t + \frac{1}{2}\Delta t)$ is the velocity of fluid particle i at time $t + \frac{1}{2}\Delta t$, and Δt is the time step.

2.4. Artificial Particle Displacement

In simulations of flow in porous media (Section 3.5), artificial particle displacement is suggested as a particle motion correction to avoid particles in the vicinity of the stagnation points of fluid flow [38] and to avoid poor particle distributions [39]. Artificial particle displacement can be expressed as

$$\delta r_i = \alpha \overline{r}_i^2 v_{\max} \Delta t \sum_{j=1}^{N} \frac{r_{ij}}{r_{ij}^3}, \tag{22}$$

where r_i is the position of particle i, α is a problem-dependent parameter that is usually set between 0.01 and 0.1, $v_{\max}$ is the maximum velocity of all particles in the computational domain, $r_{ij} = r_i - r_j$

which is the distance between particles i and j, and $\overline{r_i}$ is the average distance between the neighboring particles of particle i:

$$\overline{r_i} = \frac{1}{N} \sum_{j=1}^{N} r_{ij}. \tag{23}$$

It is noted that the problem-dependent parameter α should be selected carefully. This value should be small enough not to affect the physics of the flow but also large enough to avoid the accumulation of particles to form groups. In the present work, the value of artificial particle displacement is less than 0.1% of the physical particle displacement for a given time step, which is consistent with the magnitude in [40].

After moving the particles, the pressure and velocity components should be corrected by Taylor expansion.

2.5. Boundary Conditions

Several layers of virtual particles are used to implement the boundary condition. Similar treatments can be observed in the SPH and MPS simulations. On a flat wall, virtual particles are obtained by extending the boundary particles to the outside of the computational region, and the distribution of virtual particles is regular. The number of layers can be chosen according to the scale of the support domain. Figure 2 is a sketch of the treatment of particles near the wall. i represents the particle number, and Δx is the particle spacing.

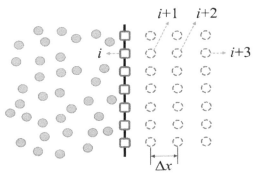

● fluid particles □ boundary particles ◌ virtual particles

Figure 2. A sketch of a simulation of an acoustic boundary using virtual particles. Fluid particles are inside the computational domain and boundary particles are fixed on the boundary.

For a flat wall, both the no-slip and free-slip boundary conditions can be implemented using virtual particles. For no-slip walls, the particle-based boundary conditions are as follows

$$p_{i+3} = p_{i+2} = p_{i+1} = p_i, \; v_{i+3} = v_{i+2} = v_{i+1} = 0, \tag{24}$$

For free-slip walls, the tangential velocity component of virtual particles is maintained the same as the boundary particles.

For a round surface, virtual particles are established based on a radial distribution inside the object domain with particle spacing Δx. The particle distribution is shown in Figure 3.

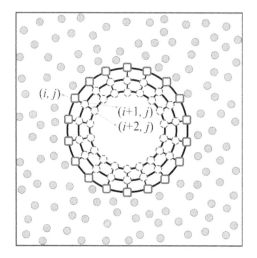

○ fluid particles □ boundary particles ◌ virtual particles

Figure 3. A sketch of a simulation of an acoustic boundary with a curved surface using virtual particles. Fluid particles are inside the computational domain and boundary particles are fixed on the boundary.

The boundary condition for a rigid wall satisfies the following equations.

$$p_{i+2} = p_{i+1} = p_i, \ v_{i+3} = v_{i+2} = v_{i+1} = 0. \tag{25}$$

Since the FDPM simulation is still based on the local support domain, most boundary conditions in the existing Lagrangian particle-based methods, such as the SPH and MPS methods, can be used directly or implemented with minor changes. The particle representation of no-slip, free-slip and superhydrophobic surfaces [41–43] can be found in [44–46].

3. Applications of the Finite Difference Particle Method

3.1. Fundamental Definition

Several test cases are simulated with the FDPM based on second-order Taylor series expansion. The numerical accuracy is evaluated by the root mean square errors ($\varepsilon_{\mathrm{RMS}}$) and the maximum errors ($\varepsilon_{\mathrm{MAX}}$), which are defined as

$$\varepsilon_{\mathrm{RMS}}(S) = \frac{\sqrt{\frac{1}{NT} \sum_{k=1}^{NT} |S_{\mathrm{num}}(k) - S_{\mathrm{ana}}(k)|^2}}{\sqrt{\frac{1}{NT} \sum_{k=1}^{NT} |S_{\mathrm{ana}}(k)|^2}}, \tag{26}$$

$$\varepsilon_{\mathrm{MAX}}(S) = \max_{1 \ll k \ll NT} |S_{\mathrm{num}}(k) - S_{\mathrm{ana}}(k)|, \tag{27}$$

where $S_{\mathrm{num}}(k)$ and $S_{\mathrm{ana}}(k)$ are the numerical and analytical results of variable k, respectively. k could be the velocity or pressure.

The convergence rate of the FDPM is evaluated based on the root mean square error convergence rate (R_{ERMS}) and the maximum error convergence rate (R_{EMAX}) as follows:

$$R_{\mathrm{ERMS}} = \left| \frac{\ln(\varepsilon_{\mathrm{RMS}}(NT_{\max})) - \ln(\varepsilon_{\mathrm{RMS}}(NT_{\min}))}{\ln(NT_{\max}) - \ln(NT_{\min})} \right|, \tag{28}$$

$$R_{\mathrm{EMAX}} = \left| \frac{\ln(\varepsilon_{\mathrm{MAX}}(NT_{\max})) - \ln(\varepsilon_{\mathrm{MAX}}(NT_{\min}))}{\ln(NT_{\max}) - \ln(NT_{\min})} \right|. \tag{29}$$

3.2. Unsteady Flow in a Pipeline

Unsteady flow field in the pipeline is simulated to verify the FDPM method. The theoretical solutions of unsteady flow in a 1D pipeline (chapter 3 in book [47]) are

$$u = \frac{2}{\gamma + 1} \frac{x}{t} + C_1, \tag{30}$$

$$p = C_3 \left(\frac{\gamma - 1}{C_3 \gamma} \left(-\frac{1}{2} \left(C_1 + \frac{2x}{(\gamma + 1)t} \right)^2 + \frac{C_1^2(\gamma + 1)}{2(\gamma - 1)} + \frac{C_2(\gamma - 3)}{\gamma + 1} t^{\frac{2-2\gamma}{\gamma+1}} + \frac{x^2}{(\gamma + 1)t^2} \right) \right)^{\gamma/(\gamma-1)}, \tag{31}$$

$$\rho = \left(\frac{\gamma - 1}{C_3 \gamma} \left(-\frac{1}{2} \left(C_1 + \frac{2x}{(\gamma + 1)t} \right)^2 + \frac{C_1^2(\gamma + 1)}{2(\gamma - 1)} + \frac{C_2(\gamma - 3)}{\gamma + 1} t^{\frac{2-2\gamma}{\gamma+1}} + \frac{x^2}{(\gamma + 1)t^2} \right) \right)^{1/(\gamma-1)}, \tag{32}$$

$$c^2 = \left(\frac{\gamma - 1}{\gamma + 1} - C_1 \right)^2 - \frac{3 - \gamma}{\gamma + 1} (\gamma - 1) C_2 t^{-2(\gamma-1)(\gamma+1)}, \tag{33}$$

where u_0 is the flow velocity distribution and x is the coordinate along the length of the pipeline. The coefficients (the unit can be obtained from dimensional analysis) $C_1 = 30.0$, $C_2 = -1.0 \times 10^6$, $C_3 = 82571.0$, and $\gamma = 1.4$; moreover, the initial time is 12.5 s, and the pipe length x is 700 m.

The FDPM algorithm with a second-order Taylor truncation is used, and the time step (Δt) is 0.0029 s. The effect of viscosity in the process of fluid motion is not considered. Dirichlet boundary conditions are used at both ends of the boundary. The velocity, pressure and density of four particles at both ends are set based on theoretical values.

The space of the initial particle (Δx) is 5.0 m, and r_s is 3.2 times Δx. The cubic spline kernel function is used in the calculations. Table 1 provides data for comparing the numerical velocity and theoretical solution of a particular particle at different times and positions. Notably, although the particle moves from $x = 2.48$ m to $x = 25.64$ m, the FDPM results agree well with the theoretical values, and this result verifies the algorithm.

Table 1. FDPM results and theoretical solutions of the position and velocity of a particle (particle number: 70) at different times.

Time (s)	Particle Method: x (m)	Theoretical Solution: u (m/s)	Particle Method: u (m/s)	Error (10⁻⁸)
0.25	2.48	30.16201462	30.16201544	2.74
0.50	10.08	30.64615858	30.64615942	2.75
0.75	17.80	31.11956025	31.11956110	2.75
1.00	25.64	31.58265402	31.58265470	2.13

The convergence verification of the FDPM method for unsteady flow in a pipeline is shown in Figure 4. The numerical error curves at three different moments are given using different Δx values, and r_s is 3.2 times Δx in the computation.

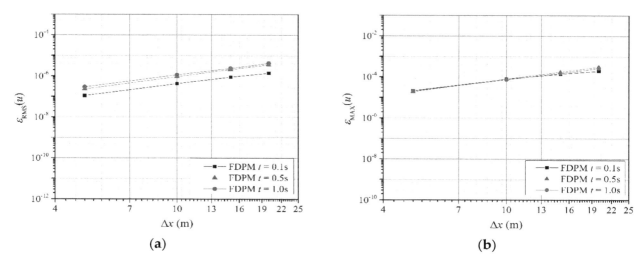

Figure 4. Convergence curves of the FDPM method for 1D unsteady flow simulation: (**a**) Root mean square error, see Equation (26), of FDPM simulation using different particle spacing Δx at different time; (**b**) maximum error, see Equation (27), of FDPM simulation using different particle spacing Δx at different time.

Figure 4 shows that the FDPM method displays good convergence at different times. When $t = 1.0$ s, the convergence curve yields a R_{ERMS} value of 1.7 and R_{EMAX} value of 1.8.

Given that the FDPM is a KGF method, the effect of the type of kernel function on this method is evaluated. Four types of kernel functions, including $\frac{1}{r^3}$, Gaussian, cubic spline and quintic spline functions, are compared through two types of errors with different r_s conditions, as shown in Figure 5. The figure shows that the maximum error of the Gaussian kernel function is larger than that of the other methods. The errors of other types of functions are similar.

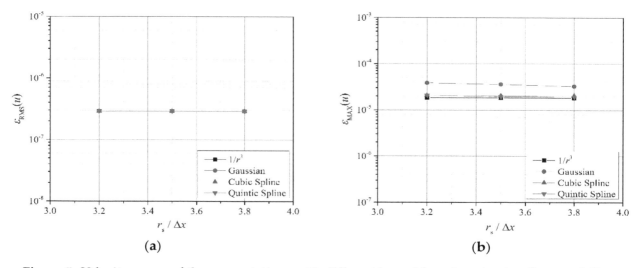

Figure 5. Velocity errors of the computations with different kernel functions in pipe flow modeling: (**a**) Root mean square error, see Equation (26), of FDPM simulation using different smoothing length and kernel functions; (**b**) maximum error, see Equation (27), of FDPM simulation using different smoothing length and kernel functions.

3.3. Poisseuille Flow

Steady, axisymmetric Poisseuille flow between two infinite plates is a classical test model in hydrodynamics. In this section, the model is used to verify the governing equations and the rigid wall boundaries. Assuming that the distance between two infinite plates is L, the volume force F is loaded on the fluid between plates in the x direction from time $t = 0$. The theoretical solution of the velocity distribution of the flow at a given time [48] is as follows.

$$u(y,t) = \frac{F}{v_k}y(y-L) + \sum_{n=0}^{\infty} \frac{4FL^2}{v_k\pi^3(2n+1)^3}\sin\left[\frac{\pi y}{L}(2n+1)\right]\exp\left[-\frac{(2n+1)^2\pi^2 v_k}{L^2}t\right]. \qquad (34)$$

The numerical simulation for Poisseuille flow is obtained under weakly compressible ($Ma = 0.0125$) conditions. Based on reference [48], the parameters of the Poisseuille field are chosen as $v_k = 10^{-6}$ m^{2}s^{-1}, $L = 10^{-3}$ m, $\rho = 10^3$ kgm^{-3}, and $F = 10^{-4}$ ms^{-2}, so the maximum velocity is 1.25×10^{-5} ms^{-1} and the Reynold number is $Re = 1.25 \times 10^{-2}$. The plate boundaries at the upper and lower ends are established using rigid walls. One layer of boundary particles and three layers of virtual particles are used. The FDPM with second-order Taylor truncation is utilized to perform the computation. The speed of sound c is taken as 0.001 m/s, as suggested in [49], and the time step Δt is 3.0×10^{-4} s. The initial particle spacings Δx and Δy are both set as 5×10^{-5} m, r_s is 3.2 times Δx, and the kernel function is selected as a cubic spline function.

A comparison of the numerical and theoretical solutions of the velocity of the flow field in the x direction at different times is shown in Figure 6. The particle velocity is obtained by bilinear interpolation. As time increases, the positions of particles gradually change until the uniformly distributed particles at the initial stage are completely mixed in disorder. At this time, the numerical solution of the particle velocity is still consistent with the theoretical solution.

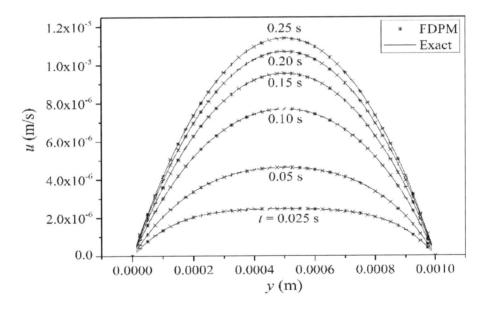

Figure 6. Velocity profiles of the FDPM results (stars) and theoretical solutions (lines) at different times along the y direction (from the bottom plate to the top plate).

During the computation, the particle velocity remains symmetrically distributed and gradually increases at different times before reaching a steady state. The velocity of particles in the middle of the two plates is the largest due to the viscous force, and the velocity is small near the plates. The FDPM solution is in good agreement with the theoretical solution.

Figure 7 shows the numerical error curves of the FDPM method at different times and is used to analyze the convergence of the FDPM method. During the computation, r_s remains 3.2 times Δx.

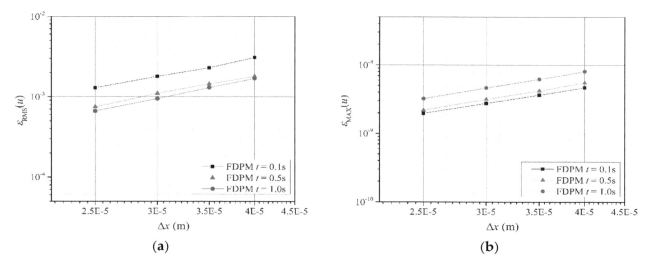

Figure 7. Convergence curves of the FDPM computation in Poisseuille flow modeling: (**a**) Root mean square error, see Equation (26), of FDPM simulation using different particle spacing Δx at different time; (**b**) maximum error, see Equation (27), of FDPM simulation using different particle spacing Δx at different time.

Figure 7 shows that ε_{RMS} and ε_{MAX} decrease as Δx decreases, which indicates that the numerical accuracy converges with the initial particle spacing at different times. ε_{RMS} is on the order of 10^{-3}, indicating that the computational results agree well with the theoretical solutions. For different error evaluation indexes, the R_{ERMS} and R_{EMAX} values of the FDPM method are approximately 1.7 and 1.8, respectively, with good convergence at $t = 1.0$ s. Since the second-order Taylor expansion-based FDPM is implemented in the test, the convergence rate is reasonable.

An error analysis of the four different types of kernel functions is conducted to analyze the sensitivity of the FDPM method to the kernel function, as shown in Figure 8.

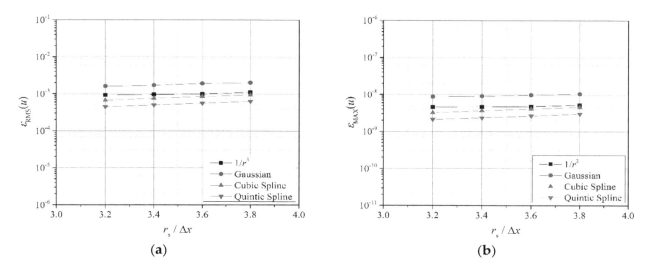

Figure 8. Velocity errors of the computations with different kernel functions in Poisseuille flow modeling: (**a**) Root mean square error, see Equation (26), of FDPM simulation using different smoothing length and kernel functions; (**b**) maximum error, see Equation (27), of FDPM simulation using different smoothing length and kernel functions.

Figure 8 shows that different types of kernel functions can be used in the FDPM method and that the differences in the calculation errors are insignificant. The calculation errors of the four types of kernel functions from large to small exhibit the following order: Gaussian, r^{-3}, cubic spline, and quantic spline.

3.4. Couette Flow

Couette flow considers the fluid flow between a stationary plate and a sliding plate. To accurately solve the flow distribution, the viscous term and boundary flow must be solved correctly. Initially, the two plates and the fluid between them remain stationary. At a constant speed, the upper plate begins to slide parallel to the lower plate. Assuming that the plate spacing is L and the sliding velocity is u_0, the theoretical solution of the flow velocity over time in the direction perpendicular to the plate [48] is as follows:

$$u(y,t) = \frac{u_0}{L}y + \sum_{n=1}^{\infty} \frac{2u_0}{n\pi}(-1)^n \sin(\frac{n\pi}{L}y)\exp(-\frac{n^2\pi^2 v_k}{L^2}t). \tag{35}$$

Couette flow is numerically simulated under weakly compressible ($Ma = 0.0125$) conditions. The parameters for Couette flow are $v_k = 10^6$ m^{2}s^{-1}, $L = 10^{-3}$ m, $\rho = 10^3$ kgm^{-3}, and $u_0 = 1.25 \times 10^{-5}$ m/s.

The plate boundaries at the upper and lower ends are obtained using rigid walls, and the upper plate is set with a constant velocity u_0. One layer of boundary particles and three layers of virtual particles are used. The FDPM with second-order Taylor truncation is utilized to perform the computation. The speed of sound c is taken as 0.001 m/s, and the time step Δt is 5.0×10^{-5} s. The initial particle spacings Δx and Δy are both set as 2.5×10^{-5} m, r_s is 3.2 times Δx, and the kernel function selected is the cubic spline function. Figure 9 shows a comparison between the FDPM method and the theoretical solution for the flow velocity at different times.

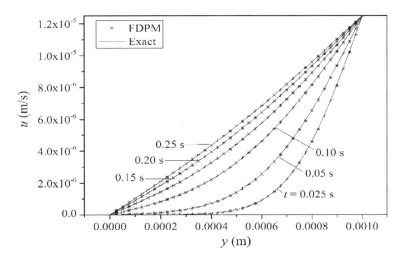

Figure 9. Comparison of the FDPM result (stars) and the theoretical solution (lines) for the flow velocity at different times along the y direction (from the stationary plate to the sliding plate).

Before reaching a steady state, the particle velocity near the upper plate rapidly increases due to the viscous force, and that near the lower plate increases in a relatively slow manner. The velocity distribution of the particles between the two plates is nonlinear.

The velocity error (ε_{RMS} and ε_{MAX}) at different times and at different Δx values is used to evaluate the convergence of the FDPM, as shown in Figure 10.

From the numerical results, both error indexes gradually decrease with decreasing Δx, which suggests that the numerical accuracy converges with the initial particle spacing at different times. ε_{RMS} is on the order of 10^{-2}, indicating that the computational results agree well with the theoretical solution. When $t = 1.0$ s, the two errors result in an R_{ERMS} value of 1.7 and R_{EMAX} value of 1.8. Since the second-order Taylor expansion-based FDPM is implemented in the test, the convergence rate is reasonable.

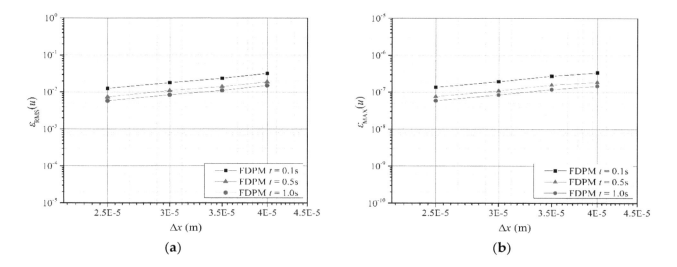

Figure 10. Convergence curves of the FDPM computations in Couette flow modeling: (**a**) Root mean square error, see Equation (26), of FDPM simulation using different particle spacing Δx at different time; (**b**) maximum error, see Equation (27), of FDPM simulation using different particle spacing Δx at different time.

To analyze the sensitivity of the FDPM method to the kernel function, four different types of kernel functions with different $r_s/\Delta x$ values are applied, as shown in Figure 11. Different types of kernel functions can be used in the FDPM method, and the calculation error of the Gaussian kernel function is larger than that of the other methods.

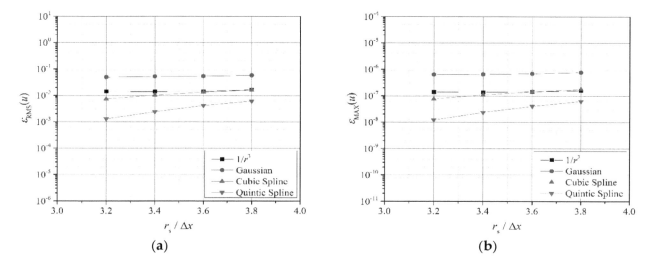

Figure 11. Velocity errors of the computations with different kernel functions in Couette flow modeling: (**a**) Root mean square error, see Equation (26), of FDPM simulation using different smoothing length and kernel functions; (**b**) maximum error, see Equation (27), of FDPM simulation using different smoothing length and kernel functions.

3.5. Flow in Porous Media

In this section, the FDPM algorithm is used to simulate the flow in a simplified model of porous media [50]. The simplified model can be seen as flow around a circular cylinder, as shown in Figure 12, and four sides of the domain are periodic boundaries.

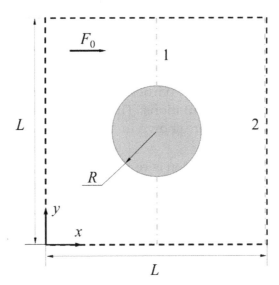

Figure 12. Simplified model of porous media. The solid circle is a circular cylinder and four sides of the domain are periodic boundaries. L is the size of the computational domain, R is the cylindrical radius, and F_0 is the volume force.

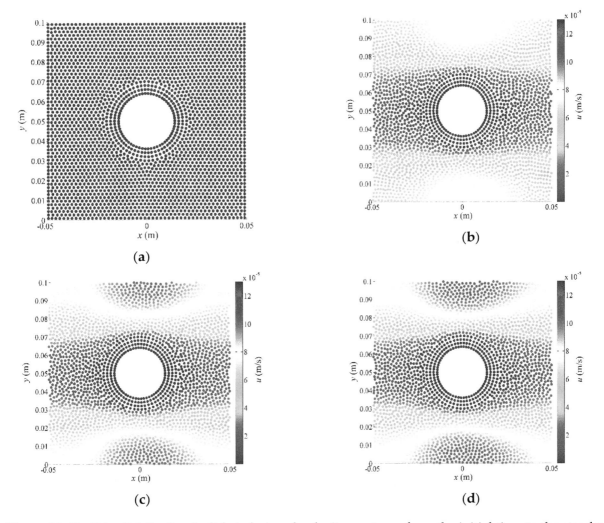

Figure 13. Particle distribution (solid circles) and velocity contours from the initial time to the steady state: (**a**) $t = 0$, (**b**) $t = 693$ s, (**c**) $t = 1386$ s and (**d**) $t = 2080$ s.

The size of the computational domain $L = 0.1$ m, the kinematic viscosity $v_k = 10^{-6}$ m^{2}s^{-1}, the cylindrical radius $R = 2 \times 10^{-2}$ m, the volume force $F_0 = 1.5 \times 10^{-7}$ ms^{-2}, and the speed of sound $c = 5.77 \times 10^{-4}$ ms^{-1}. Δx and Δy are 0.003 m, r_s is 3.2 times Δx, $\Delta t = 1.04$ s with 2000 steps, and the coefficient of artificial particle displacement is 0.05. A rigid wall boundary is used for the cylindrical boundary, and a periodic boundary is used on the four sides of the computational domain. One layer of boundary particles and three layers of virtual particles are used. The FDPM with second-order Taylor truncation is utilized to perform the computations. The particle distribution and velocity contours at the initial time and the final steady state are shown in Figure 13.

At the initial time, the particle distribution is regular. Then, the fluid begins to flow, and particles are gradually scattered and evenly distributed in the computational domain.

The velocity distributions along lines 1 and 2 (dotted-dashed lines in Figure 12) are shown in Figure 14. Both the FDPM results and finite element method (FEM) results are given to evaluate the accuracy of the numerical method. The FEM results come from the data of figure 6 in [48].

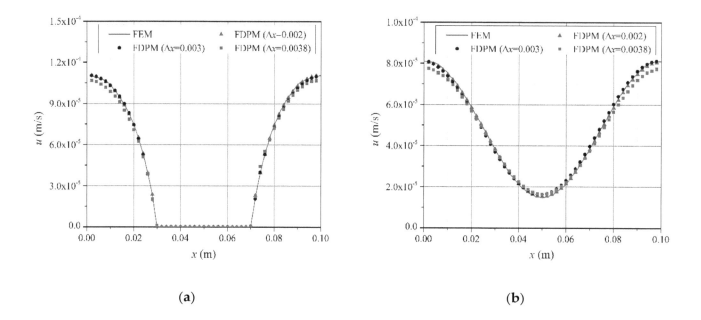

(a)   (b)

Figure 14. Velocity distributions along observation lines 1 and 2 (dotted-dashed lines in Figure 12): (a) Observation line 1 and (b) observation line 2. Lines are FEM results and solid points are FDPM results with different particle spacing Δx.

When $\Delta x > 0.004$, the computation is not sufficiently stable and can easily collapse, so this condition is not shown in Figure 14. When $\Delta x = 0.0038$ m, the FDPM calculation results and the FEM results at $x = 0$ and 0.1 m produce significant differences. When $\Delta x = 0.002$ m, the FDPM results are similar to the FEM results. After convergence is obtained, the results of the FDPM method are consistent with the FEM results, which verifies the correctness of the numerical method and the boundary conditions.

A comparison of the FDPM results ($\Delta x = 0.002$ m) with different kernel functions and the FEM reference results is shown in Figure 15. The figure shows that the FDPM results with different types of kernel functions are comparable to the reference results and exhibit only minor differences.

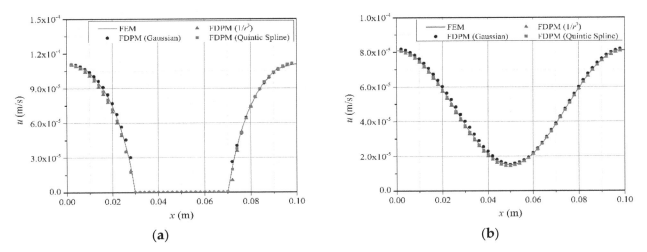

Figure 15. Comparison along observation lines 1 and 2 (dotted-dashed lines in Figure 12) between the FDPM results (solid points) with different kernel functions and the FEM results (lines): (**a**) Observation line 1 and (**b**) observation line 2.

3.6. Lid-Driven Cavity Flow

The lid-driven cavity flow is widely used as a benchmark test case and the model in [51] is used to verify the method. The case is in a square cavity with a sliding plate on the upper side and three fixed rigid walls around, as shown in Figure 16. Initially, the fluid in the cavity remain stationary. At a constant speed, the upper plate begins to slide horizontally and the simulation is at $Re = 100$.

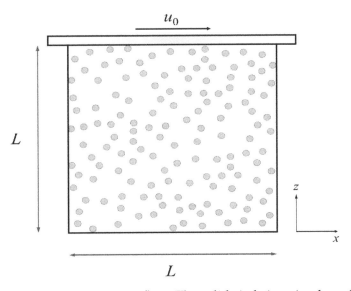

Figure 16. Schematic of the lid-driven cavity flow. The solid circle is a circular cylinder and four sides of the domain are periodic boundaries. L is the size of the computational domain, R is the cylindrical radius, and F_0 is the volume force.

Rigid wall boundary condition is used for four sides of the cavity, and the upper plate is set with a constant velocity u_0. The size of the cavity $L = 1.0$ m, the kinematic viscosity $v_k = 0.01$ m^{2}s^{-1}, the sliding velocity $u_0 = 1.0$ m/s, and the speed of sound $c = 10.0$ ms^{-1}. Δx and Δy are 0.025 m, r_s is 2.7 times Δx, $\Delta t = 0.001$ s with 3000 steps, the coefficient of artificial particle displacement is 0.05, and the kernel function selected is the cubic spline function. One layer of boundary particles and three layers of virtual particles are used. The FDPM with second-order Taylor truncation is utilized to perform the computations. The particle distribution and velocity contours at the initial time and the final steady state are shown in Figure 17.

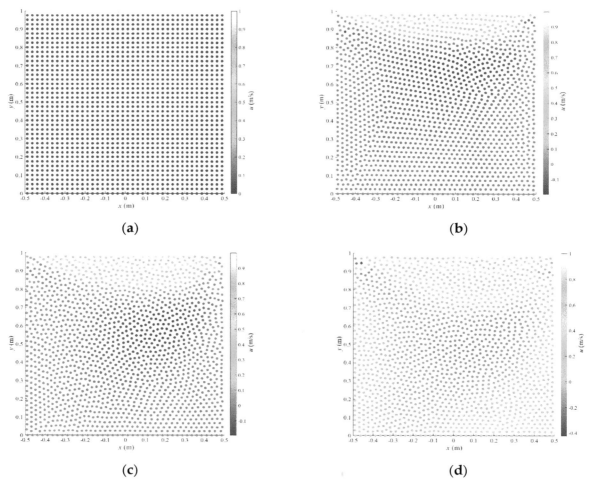

Figure 17. Particle distribution (solid circles) and velocity contours from the initial time to the steady state: (**a**) $t = 0$, (**b**) $t = 1.0$ s, (**c**) $t = 2.0$ s and (**d**) $t = 3.0$ s.

At the initial time, the particle distribution is regular. Then, the fluid begins to flow, and particles are gradually scattered and evenly distributed in the computational domain.

Horizontal velocity component profiles along horizontal and vertical geometric centerlines at $t = 3.0$ s, respectively, are shown in Figure 18. Although the present FDPM computation employed only 2/5 particles in the work [51], these profiles are in good agreement with the reference results. The particle distribution shows the method works well in geometries with corners.

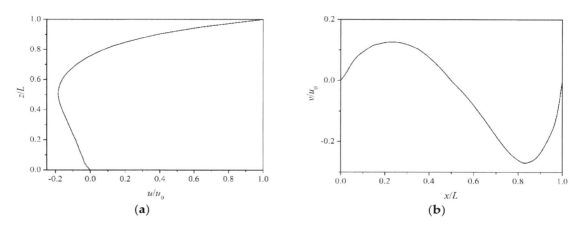

Figure 18. Horizontal Velocity distributions along horizontal and vertical geometric centerlines at $t = 3.0$ s: (**a**) Along vertical geometric centerlines and (**b**) along horizontal geometric centerlines.

4. Conclusions

In this paper, a particle method based on the GFD scheme is proposed to simulate weakly compressible viscous flow. This approach represents the partial differential terms in the Navier-Stokes equations as the solution of a symmetric system of linear equations. The convergence and accuracy of the symmetric particle-based method are tested by modeling flow in a pipeline, Poisseuille flow, Couette flow, flow in porous media, and lid-driven cavity flow. The numerical results exhibit close agreement with the theoretical solutions and finite element results. The particle method utilizes the kernel function itself instead of its gradient, which avoids using artificial parameters to solve for the viscous term and reduces the limitations on the choice of kernel function. Moreover, the order of the Taylor series expansion can easily be improved in the meshless algorithm. The convergence rate of the particle-based calculations with second-order Taylor truncation is approximately 1.7 in the tests, and four different kernel functions are tested and determined to be reliable.

Author Contributions: These authors contributed equally to this work.

References

1. Kim, K.S. A Mesh-Free Particle Method for Simulation of Mobile-Bed Behavior Induced by Dam Break. *Appl. Sci.* **2018**, *8*, 1070. [CrossRef]

2. De Padova, D.; Mossa, M.; Sibilla, S. SPH numerical investigation of characteristics of hydraulic jumps. *Environ. Fluid Mech.* **2018**, *18*, 849–870. [CrossRef]

3. Zuo, J.; Tian, W.; Chen, R.; Qiu, S.; Su, G. Two-dimensional numerical simulation of single bubble rising behavior in liquid metal using moving particle semi-implicit method. *Prog. Nucl. Energy* **2013**, *64*, 31–40. [CrossRef]

4. Zhang, A.; Sun, P.; Ming, F. An SPH modeling of bubble rising and coalescing in three dimensions. *Comput. Methods Appl. Mech. Eng.* **2015**, *294*, 189–209. [CrossRef]

5. Wang, X.; Ban, X.; He, R.; Wu, D.; Liu, X.; Xu, Y. Fluid-Solid Boundary Handling Using Pairwise Interaction Model for Non-Newtonian Fluid. *Symmetry* **2018**, *10*, 94. [CrossRef]

6. Liu, S.; Ban, X.; Wang, B.; Wang, X. A Symmetric Particle-Based Simulation Scheme towards Large Scale Diffuse Fluids. *Symmetry* **2018**, *10*, 86. [CrossRef]

7. Lucy, L.B. A numerical approach to the testing of the fission hypothesis. *Astron. J.* **1977**, *82*, 1013–1024. [CrossRef]

8. Gingold, R.A.; Monaghan, J.J. Smoothed particle hydrodynamics: Theory and application to non-spherical stars. *Mon. Not. R. Astron. Soc.* **1977**, *181*, 375–389. [CrossRef]

9. Monaghan, J. Smoothed Particle Hydrodynamics and Its Diverse Applications. *Annu. Rev. Fluid Mech.* **2012**, *44*, 323–346. [CrossRef]

10. Violeau, D.; Rogers, B.D. Smoothed particle hydrodynamics (SPH) for free-surface flows: Past, present and future. *J. Hydraul. Res.* **2016**, *54*, 1–26. [CrossRef]

11. Wang, Z.-B.; Chen, R.; Wang, H.; Liao, Q.; Zhu, X.; Li, S.-Z. An overview of smoothed particle hydrodynamics for simulating multiphase flow. *Appl. Math. Model.* **2016**, *40*, 9625–9655. [CrossRef]

12. Zhang, A.-M.; Sun, P.-N.; Ming, F.-R.; Colagrossi, A. Smoothed particle hydrodynamics and its applications in fluid-structure interactions. *J. Hydrodyn.* **2017**, *29*, 187–216. [CrossRef]

13. Liu, G.R.; Liu, M.B. *Smoothed Particle Hydrodynamics—A Meshfree Particle Method*; World Scientific Publishing: Singapore, 2003; pp. 103–176.

14. Liu, M.B.; Liu, G.R. Smoothed Particle Hydrodynamics (SPH): An Overview and Recent Developments. *Arch. Comput. Methods Eng.* **2010**, *17*, 25–76. [CrossRef]

15. Cleary, P.W.; Prakash, M.; Ha, J.; Stokes, N.; Scott, C. Smooth particle hydrodynamics: Status and future potential. *Prog. Comput. Fluid Dyn. Int. J.* **2007**, *7*, 70–90. [CrossRef]

16. Cummins, S.J.; Rudman, M. An SPH projection method. *J. Comput. Phys.* **1999**, *152*, 584–607. [CrossRef]

17. Chen, J.K.; Beraun, J.E.; Jih, C.J. An improvement for tensile instability in smoothed particle hydrodynamics. *Comput. Mech.* **1999**, *23*, 279–287. [CrossRef]

18. Chen, J.K.; Beraun, J.E.; Carney, T.C. A corrective smoothed particle method for boundary value problems in heat conduction. *Int. J. Numer. Methods Eng.* **1999**, *46*, 231–252. [CrossRef]

19. Zhang, G.M.; Batra, R.C. Modified smoothed particle hydrodynamics method and its application to transient problems. *Comput. Mech.* **2004**, *34*, 137–146. [CrossRef]

20. Liu, M.; Liu, G.; Lam, K. Constructing smoothing functions in smoothed particle hydrodynamics with applications. *J. Comput. Appl. Math.* **2003**, *155*, 263–284. [CrossRef]

21. Huang, C.; Lei, J.M.; Liu, M.B.; Peng, X.Y. A kernel gradient free (KGF) SPH method. *Int. J. Numer. Methods Fluids* **2015**, *78*, 691–707. [CrossRef]

22. Lei, J.M.; Peng, X.Y. Improved kernel gradient free-smoothed particle hydrodynamics and its applications to heat transfer problems. *Chin. Phys. B* **2015**, *25*, 020202. [CrossRef]

23. Huang, C.; Lei, J.M.; Liu, M.B.; Peng, X.Y. An improved KGF-SPH with a novel discrete scheme of Laplacian operator for viscous incompressible fluid flows. *Int. J. Numer. Methods Fluids* **2016**, *81*, 377–396. [CrossRef]

24. Koh, C.G.; Gao, M.; Luo, C. A new particle method for simulation of incompressible free surface flow problems. *Int. J. Numer. Meth. Eng.* **2012**, *89*, 1582–1604. [CrossRef]

25. Luo, M.; Koh, C.G.; Bai, W.; Gao, M. A particle method for two-phase flows with compressible air pocket. *Int. J. Numer. Methods Eng.* **2016**, *108*, 695–721. [CrossRef]

26. Jensen, P.S. Finite difference techniques for variable grids. *Comput. Struct.* **1972**, *2*, 17–29. [CrossRef]

27. Perrone, N.; Kao, R. A general finite difference method for arbitrary meshes. *Comput. Struct.* **1975**, *5*, 45–57. [CrossRef]

28. Liszka, T.; Orkisz, J. The finite difference method at arbitrary irregular grids and its application in applied mechanics. *Comput. Struct.* **1980**, *11*, 83–95. [CrossRef]

29. Ding, H.; Shu, C.; Yeo, K.; Xu, D. Simulation of incompressible viscous flows past a circular cylinder by hybrid FD scheme and meshless least square-based finite difference method. *Comput. Methods Appl. Mech. Eng.* **2004**, *193*, 727–744. [CrossRef]

30. Li, P.-W.; Chen, W.; Fu, Z.-J.; Fan, C.-M. Generalized finite difference method for solving the double-diffusive natural convection in fluid-saturated porous media. *Eng. Anal. Bound. Elem.* **2018**, *95*, 175–186. [CrossRef]

31. Gavete, L.; Ureña, F.; Benito, J.; García, A.; Ureña, M.; Salete, E.; Corvinos, L.A.G. Solving second order non-linear elliptic partial differential equations using generalized finite difference method. *J. Comput. Appl. Math.* **2017**, *318*, 378–387. [CrossRef]

32. Benito, J.; Ureña, F.; Gavete, L. Influence of several factors in the generalized finite difference method. *Appl. Math. Model.* **2001**, *25*, 1039–1053. [CrossRef]

33. Gavete, L.; Gavete, M.; Benito, J. Improvements of generalized finite difference method and comparison with other meshless method. *Appl. Math. Model.* **2003**, *27*, 831–847. [CrossRef]

34. Benito, J.; Ureña, F.; Gavete, L. Solving parabolic and hyperbolic equations by the generalized finite difference method. *J. Comput. Appl. Math.* **2007**, *209*, 208–233. [CrossRef]

35. Alamri, S.Z.; Ellahi, R.; Shehzad, N.; Zeeshan, A. Convective radiative plane Poiseuille flow of nanofluid through porous medium with slip: An application of Stefan blowing. *J. Mol. Liq.* **2019**, *273*, 292–304. [CrossRef]

36. Ellahi, R.; Shivanian, E.; Abbasbandy, S.; Hayat, T. Numerical study of magnetohydrodynamics generalized Couette flow of Eyring-Powell fluid with heat transfer and slip condition. *Int. J. Numer. Methods Heat Fluid Flow* **2016**, *26*, 1433–1445. [CrossRef]

37. Shehzad, N.; Zeeshan, A.; Ellahi, R.; Rashidi, S. Modelling Study on Internal Energy Loss Due to Entropy Generation for Non-Darcy Poiseuille Flow of Silver-Water Nanofluid: An Application of Purification. *Entropy* **2018**, *20*, 851. [CrossRef]

38. Bašić, J.; Degiuli, N.; Werner, A. Simulation of water entry and exit of a circular cylinder using the ISPH method. *Trans. Famena* **2014**, *38*, 45–62.

39. Nestor, R.M.; Basa, M.; Lastiwka, M.; Quinlan, N.J. Extension of the finite volume particle method to viscous flow. *J. Comput. Phys.* **2009**, *228*, 1733–1749. [CrossRef]

40. Ozbulut, M.; Yildiz, M.; Goren, O. A numerical investigation into the correction algorithms for SPH method in modeling violent free surface flows. *Int. J. Mech. Sci.* **2014**, *79*, 56–65. [CrossRef]

41. Rothstein, J.P. Slip on superhydrophobic surfaces. *Annu. Rev. Fluid Mech.* **2010**, *42*, 89–109. [CrossRef]

42. Gentili, D.; Bolognesi, G.; Giacomello, A.; Chinappi, M.; Casciola, C.M. Pressure effects on water slippage over silane-coated rough surfaces: Pillars and holes. *Microfluid. Nanofluidics* **2014**, *16*, 1009–1018. [CrossRef]

43. Bolognesi, G.; Cottin-Bizonne, C.; Pirat, C. Evidence of slippage breakdown for a superhydrophobic microchannel. *Phys. Fluids* **2014**, *26*, 082004. [CrossRef]

44. Marrone, S.; Antuono, M.; Colagrossi, A.; Colicchio, G.; Le Touzé, D.; Graziani, G. δ-SPH model for simulating violent impact flows. *Comput. Methods Appl. Mech. Eng.* **2011**, *200*, 1526–1542. [CrossRef]

45. Adami, S.; Hu, X.; Adams, N.; Adams, N. A generalized wall boundary condition for smoothed particle hydrodynamics. *J. Comput. Phys.* **2012**, *231*, 7057–7075. [CrossRef]

46. Sun, P.; Ming, F.; Zhang, A. Numerical simulation of interactions between free surface and rigid body using a robust SPH method. *Ocean Eng.* **2015**, *98*, 32–49. [CrossRef]

47. Von Mises, R.; Geiringer, H.; Ludford, G.S.S. *Mathematical Theory of Compressible Fluid Flow*; Chapter 3; Academic Press: New York, NY, USA, 2004.

48. Morris, J.P.; Fox, P.J.; Zhu, Y. Modeling Low Reynolds Number Incompressible Flows Using SPH. *J. Comput. Phys.* **1997**, *136*, 214–226. [CrossRef]

49. Liu, M.; Xie, W.; Liu, G. Modeling incompressible flows using a finite particle method. *Appl. Math. Model.* **2005**, *29*, 1252–1270. [CrossRef]

50. Fang, J.; Parriaux, A. A regularized Lagrangian finite point method for the simulation of incompressible viscous flows. *J. Comput. Phys.* **2008**, *227*, 8894–8908. [CrossRef]

51. Khorasanizade, S.; Sousa, J.M.M.; De Sousa, J.M. A detailed study of lid-driven cavity flow at moderate Reynolds numbers using Incompressible SPH. *Int. J. Numer. Methods Fluids* **2014**, *76*, 653–668. [CrossRef]

11

Modeling and Optimization of Gaseous Thermal Slip Flow in Rectangular Microducts using a Particle Swarm Optimization Algorithm

Nawaf N. Hamadneh [1], Waqar A. Khan [2], Ilyas Khan [3,*] and Ali S. Alsagri [4]

[1] Department of Basic Sciences, College of Science and Theoretical Studies, Saudi Electronic University, Riyadh 11673, Saudi Arabia; nhamadneh@seu.edu.sa
[2] Department of Mechanical Engineering, College of Engineering, Prince Mohammad Bin Fahd University, Al Khobar 31952, Saudi Arabia; wkhan@pmu.edu.sa
[3] Faculty of Mathematics and Statistics, Ton Duc Thang University, Ho Chi Minh City 72915, Vietnam
[4] Mechanical Engineering Department, Qassim University, Buraydah 51431, Saudi Arabia; a.alsagri@qu.edu.sa
* Correspondence: ilyaskhan@tdtu.edu.vn

Abstract: In this study, pressure-driven flow in the slip regime is investigated in rectangular microducts. In this regime, the Knudsen number lies between 0.001 and 0.1. The duct aspect ratio is taken as $0 \leq \varepsilon \leq 1$. Rarefaction effects are introduced through the boundary conditions. The dimensionless governing equations are solved numerically using MAPLE and MATLAB is used for artificial neural network modeling. Using a MAPLE numerical solution, the shear stress and heat transfer rate are obtained. The numerical solution can be validated for the special cases when there is no slip (continuum flow), $\varepsilon = 0$ (parallel plates) and $\varepsilon = 1$ (square microducts). An artificial neural network is used to develop separate models for the shear stress and heat transfer rate. Both physical quantities are optimized using a particle swarm optimization algorithm. Using these results, the optimum values of both physical quantities are obtained in the slip regime. It is shown that the optimal values ensue for the square microducts at the beginning of the slip regime.

Keywords: forced convection; microducts; Knudsen number; Nusselt number; artificial neural networks; particle swarm optimization

1. Introduction

1.1. Rarefied Gas Flows

Flows in microducts are found in microelectromechanical systems, nanotechnology applications, therapeutic and superhydrophobic microchannels, low-pressure environments, biochemical applications and cryogenics. The rarefaction effect can be found in microchannels and can be expressed in terms of the Knudsen number. In this case, the deviations from continuum behavior are smaller. Therefore, the Navier–Stokes equations can be employed with slip boundary conditions. The difference between a fully developed flow in a rectangular duct and in a microchannel is that a rectangular microduct needs 2D analysis, whereas a microchannel entail a one-dimensional analysis.

Generally, the Navier–Stokes equations can be solved with slip boundary conditions at the microduct walls [1,2]. Previous studies [1,2] found that the solutions were validated by the experimental data in several microchannel flows. Otherwise, significant departures were observed between the numerical and experimental results without applying the slip conditions. Ameel et al. [3] confirmed the statement of Reference [1,2]. Zade et al. [4] considered variable physical properties and investigated thermal features of developing flows in rectangular microchannels. They employed a collocated finite volume method and studied several channel dimensions for different values of slip parameters. They found that variable properties show substantial changes in flow behavior. Ghodoossi and Eğrican [5] determined the heat transfer rate using an integral method. They predicted thermal features for isothermal boundary conditions. They noticed that, for any aspect ratio, the rarefaction had obvious effects on the heat transfer.

The effects of rarefaction on the Nusselt numbers were investigated by Hettiarachchi et al. [6] for thermally developing flows. They noticed that these Nusselt numbers increased with the slip velocity, and decline with the thermal jump boundary condition. The slip flow forced convection in rectangular microchannels was also considered by Yu and Ameel [7]. It was found that the velocity slip and aspect ratio enhance heat transfer. For rarefied gas flows, Renksizbulut et al. [8] observed exceptionally large reductions in the friction and heat transfer coefficients. They found that these coefficients are insensible to rarefaction effects in corner-dominated flows. Tamayol and Hooman [9] analyzed microchannels of polygonal, rectangular, and rhombic cross-sections and showed that the Poiseuille numbers decrease with increasing aspect ratios and Knudsen numbers for rectangular channels. Hooman [10] investigated forced convection in microducts using a superposition approach for both thermal boundary conditions and demonstrated that their results were valid for several cross-sections. Sadeghi and Saidi [11] considered the rarefaction effects in two different microgeometries and demonstrated that for both geometries, the Brinkman number reduced the heat transfer rate. In fact, the Brinkman number measures the viscous heating effects with reference to the heat conduction. They also developed a correlation for the Nusselt number.

Yovanovich and Khan [12] developed several slip flow models for microchannels of different cross-sections. The slip flow in circular and other cross-sections was investigated by a number of authors, including References [13–16]. They developed different models for envisaging the friction factor under both developing and fully developed conditions. Duan and Muzychka [13,14] proposed a simple model to predict the Poiseuille numbers in these microchannels. Duan and Muzychka [15] considered slip flow in the entrance of circular and parallel plate microchannels and noticed that the Poisseuille number depends on the Knudsen number. Yovanovich and Khan [16] modeled the Poiseuille flow in long channels of different cross-sections. Ebert and Sparrow [17] analyzed abstemiously rarefied gas flows in different ducts and noticed that the velocity flattened and the friction lessened due to the slip parameter.

Baghani et al. [18] treated rarefaction effects in microducts of different cross-sections and obtained Nusselt numbers under isothermal boundary condition. They tabulated Nusselt numbers for the entire slip flow range of the Knudsen number. Wang [19] considered the slip flow under an isothermal boundary condition in different ducts. It was found that both hydrodynamic and thermal boundary conditions have substantial effects on both the Poiseuille and Nusselt numbers. Also, both parameters showed an opposite trend with an increasing velocity slip. Beskok and Karniadakis [20] considered internal rarefied gas flows and presented mathematical models for different regimes. To account for heat transfer from the surface, they suggested a new boundary condition and demonstrated that it is applicable in all flow regimes. Yovanovich and Khan [21] proposed compact models for isothermal, incompressible slip flows in different microchannels. They used the principle of superposition in the analysis and introduced new flow parameters. They compared their results with previous numerical results for different cross-sections and found them in a very good agreement. Klášterka et al. [22] obtained analytical and numerical solutions for the flow variables and demonstrated the effect of slip parameters on the friction and heat transfer coefficients.

The above literature survey reveals that there are only a few studies related to rectangular microducts. However, there has been no study where the numerical models are developed using an artificial neural network (ANN) and are optimized for the duct geometry. This fact inspires us to model our numerical results using ANN and to acquire optimized values of friction factors and Nusselt numbers using a particle swarm optimization algorithm.

1.2. Artificial Neural Networks

Artificial neural networks (ANNs) are mathematical models, which are inspired by biological nervous systems [23,24]. The multilayer perceptron neural network (MLPNN) is one of the widely known feedforward neural networks. MLPNNs have three types of layers, which are the input layer, the hidden layers, and the output layer. Figure 1 shows the structure of MLPNN [25,26].

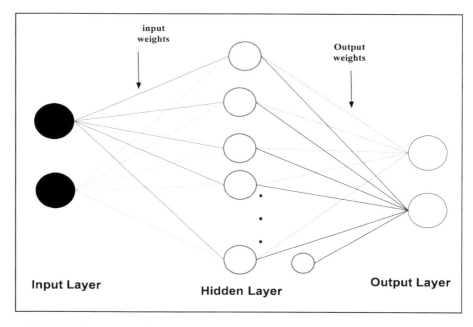

Figure 1. Structure of a multilayer perceptron neural network (MLPNN).

The activation function is defined as an output of a node or a neuron and can be written as:

$$Y = \frac{1}{1 + e^{-x}} \tag{1}$$

where x is an input value from the input layer. Supervised learning of ANNs is useful for improving the performance of ANN models. Several algorithms are used for the training to find the best parameters of the neural networks [27,28]. The optimization algorithms are used to find the best values for inputs and output weights that are parameters of MLPNNs. In this study, the particle swarm optimization (PSO) algorithm is used to train the neural networks, because it is one of the most effective meta-heuristic algorithms used for optimization problems. Equation (2) is the sum of squared error (SSE) function, which was used as a fitness function for testing the performance of the ANNs [25].

$$SEE = \sum (Actual\ output - target\ output)^2 \tag{2}$$

where the *target output* values are determined by Equation (3), while the *Actual output* values are determined by the neural network (NN) output values.

$$Output(y_k) = \sum_{j=1}^{m} \sum_{i=1}^{n} w_{jk} \sum_{k=1}^{s} w'_{ki} \frac{1}{1 + e^{-x_j}} \tag{3}$$

where n, m, k, w_{jk}, and w'_{ki} represent the number of neurons in the hidden layer, number of input data, number of neurons in input layer, number of neurons in output layer, input weight—which is between the input neurons and hidden neurons—and output weight, which is between the hidden neurons and output neurons, respectively.

1.3. Particle Swarm Optimization Algorithm

The principle of the particle swarm optimization (PSO) algorithm came from the ideas of the social behavior of bird flocks, as well as from the shoaling behavior of fish [27–29]. All solution members of the algorithm are called particles. The particle flies by updating its velocity vectors v and position x, by using Equations (4) and (5), respectively [30].

$$v_{ik}(t+1) = v_{ik}(t) + c_1 * r_{1k}(t) * (y_{ik}(t) - x_{ik}(t)) + c_2 \atop *r_{2k}(t) * (\hat{y}_k(t) - x_{ik}(t)) \tag{4}$$

$$x_i(t+1) = x_i(t) + v_i(t+1) \tag{5}$$

where $x_i(t)$, $v_i(t)$, $y_i(t)$, and $\hat{y}_k(t)$ represent the position at time t, the velocity at time t, the best personal position (p^{best}) at time t, and the global best position (g^{best}) of population at time t respectively, whereas c_1 and c_2 are the learning factors of (p^{best}) in interval between 0 and 2 and r_1 and r_2 are the random numbers in the interval 0 to 1.

It is important to note that the other particles follow the performance of the best particle [31–34]. In addition, each particle keeps the best position that it has achieved so far. The flowchart of Figure 2 shows the steps of the present PSO algorithm [27,32,35].

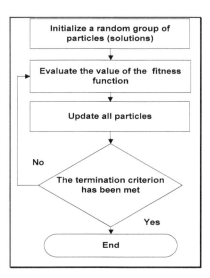

Figure 2. Structure of the particle swarm optimization (PSO) algorithm.

Thus, the particles should move towards the best directions, and utilize useful information from other particles along with the best particles [30,35].

In this study, the algorithm was used to find the optimal values of the Nusselt number (Nu) and Poiseuille number (Po) corresponding to the aspect ratio and the Knudsen number in the slip flow regime. The number of the iterations was set as 50, with initial populations 60, and $c_1 = c_2 = 0.05$ [30].

To the best of our knowledge, most of the existing analytical solutions for slip flows are limited to fluid flow in rectangular ducts. There is a lack of literature related to heat transfer in microducts. Furthermore, there is no study related to ANN modeling and optimization of pressure drop and heat transfer in microducts. The aim of this paper was to cover these aspects.

2. Mathematical Model

2.1. Hydrodynamic Analysis

Let us consider a fully developed flow in a rectangular duct, as shown in Figure 3. The semi-major and -minor axes are taken as a and b along the x- and y-axes, respectively. The aspect ratio is $\varepsilon = b/a$.

The gas is assumed to flow along the z-axis. The momentum equation for the flow in a microduct can be written as:

$$\frac{\partial^2 u}{\partial x^2} + \frac{\partial^2 u}{\partial y^2} = -\frac{1}{\mu}\frac{dP}{dz} \tag{6}$$

which is a 2D equation for the axial velocity u, uniform gas viscosity μ and constant pressure gradient dP/dz. Due to double symmetry, only the OABCO is selected for the simulation. The velocity slip and symmetry boundary conditions for this problem can be written as:

$$u = -\lambda\frac{2-\sigma_v}{\sigma_v}\frac{\partial u}{\partial y}\bigg|_{y=b} \quad u = -\lambda\frac{2-\sigma_v}{\sigma_v}\frac{\partial u}{\partial x}\bigg|_{x=a} \tag{7}$$

and:

$$\frac{\partial u}{\partial y}\bigg|_{y=0} = 0; \frac{\partial u}{\partial x}\bigg|_{x=0} = 0 \tag{8}$$

where σ_v is the tangential–momentum accommodation coefficient, which lies between 0.9 and 1, and λ is the mean free path. If A is the area of cross-section and P is the perimeter of the duct, then the hydraulic diameter, D_h can be written as:

$$D_h = 4A/P = 4b/(1+\varepsilon) \tag{9}$$

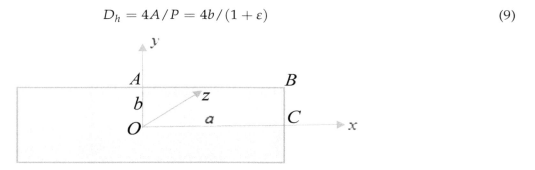

Figure 3. Schematic diagram of rectangular microduct.

At this stage, it is convenient to define the Knudsen number (the measure of rarefaction effects) as follows:

$$Kn = \lambda/D_h \Rightarrow \lambda = 4bKn/1+\varepsilon \tag{10}$$

Following Ebert and Sparrow [17], Equations (6)–(8) can be non-dimensionalized as follows. Let:

$$\xi = x/a, \eta = y/b \tag{11}$$

The dimensionless momentum equation can be written as:

$$\varepsilon^2\frac{\partial^2 u}{\partial \xi^2} + \frac{\partial^2 u}{\partial \eta^2} = -1 \tag{12}$$

where the axial velocity u is normalized with $u_o = \left(\frac{b^2}{\mu}\right)\left|\frac{dP}{dz}\right|$. Using Equation (11), dimensionless boundary conditions can be written as:

$$\begin{aligned} u &= -\frac{4Kn}{1+\varepsilon}\frac{2-\sigma_v}{\sigma_v}\frac{\partial u}{\partial \eta}\bigg|_{\eta=1}; 0 \le \xi < 1 \\ u &= -\frac{4\varepsilon Kn}{1+\varepsilon}\frac{2-\sigma_v}{\sigma_v}\frac{\partial u}{\partial \xi}\bigg|_{\xi=1}; 0 \le \eta < 1 \\ \frac{\partial u}{\partial \eta}\bigg|_{\eta=0} &= 0; 0 \le \xi < 1; \frac{\partial u}{\partial x}\bigg|_{\xi=0} = 0; 0 \le \eta < 1 \end{aligned} \tag{13}$$

The mean velocity in the duct is defined as:

$$\bar{u} = u_0 \int_0^1 \int_0^1 u(\xi, \eta) \cdot d\xi \cdot d\eta \tag{14}$$

The Poiseuille number Po is defined as:

$$Po = c_f \cdot \mathrm{Re}_{D_h} \tag{15}$$

where $c_f = \bar{\tau}_w / 0.5\rho\bar{u}^2$ is the coefficient of friction and $\mathrm{Re}_{D_h} = \bar{u}D_h/\nu$ is the Reynolds number, based on mean velocity. For a fully developed flow, the force balance on a control volume of width dz gives:

$$-A \cdot dp = \bar{\tau}_w P dz \;\Rightarrow\; \bar{\tau}_w = (dp/dz)(D_h/4) \tag{16}$$

Thus, the dimensionless shear stress can be expressed in terms of the Poiseuille number Po:

$$Po = -\frac{32}{(1+\varepsilon)^2} \cdot \frac{1}{(\bar{u}/u_0)} \tag{17}$$

where $\bar{u}/u_0$ is the dimensionless average axial velocity in the microduct and can be determined from Equation (14).

2.2. Thermal Analysis

For fully thermally developed and steady flow in the rectangular microducts, the energy equation is given by:

$$\frac{k_f}{\rho c_p}\left(\frac{\partial^2 T}{\partial x^2} + \frac{\partial^2 T}{\partial y^2}\right) = u\frac{\partial T}{\partial z} \tag{18}$$

with temperature jump and symmetry boundary conditions:

$$T - T_w = -\frac{\lambda}{\mathrm{Pr}}\frac{2-\sigma_t}{\sigma_t}\frac{2\gamma}{1+\gamma}\frac{\partial T}{\partial y}\Big|_{y=b}$$
$$T - T_w = -\frac{\lambda}{\mathrm{Pr}}\frac{2-\sigma_t}{\sigma_t}\frac{2\gamma}{1+\gamma}\frac{\partial T}{\partial x}\Big|_{x=a} \tag{19}$$

and:

$$\frac{\partial T}{\partial y}\Big|_{y=0} = 0 \quad \frac{\partial T}{\partial x}\Big|_{x=0} = 0 \tag{20}$$

where Pr is the Prandtl number for the selected gas, σ_t is the energy accommodation coefficient, and γ is the ratio of specific heats of gas.

Using Equation (11), the energy balance in the flow direction can be written as:

$$\frac{k_f}{\rho c_p}\left(\varepsilon^2\frac{\partial^2 T}{\partial \xi^2} + \frac{\partial^2 T}{\partial \eta^2}\right) = u\frac{b^2 Q}{4ab\rho c_p\bar{u}} \tag{21}$$

or:

$$\varepsilon^2\frac{\partial^2 T}{\partial \xi^2} + \frac{\partial^2 T}{\partial \eta^2} = \frac{u}{\bar{u}}\frac{\varepsilon}{4}\frac{Q}{k_f} \tag{22}$$

where Q is the total energy per unit length of the duct and is assumed to be constant. Using $\theta = (T - Tw)/(Q/k_f)$, Equation (22) can be written as:

$$\varepsilon^2\frac{\partial^2 \theta}{\partial \xi^2} + \frac{\partial^2 \theta}{\partial \eta^2} = \frac{u}{\bar{u}}\frac{\varepsilon}{4} \tag{23}$$

with transformed boundary conditions [17]:

$$
\left.
\begin{aligned}
\theta &= -\frac{4}{1+\varepsilon}\frac{Kn}{Pr}\frac{2-\sigma_t}{\sigma_t}\frac{2\gamma}{1+\gamma}\frac{\partial\theta}{\partial\eta}\Big|_{\eta=1}; 0 \le \xi < 1 \\
\theta &= -\frac{4\varepsilon}{1+\varepsilon}\frac{Kn}{Pr}\frac{2-\sigma_t}{\sigma_t}\frac{2\gamma}{1+\gamma}\frac{\partial\theta}{\partial\xi}\Big|_{\xi=1}; 0 \le \eta < 1 \\
\frac{\partial\theta}{\partial\eta}\Big|_{\eta=0} &= 0; 0 \le \xi < 1; \quad \frac{\partial\theta}{\partial\xi}\Big|_{\xi=0} = 0; 0 \le \eta < 1
\end{aligned}
\right\}
\tag{24}
$$

For a constant temperature jump condition, the Nusselt number can be defined as:

$$
Nu_T = \frac{\partial T/\partial y|_{\text{wall}}}{T - T_{\text{wall}}}
\tag{25}
$$

2.3. Solution Procedure

The solution procedure comprises three steps. In the first step, the numerical values of the Nu and Po numbers are determined, corresponding to different values of ε and Kn by solving the governing Equations (12) and (23) by a finite difference method using MAPLE, and numerical data was obtained. In the second step, the back-error-propagation algorithm is employed to train a multi-layer perceptron neural network (MLPNN) using the neural network toolbox in MATLAB. There are two inputs, namely, the aspect ratio ε and the Knudsen number Kn. The corresponding outputs are the predicted values of the Nu and Po numbers, where each output is represented by a separate model. In each model, different input, hidden and output neurons are used. Finally, the results are optimized using PSO algorithm in order to find the optimal values of Nu and Po without gaining to the numerical values. Accordingly, the dimensionless partial differential equations will not use again in any stage of our work. As a result, the PSO algorithm uses neural network models to obtain the optimum values of Nu and Po.

3. Results and Discussion

The numerical values of Poiseuille numbers for forced convection in microducts were compared with the available data for different values of Kn in Tables 1–3. The comparison showed excellent agreement with the available data. The predicted values of the Nusselt numbers are also provided in Table 4 for different values of ε and Kn. To build the NN models, 70% of data was used as the training data, 15% as the testing data, and 15% as the validation data. In NN MATLAB tools, the back-error-propagation algorithm (an optimization algorithm) was used to determine the best models that have the optimal errors. These models were based on the mean squared error function of MSE = 0.00020768 and MSE = 0.00022112 for the two models, respectively, as shown in Figures 4 and 5. The figures show that the best validation performance, in terms of the mean squared error function, were found after 187 and 452 iterations, respectively.

Table 1. Comparison of the Poiseuille numbers for forced convection when $Kn = 0.1$.

ε	Morini et al. [36]	Sadeghi et al. [37]	Present Results
0.2	9.46	9.464	9.519
0.4	8.65	8.654	8.721
0.6	8.25	8.248	8.331
0.8	8.08	8.076	8.178
1	8.04	8.033	8.159

Table 2. Comparison of numerical and exact values of Po when $Kn = 0.001$.

ε	Numerical	Exact [38]	% Difference
0.001	24.031	23.7	1.3774
0.01	23.743	23.41	1.4025
0.02	23.432	22.24	5.0871
0.1	21.259	20.95	1.4535
0.2	19.182	18.89	1.5223
0.3	17.641	17.36	1.5929
0.4	16.514	16.24	1.6592
0.5	15.712	15.43	1.7948
0.6	15.164	14.87	1.9388
0.7	14.811	14.5	2.0998
0.8	14.608	14.28	2.2453
0.9	14.519	14.17	2.4037
1	14.514	14.14	2.5768

Table 3. Comparison of numerical and exact values of Po when $Kn = 0.1$.

ε	Numerical	Exact [38]	% Difference
0.001	10.944	10.9	0.402
0.01	10.862	10.82	0.3867
0.02	10.773	10.47	2.8126
0.1	10.139	10.09	0.4833
0.2	9.519	9.46	0.6198
0.3	9.056	9	0.6184
0.4	8.721	8.66	0.6995
0.5	8.487	8.41	0.9073
0.6	8.331	8.25	0.9723
0.7	8.233	8.14	1.1296
0.8	8.178	8.08	1.1983
0.9	8.157	8.04	1.4344
1	8.159	8.04	1.4585

Table 4. Numerical values of Nu for different values of Kn.

ε	$Kn = 0.001$	$Kn = 0.05$	$Kn = 0.1$
0.001	6.1658	1.4213	0.7164
0.25	5.1047	1.7625	1.0819
0.35	4.6786	1.8996	1.2287
0.5	4.0394	2.1051	1.4488
0.65	3.8048	2.1830	1.5590
0.75	3.6485	2.2348	1.6325
0.85	3.6156	2.2745	1.6816
1	3.5662	2.3341	1.7552

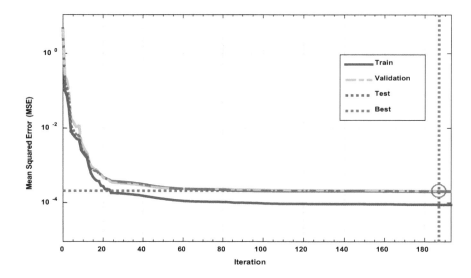

Figure 4. Best Validation Performance is 0.00020768 at iteration 187 of using MATLAB to generate the NN model of Nu.

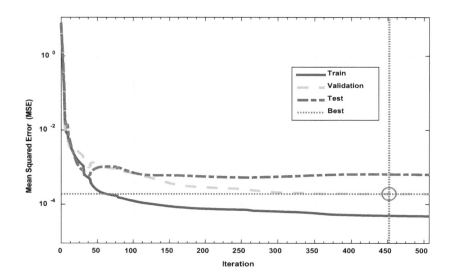

Figure 5. Best Validation Performance is 0.00022112 at iteration 452 using MATLAB to generate the neural network (NN) model of *Po*.

The correlation coefficients of the models are shown in Figures 6 and 7. As expected, the correlation coefficient of the models was greater than 0.95. In addition, Figures 6 and 7 show a significant convergence between the real values and the corresponding values that resulted from models. Accordingly, there was enough support to believe that the models were suitable to find the estimated values of the *Nu* and *Po* numbers. In addition, the coefficient of determination of the models was more than 0.99, which meant that both models could represent whole target data.

For optimization, the PSO algorithm was used to find the optimal values of *Po* and *Nu*, based on NN models. The optimal values of *Po* and *Nu* were achieved when $\varepsilon = 1$, and $Kn = 0.001$. The behavior of *Po* and *Nu* values in the PSO algorithm is shown in Figures 8 and 9. The optimal values of the curves were determined when the curves became stable. Accordingly, the optimal values for *Nu* and *Po* were found in iteration number 4 and 10, respectively. Note that, the data set values were converted into z-score values for use in the neural network toolbox in MATLAB.

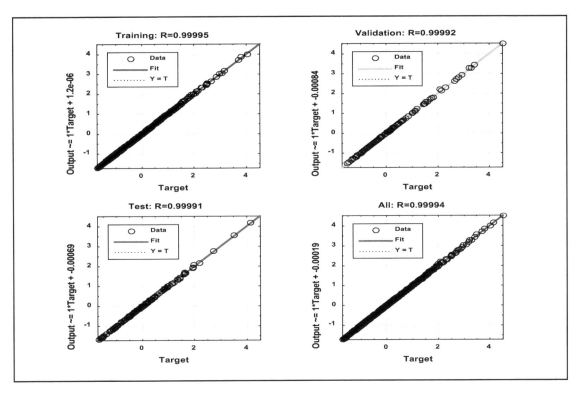

Figure 6. Correlation coefficient and the regression of the NN model values of Nu and the desired values.

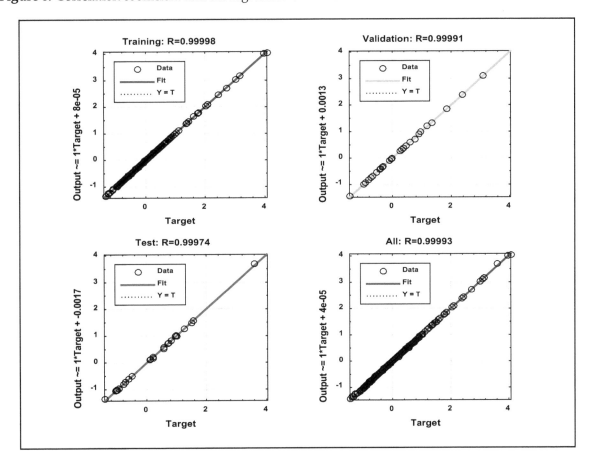

Figure 7. Correlation coefficient and the regression of the NN model values of Po and the desired values.

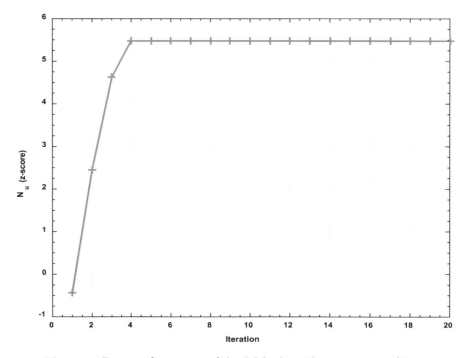

Figure 8. Best performance of the PSO algorithm in terms of Nu.

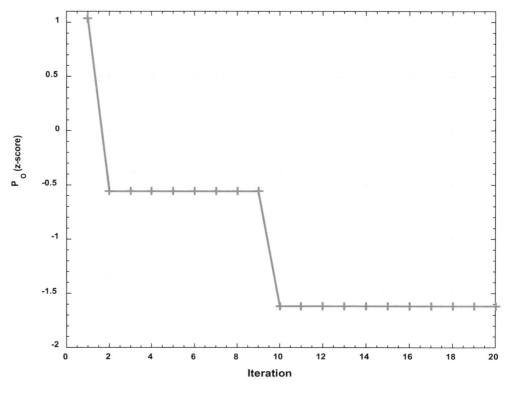

Figure 9. Best performance of the PSO algorithm in terms of Po.

4. Conclusions

In this study, the slip flow and heat transfer were investigated in rectangular microducts. The dimensionless governing partial differential equations with slip boundary conditions were solved numerically using MATLAB. The numerical data were used successfully for building ANN models. The outputs of the models were the corresponding Poiseuille Po and Nusselt Nu numbers to the input values ε and Kn. The main outcomes of the study were:

1. The small tolerance values for the mean squared error function values and a very close correlation coefficient of 1 in the NN models proved that the NN models were suitable to use for predicting the Poiseuille Po and Nusselt Nu numbers.

2. Without going back to the numerical data, the proposed models were used with the PSO algorithm to find the optimal values of Po and Nu.

3. The optimal values of Po and Nu are found when $\varepsilon = 1$ and $Kn = 0.001$.

4. The correlation coefficient of the models was greater than 0.95.

5. The optimal values of the curves were determined when the curves became stable.

Author Contributions: Optimization work: N.N.H.; Results and discussion: N.N.H. and I.K.; Introduction and literature review: A.S.A. and I.K.; Abstract, Conclusios and final revision: W.A.K. and A.S.A.

Acknowledgments: The first author is grateful to the Deanship of Scientific Research at Saudi Electronic University for providing great assistance in completing this research.

References

1. Liu, J.; Tai, Y.-C.; Pong, C.-M.H.-C. MEMS for pressure distribution studies of gaseous flows in microchannels. In Proceedings of the IEEE Micro Electro Mechanical Systems, Amsterdam, The Netherlands, 29 January–2 February 1995; p. 209.

2. Arkilic, E.B. Gaseous Flow in Micro-channels, Application of Microfabrication to Fluid Mechanics. *AsmeFed* **1994**, *197*, 57–66.

3. Ameel, T.A.; Wang, X.; Barron, R.F.; Warrington, R.O. Laminar forced convection in a circular tube with constant heat flux and slip flow. *Microscale Thermophys. Eng.* **1997**, *1*, 303–320.

4. Zade, A.Q.; Renksizbulut, M.; Friedman, J. Heat transfer characteristics of developing gaseous slip-flow in rectangular microchannels with variable physical properties. *Int. J. Heat Fluid Flow* **2011**, *32*, 117–127. [CrossRef]

5. Ghodoossi, L. Prediction of heat transfer characteristics in rectangular microchannels for slip flow regime and H1 boundary condition. *Int. J. Therm. Sci.* **2005**, *44*, 513–520. [CrossRef]

6. Hettiarachchi, H.M.; Golubovic, M.; Worek, W.M.; Minkowycz, W. Three-dimensional laminar slip-flow and heat transfer in a rectangular microchannel with constant wall temperature. *Int. J. Heat Mass Transf.* **2008**, *51*, 5088–5096. [CrossRef]

7. Yu, S.; Ameel, T.A. Slip-flow heat transfer in rectangular microchannels. *Int. J. Heat Mass Transf.* **2001**, *44*, 4225–4234. [CrossRef]

8. Renksizbulut, M.; Niazmand, H.; Tercan, G. Slip-flow and heat transfer in rectangular microchannels with constant wall temperature. *Int. J. Therm. Sci.* **2006**, *45*, 870–881. [CrossRef]

9. Tamayol, A.; Hooman, K. Slip-flow in microchannels of non-circular cross sections. *J. Fluids Eng.* **2011**, *133*, 091202. [CrossRef]

10. Hooman, K. A superposition approach to study slip-flow forced convection in straight microchannels of uniform but arbitrary cross-section. *Int. J. Heat Mass Transf.* **2008**, *51*, 3753–3762. [CrossRef]

11. Sadeghi, A.; Saidi, M.H. Viscous dissipation and rarefaction effects on laminar forced convection in microchannels. *J. Heat Transf.* **2010**, *132*, 072401. [CrossRef]

12. Yovanovich, M.; Khan, W.A. Compact Slip Flow Models for Gas Flows in Rectangular, Trapezoidal and Hexagonal Microchannels. In Proceedings of the ASME 2015 International Technical Conference and Exhibition on Packaging and Integration of Electronic and Photonic Microsystems, San Francisco, CA, USA, 6–9 July 2015.

13. Duan, Z.; Muzychka, Y. Slip flow in elliptic microchannels. *Int. J. Therm. Sci.* **2007**, *46*, 1104–1111. [CrossRef]

14. Duan, Z.; Muzychka, Y. Slip flow in non-circular microchannels. *Microfluid. Nanofluid.* **2007**, *3*, 473–484. [CrossRef]

15. Duan, Z.; Muzychka, Y. Slip flow in the hydrodynamic entrance region of circular and noncircular microchannels. *J. Fluids Eng.* **2010**, *132*, 011201. [CrossRef]

16. Yovanovich, M.M.; Khan, W.A. Friction and Heat Transfer in Liquid and Gas Flows in Micro-and Nanochannels. *Adv. Heat Transf.* **2015**, *47*, 203–307.

17. Ebert, W.; Sparrow, E.M. Slip flow in rectangular and annular ducts. *J. Basic Eng.* **1965**, *87*, 1018–1024. [CrossRef]

18. Baghani, M.; Sadeghi, A.; Baghani, M. Gaseous slip flow forced convection in microducts of arbitrary but constant cross section. *Nanoscale Microscale Thermophys. Eng.* **2014**, *18*, 354–372. [CrossRef]

19. Wang, C. Benchmark solutions for slip flow and H1 heat transfer in rectangular and equilateral triangular ducts. *J. Heat Transf.* **2013**, *135*, 021703. [CrossRef]

20. Beskok, A.; Karniadakis, G.E. Report: A model for flows in channels, pipes, and ducts at micro and nano scales. *Microscale Thermophys. Eng.* **1999**, *3*, 43–77.

21. Yovanovich, M.; Khan, W. Similarities of rarefied gas flows in elliptical and rectangular microducts. *Int. J. Heat Mass Transf.* **2016**, *93*, 629–636. [CrossRef]

22. Klášterka, H.; Vimmr, J.; Hajžman, M. Contribution to the gas flow and heat transfer modelling in microchannels. *Appl. Comput. Mech.* **2009**, *12*, 63–74.

23. Hamadneh, N.; Tilahun, S.; Sathasivam, S.; Choon, O.H. Prey-predator algorithm as a new optimization technique using in radial basis function neural networks. *Res. J. Appl. Sci.* **2013**, *8*, 383–387.

24. Rakhshandehroo, G.R.; Vaghefi, M.; Aghbolaghi, M.A. Forecasting groundwater level in Shiraz plain using artificial neural networks. *Arab. J. Sci. Eng.* **2012**, *37*, 1871–1883. [CrossRef]

25. Yeung, D.S.; Li, J.-C.; Ng, W.W.; Chan, P.P. MLPNN training via a multiobjective optimization of training error and stochastic sensitivity. *IEEE Trans. Neural Netw. Learn. Syst.* **2016**, *27*, 978–992. [CrossRef] [PubMed]

26. Mefoued, S. Assistance of knee movements using an actuated orthosis through subject's intention based on MLPNN approximators. In Proceedings of the 2013 International Joint Conference on Neural Networks, Dallas, TX, USA, 4–9 August 2013; pp. 1–6.

27. Shi, Y.; Eberhart, R.C. Empirical study of particle swarm optimization. In Proceedings of the 1999 Congress on Evolutionary Computation, Washington, DC, USA, 6–9 July 1999; pp. 1945–1950.

28. Trelea, I.C. The particle swarm optimization algorithm: Convergence analysis and parameter selection. *Inf. Process. Lett.* **2003**, *85*, 317–325. [CrossRef]

29. Liu, X. Radial basis function neural network based on PSO with mutation operation to solve function approximation problem. In Proceedings of the International Conference in Swarm Intelligence, Brussels, Belgium, 8–10 September 2010; pp. 92–99.

30. Eberhart, R.; Kennedy, J. A new optimizer using particle swarm theory. In Proceedings of the Sixth International Symposium on Micro Machine and Human Science, Nagoya, Japan, 4–6 October 1995; pp. 39–43.

31. Shi, Y. Particle swarm optimization: Developments, applications and resources. In Proceedings of the 2001 Congress on Evolutionary Computation, Seoul, Korea, 27–30 May 2001; pp. 81–86.

32. Marinke, R.; Araujo, E.; Coelho, L.S.; Matiko, I. Particle swarm optimization (PSO) applied to fuzzy modeling in a thermal-vacuum system. In Proceedings of the Fifth International Conference on Hybrid Intelligent Systems, Rio de Janeiro, Brazil, 6–9 November 2005; p. 6.

33. Nenortaite, J.; Butleris, R. Application of particle swarm optimization algorithm to decision making model incorporating cluster analysis. In Proceedings of the 2008 Conference on Human System Interactions, Krakow, Poland, 25–27 May 2008; pp. 88–93.

34. Hamadneh, N.; Khan, W.A.; Sathasivam, S.; Ong, H.C. Design optimization of pin fin geometry using particle swarm optimization algorithm. *PLoS ONE* **2013**, *8*, e66080. [CrossRef] [PubMed]

35. Bai, Q. Analysis of particle swarm optimization algorithm. *Comput. Inf. Sci.* **2010**, *3*, 180. [CrossRef]

36. Morini, G.L.; Spiga, M.; Tartarini, P. The rarefaction effect on the friction factor of gas flow in microchannels. *Superlattices Microstruct.* **2004**, *35*, 587–599. [CrossRef]

37. Sadeghi, M.; Sadeghi, A.; Saidi, M.H. Gaseous slip flow mixed convection in vertical microducts of constant but arbitrary geometry. *J. Thermophys. Heat Transf.* **2014**, *28*, 771–784. [CrossRef]

38. Shah, R.; London, A. *Laminar Flow Forced Convection in Ducts, Advances in Heat Transfer*; Academic Press: New York, NY, USA, 1978.

12

Heat Transfer of Oil/MWCNT Nanofluid Jet Injection Inside a Rectangular Microchannel

Esmaeil Jalali [1]**, Omid Ali Akbari** [2]**, M. M. Sarafraz** [3]**, Tehseen Abbas** [4] **and Mohammad Reza Safaei** [5,6,*]

[1]   Department of Mechanical Engineering, Najafabad Branch, Islamic Azad University, Najafabad, Iran; esmaiil.j66@yahoo.com
[2]   Young Researchers and Elite Club, Khomeinishahr Branch, Islamic Azad University, Khomeinishahr, Iran; Akbariomid11@gmail.com
[3]   Centre for Energy Technology, School of Mechanical Engineering, The University of Adelaide, South Australia, Australia; Mohammadmohsen.sarafraz@adelaide.edu.au
[4]   Department of Mathematics, University of Education Lahore, Faisalabad Campus, Faisalabad, Pakistan; tehseen.abbas@ue.edu.pk
[5]   Division of Computational Physics, Institute for Computational Science, Ton Duc Thang University, Ho Chi Minh City, Vietnam
[6]   Faculty of Electrical and Electronics Engineering, Ton Duc Thang University, Ho Chi Minh City, Vietnam
*   Correspondence: cfd_safaei@tdtu.edu.vn

Abstract: In the current study, laminar heat transfer and direct fluid jet injection of oil/MWCNT nanofluid were numerically investigated with a finite volume method. Both slip and no-slip boundary conditions on solid walls were used. The objective of this study was to increase the cooling performance of heated walls inside a rectangular microchannel. Reynolds numbers ranged from 10 to 50; slip coefficients were 0.0, 0.04, and 0.08; and nanoparticle volume fractions were 0–4%. The results showed that using techniques for improving heat transfer, such as fluid jet injection with low temperature and adding nanoparticles to the base fluid, allowed for good results to be obtained. By increasing jet injection, areas with eliminated boundary layers along the fluid direction spread in the domain. Dispersing solid nanoparticles in the base fluid with higher volume fractions resulted in better temperature distribution and Nusselt number. By increasing the nanoparticle volume fraction, the temperature of the heated surface penetrated to the flow centerline and the fluid temperature increased. Jet injection with higher velocity, due to its higher fluid momentum, resulted in higher Nusselt number and affected lateral areas. Fluid velocity was higher in jet areas, which diminished the effect of the boundary layer.

Keywords: Oil/MWCNT nanofluid; heat transfer; finite volume method; laminar flow; slip coefficient; microchannel

1. Introduction

In recent years, industrial developments have led scientists to search for methods to improve heat transfer in heat exchangers and industrial equipment. Therefore, a new generation of cooling fluids, called nanofluids, is used in industrial and commercial applications. Cooling fluid jet is used in turbine blade cooling and indirect access surfaces. Metal and non-metal nanoparticles have a higher thermal conductivity coefficient than water and lead to higher conductive heat transfer coefficients of fluid, as well as improve the temperature distribution of the nanofluid. Experimental results show that adding nanoparticles to the base fluid increases the heat transfer coefficient of nanofluids. Using nanofluids is one of the novel heat transfer improvement methods with high efficiency [1–4].

Numerous researchers have investigated the thermal or hydrodynamic performance of nanofluids in different microchannel heat sinks [5–7]. Hang et al. [8] numerically investigated the heat transfer performance of a microchannel heat sink with different nanofluids. Akbari et al. [9] studied laminar flow and heat transfer parameters of water/Al_2O_3 nanofluid with different nanoparticle volume fractions inside a rectangular microchannel and found that using rough surfaces in microchannel leads to higher heat transfer. Behnampour et al. [10] numerically investigated laminar flow and heat transfer parameters of water/AgO nanofluid with different nanoparticle volume fractions in a rectangular microchannel and showed that by increasing fluid velocity, an optimized trade-off can be obtained between heat transfer, hydrodynamic behavior of nanofluid, and the performance evaluation criteria (PEC) variations. Geravandian et al. [11] numerically simulated the laminar heat transfer of nanofluid flow in a rectangular microchannel and revealed that by increasing TiO_2 nanoparticles, heat transfer, friction coefficient, PEC, and pressure drop increase. Studies on the effect of using cooling fluid jet injection on heated surfaces [12] and other methods to increase heat transfer, such as using dimples, rough surfaces [13,14], and twisted tapes [15,16], have been conducted for different industrial and experimental geometries. These studies show that by creating vortexes, uniform temperature distribution can be obtained. Fluid jet plays an important role in cooling technologies and by creating better mixtures of cooling fluid flow, thermal performance can be enhanced [17–19]. Chen et al. [20] numerically and experimentally investigated the forced convection heat transfer inside a rectangular channel for determining fluid flow and heat transfer properties. They also compared the performance of heat sinks with solid and perforated pins and showed that by increasing the number of perforations and their diameter, pressure drop decreases and the Nusselt number increases. In their experiments, thermal performance of the heat sink with perforated pins was better than with solid pins. Nafon et al. [21] experimentally studied the effects of inlet temperature, Reynolds number, and heat flux on heat transfer properties of a water/TiO_2 nanofluid jet in a semi-rectangular heat sink, and showed that the average heat transfer coefficient of nanofluid is higher than base fluid, and the pressure drop increases by increasing the nanoparticle volume fraction. Jasperson et al. [22] studied the thermal and hydrodynamic performance of a copper microchannel and a pin fin microchannel and showed that by increasing the volume rate of flow, the thermal resistance of a pin fin heat sink decreases. Zhuwang et al. [23] studied heat transfer inside a microchannel with fluid jet and different coolants and showed that using fluid jet results in higher heat transfer compared with ordinary parallel flows.

Few studies have investigated the thermal and hydrodynamic performance of laminar nanofluid flow in a rectangular microchannel with cooling fluid jet injection. Therefore, the main purpose of this study is to investigate heat transfer improvement methods, such as using nanofluid and direct fluid jet injection. We also investigated the effect of the hydrodynamic velocity boundary condition on the cooling performance of the microchannel. In the present numerical study, laminar flow and heat transfer of oil/MWCNT nanofluid inside a two-dimensional microchannel with nanofluid jet injection were simulated using a finite volume method. Cooling nanofluid jet injection in a microchannel disturbs the thermal boundary layer and increases the heat transfer rate. The use of fluid jet injection in various sections of the microchannel is one of the novelties of this study. Fluid flow and heat transfer behavior are separately simulated in cases with no jets and with one, two, and three jets. The effect of applying slip and no-slip boundary conditions on solid walls of microchannel on the flow and heat transfer were studied.

2. Problem Definition

In the present numerical study, oil/MWCNT nanofluid flow was simulated in 0–4% nanoparticle volume fractions and Reynolds numbers of 10 and 50. We considered different numbers of fluid jets on the insulated bottom wall of the microchannel. The top heated wall of the microchannel had a constant temperature of $T_h = 303$ K. The inlet cold fluid entered from the left side of the microchannel with the temperature of $T_c = 293$ K. Figure 1 shows the microchannel in the present study.

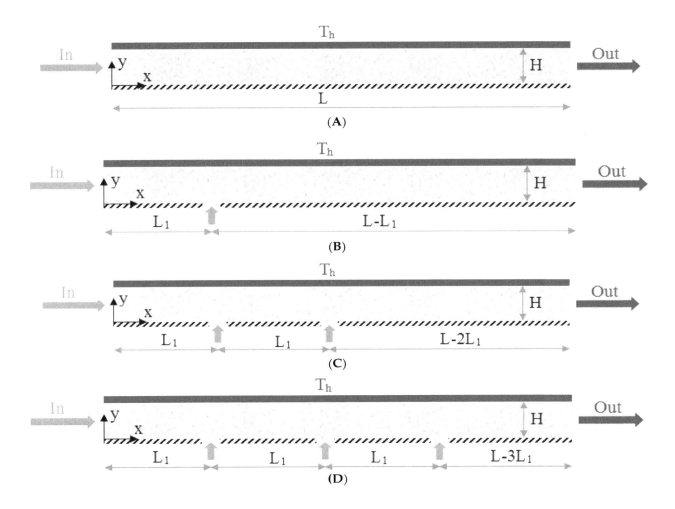

Figure 1. Schematic of the studied geometry in the present numerical study. (**A**) Case 1; (**B**) Case 2; (**C**) Case 3; (**D**) Case 4.

The length of studied microchannel was L = 2500 μm, the width of inlet jet entrance was 10 μm, the height of microchannel was H = 50 μm, and the jet pitch was L_1 = 500 μm. The velocity of the inlet nanofluid jet at all studied Reynolds numbers was constant (Re = 10). Fluid flow and heat transfer inside the microchannel were separately simulated in cases with no jet and with 1, 2, and 3 jets. Nanofluids with high velocity through the lower wall were injected into the micro-channel. As the cooling fluid jet flowed on the surface, the temperature of this surface decreased. Also, in this investigation, the effect of applying slip and no-slip boundary conditions on solid walls of the microchannel in slip coefficients of B = 0.0, 0.04, and 0.08 were investigated. The flow was laminar, forced, Newtonian, single-phase, and incompressible; the nanofluid was homogeneous and uniform; and also, the radiation effects were neglected. The properties of the base fluid and nanofluid in different nanoparticle volume fractions are presented in Table 1 [24].

Table 1. Thermophysical properties of base fluid and solid nanoparticles.

	Oil	MWCNT	ϕ = 0.02	ϕ = 0.04
c_p (J/kg K)	2032	1700	2012.9	1995.1
ρ (kg/m³)	867	2600	901.66	936.32
k (W/m K)	0.133	3000	0.5255	0.7912
μ (Pa s)	0.0289	-	0.0305	0.0321

3. Governing Equations

Dimensionless Navier–Stokes equations for forced, laminar, steady, and single-phase nanofluid are presented for 2-D space, which are continuity, momentum, and energy equations [25]:

$$\frac{\partial U}{\partial X} + \frac{\partial V}{\partial Y} = 0 \tag{1}$$

$$U\frac{\partial U}{\partial X} + V\frac{\partial U}{\partial Y} = -\frac{\partial P}{\partial X} + \frac{\mu_{nf}}{\rho_{nf}\nu_f}\frac{1}{Re}\left(\frac{\partial^2 U}{\partial X^2} + \frac{\partial^2 U}{\partial Y^2}\right) \tag{2}$$

$$U\frac{\partial V}{\partial X} + V\frac{\partial V}{\partial Y} = -\frac{\partial P}{\partial Y} + \frac{\mu_{nf}}{\rho_{nf}\nu_f}\frac{1}{Re}\left(\frac{\partial^2 V}{\partial X^2} + \frac{\partial^2 V}{\partial Y^2}\right) \tag{3}$$

$$U\frac{\partial \theta}{\partial X} + V\frac{\partial \theta}{\partial Y} = \frac{\alpha_{nf}}{\alpha_f}\frac{1}{RePr}\left(\frac{\partial^2 \theta}{\partial X^2} + \frac{\partial^2 \theta}{\partial Y^2}\right) \tag{4}$$

The dimensionless equations used this study are [26]:

$$X = \frac{x}{H}, \ Y = \frac{y}{H}, \ U = \frac{u}{u_c}, \ V = \frac{v}{u_c}, \ P = \frac{\bar{P}}{\rho_{nf}u_c^2}$$
$$\theta = \frac{T-T_c}{\Delta T}, \ Re = \frac{u_c \times H}{\nu_f}, \ Pr = \frac{\nu_f}{\alpha_f} \tag{5}$$

For calculating local and average Nusselt number along the microchannel walls, the following equations are used:

$$Nu(X) = \frac{k_{eff}}{k_f}\left(\frac{\partial \theta}{\partial Y}\right)_{Y=0} \tag{6}$$

$$Nu_{ave} = \frac{1}{L}\int_0^L Nu_s(X)\,dX \tag{7}$$

The dimensionless boundary conditions are:

$$U = 1, V = 0 \quad \text{and} \quad \theta = 0 \quad \text{for} \quad X = 0 \quad \text{and} \quad 0 \leq Y \leq 1 \tag{8}$$

$$V = 0 \quad \text{and} \quad \frac{\partial U}{\partial X} = 0 \quad \text{for} \quad X = 50 \text{ and } 0 \leq Y \leq 1 \tag{9}$$

$$U_w = -B\frac{\partial U}{\partial Y}, V = 0 \quad \text{and} \quad \theta = 1 \quad \text{for} \quad Y = 1 \quad \text{and} \quad 0 \leq X \leq 50 \tag{10}$$

$$U_w = B\frac{\partial U}{\partial Y}, V = 0 \quad \text{and} \quad \frac{\partial \theta}{\partial Y} = 0 \text{ for } Y = 0 \quad \text{and} \quad 0 \leq X \leq 50 \tag{11}$$

4. Numerical Details

Finite volume method [27,28] and the second-order upwind discretization method with maximum residual of 10^{-6} [29,30] was used. For coupling velocity–pressure equations, the SIMPLEC [31,32] algorithm was used. Also, the effect of nanoparticle volume fraction, Reynolds number, and jet number on flow and heat transfer parameters were investigated; and flow parameters, temperature, and streamlines contours are the presented.

The meshes in this study were rectangular and regular grids ranging from 25,000 to 100,000 grids were used for Reynolds numbers of 10 and 50, 2% volume fraction, and no-slip boundary conditions (Table 2). In order to ensure the independence of flow and heat transfer parameters from mesh number, Nusselt number and average velocity on flow centerline were studied for different mesh numbers. The 850×50 mesh number has better performance than 900×90, therefore, the former was used in the present simulations.

Table 2. Grid independence test.

Grid Size	500 × 50	750 × 75	850 × 80	900 × 90	1000 × 100
Re = 10					
Nu_{ave}	0.3698	0.3401	0.3202	0.2922	0.3401
$U_{out(Y=H/2)}$	3.806	3.7740	3.7532	3.6178	3.7740
Re=50					
Nu_{ave}	0.69719	0.6561	0.6178	0.5541	0.6561
$U_{out(Y=H/2)}$	3.972	3.9616	3.9578	3.9257	3.9616

Figures 2 and 3 show the validation of velocity and dimensionless flow temperature at the outlet section of microchannel against the results of Raisi et al. [33], who investigated laminar and forced flow of water/CuO nanofluid with different nanoparticle volume fractions in a two-dimensional rectangular microchannel using a finite volume method. The results of Raisi et al. [33] for 3% nanoparticle volume fraction, Re = 100, and different slip velocity coefficients were used for the validation. There is good agreement between the results of the present study and Raisi et al. [33]. Therefore, the boundary conditions and assumptions of this study are correct.

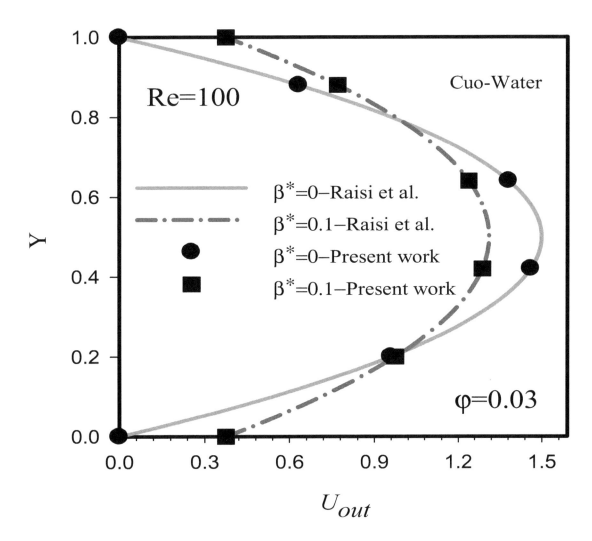

Figure 2. Validation of the present numerical study with Raisi et al. [33].

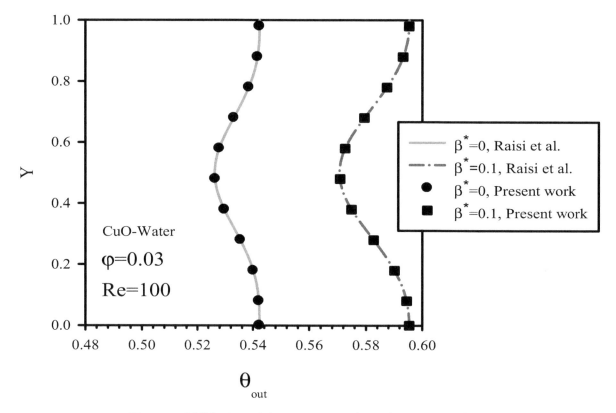

Figure 3. Validation of the present study with Raisi et al. [33].

5. Results and Discussion

In the present numerical study, the effect of nanofluid jet injection on laminar flow and heat transfer of oil/MWCNT nanofluid for 0–4% nanoparticle volume fractions, at Reynolds numbers of 10 and 50, and different slip velocity coefficients was numerically studied. The effects of the variations of jet number, nanoparticle volume fractions, and velocity boundary layer on solid walls of microchannel were also studied. Reynolds number, nanoparticle volume fraction, and slip velocity coefficient were input variables and their effect on Nusselt number, temperature domain, velocity domain, flow parameters, and temperature contours were investigated and compared based on their local and average values.

5.1. Streamlines and Isothermal Contours

In Figure 4, streamline contours (right side) and constant temperature contours (left side) are shown for Re = 10, B = 0.04, and φ = 2% for cases 1, 2, 3, and 4. In these contours, the effect of fluid motion with minimum velocity along the microchannel at different nanofluid jet injection on the heated surface was investigated. The temperature difference between inlet cold fluid and the heated wall improves heat transfer and creates temperature gradients. Fluid motion and its impact with the solid walls of the microchannel creates velocity gradients in areas close to the wall. In general, fluid motion along the microchannel leads to the creation of velocity and thermal boundary layers. Nanofluid jet injection on fluid direction results in the elimination of velocity and thermal boundary layers. In all contours, by increasing the number of fluid jets, areas with eliminated boundary layers become larger. Areas with a high number of jet injections have smaller velocity boundary layers. In the cases with more jets, due to the high volume of crossing fluid, axial velocity parameters of fluid are higher than the cases with fewer jets.

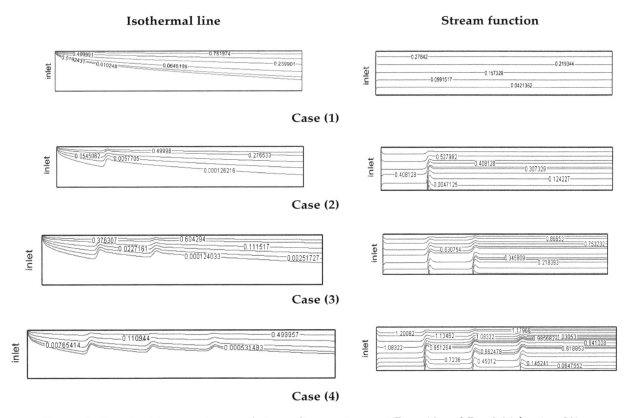

Figure 4. Constant temperature and streamline contours at Re = 10 and B = 0.04 for ϕ = 2%.

Streamline contours (right side) and constant temperature contours (left side) at Re = 50, B = 0.04, and ϕ = 2% are presented in Figure 5. The effect of increasing jet number on the heated surface on these contours is shown for cases 1, 2, 3, and 4. In the inlet areas of the microchannel and jet injection areas is 5 times higher, contrary to Figure 4. Increasing fluid velocity decreases the thermal boundary layer thickness. Increasing the fluid momentum results in higher crossing fluid velocity. At Re = 50, by increasing the number of fluid jets, due to the better mixture of fluid and reduction of temperature curve slope, a significant reduction of the thermal boundary layer thickness is observed. According to the streamline contours, nanofluid jet injection with higher velocity results in better mixture of flow and creation of local vortices. By increasing Reynolds number, especially in the last jet, the effect of vertical jet injection decreases due to the higher of volumetric rate of crossing fluid and creation of a smaller vortex behind each jet.

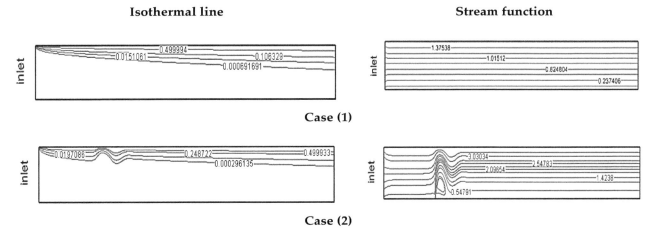

Figure 5. *Cont.*

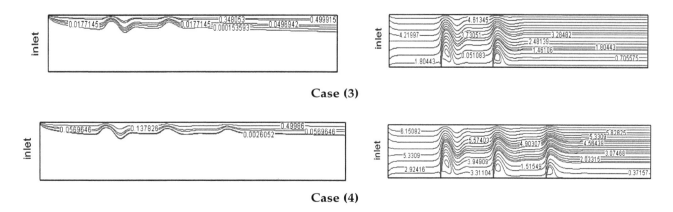

Figure 5. Streamlines and constant temperature contours at Re = 50 and B = 0.04 for φ = 2%.

5.2. Local Nusselt Number

In Figure 6, the local Nusselt number on the heated wall of the microchannel is shown for Re = 10, B = 0.04, and different nanoparticles volume fractions. In this figure, we have different jet numbers with minimum fluid velocity. Nanoparticle volume fractions were 0–4%. Heat transfer occurs because of temperature differences between fluid and the top heated wall. The value of heat transfer has a direct relationship with Nusselt number and temperature line slope. Uniform temperature distribution and removal of temperature gradients are desirable and lead to higher Nusselt numbers. The local jump in the Nusselt number in Figure 6 was because of the low temperature of the injected fluid jet on the heated surface. In these areas, the heat transfer distribution was more uniform and the Nusselt number increased. In all curves for Figure 6, by adding nanoparticle volume fraction, the Nusselt number and heat transfer were significantly increased.

In Figure 7, the local Nusselt numbers at Re = 50, B = 0.04, and different nanoparticle volume fractions are shown. In this figure, the effect of the increasing fluid velocity on Nusselt number is compared with Figure 6. Increasing φ and the number of jets resulted in the higher Nusselt number graphs in Figure 7. Figures 6 and 7 show the elimination of thermal and velocity boundary layers due to the fluid jet injection, which is more significant at higher Reynolds numbers. Jet injection with higher velocity leads to the increase of the Nusselt number due to the higher fluid momentum which affects the lateral regions. Increasing the nanoparticle volume fraction at Re = 50, compared to Re = 10, resulted in a significant rise of the Nusselt number. In general, the local Nusselt number on heated walls showed that using fluid jet injection with low temperature can increase the heat transfer rate. Compared to the primary jets, the heat transfer performance of the last jets was low, because by adding more jets, the effects of fluid momentum and crossing fluid become more significant, which diminishes the lateral flow by vertical jets as they need higher momentum. Therefore, in cases 2 and 3, the lateral cooling areas were narrower.

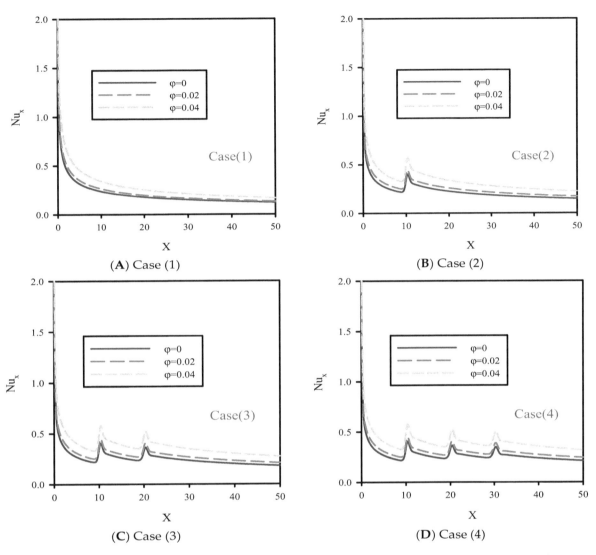

Figure 6. Local Nusselt number graphs at Re = 10 and B = 0.04 for different nanoparticle volume fractions.

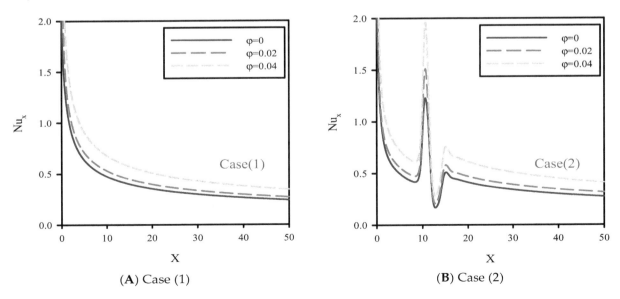

Figure 7. *Cont.*

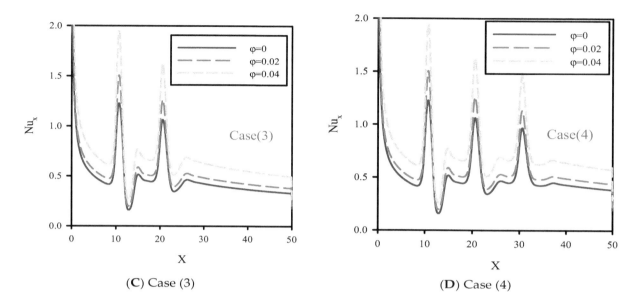

(**C**) Case (3)   (**D**) Case (4)

Figure 7. Local Nusselt number at Re = 50 and B = 0.04 for different nanoparticle volume fractions.

5.3. Temperature along the Symmetry Plane

In Figure 8, local dimensionless temperature distribution on flow centerline for Re = 10, B = 0.04, and volume fractions of 0–4% is presented. The dimensionless temperature distribution in the central section of the microchannel was influenced by the reduction of fluid velocity and jet number for cases 1, 2, 3, and 4, and different nanoparticle volume fractions. Because of the temperature difference between the fluid and heated surface, the penetration of the thermal boundary layer affected all areas of the microchannel, especially the areas in the central line of flow. Figure 8 shows that increasing ϕ resulted in higher thermal conductivity of the microchannel and uniform temperature distribution in all areas of the microchannel, especially in the heated areas. Due to the higher thermal conductivity of cooling fluid in higher nanoparticle volume fractions, temperature gradients in heated areas were diminished, and also, the conductivity of the fluid layers was higher. Therefore, by increasing ϕ, heat transfer penetrated the flow centerline, and because of the higher conductivity of the fluid layers, the fluid temperature rose.

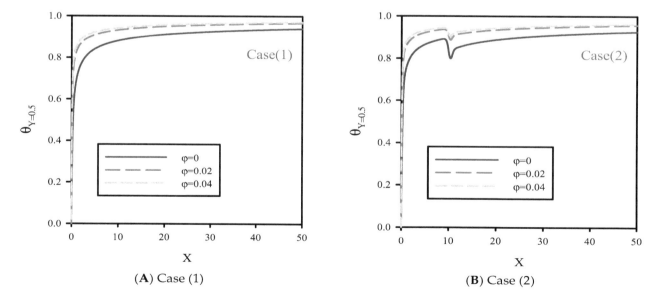

(**A**) Case (1)   (**B**) Case (2)

Figure 8. *Cont.*

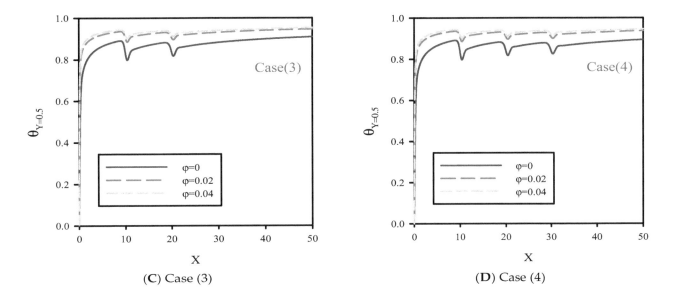

Figure 8. Temperature distribution at Re = 10, B = 0.04, and different nanoparticle volume fractions.

In Figure 9, local dimensionless temperature distribution at Re = 50, B = 0.04, and 0–4% nanoparticle volume fractions along the flow centerline are shown for cases 1, 2, 3, and 4. Increasing the number of jets resulted in lower temperature of the fluid, and thus the dimensionless temperature graphs in the microchannel centerlines were decreased, which was due to the reduction of temperature associated with increased fluid velocity and transferred temperature to the flow centerline. By increasing the velocity of injected cold fluid jet, the cold and hot fluids were better mixed. Consequently, the fluid temperature decreased significantly. According to Figures 8 and 9, by decreasing the fluid velocity, the fluid becomes thermally better developed, which is due to the penetration of heat to the inlet section of the microchannel. By increasing the fluid velocity, the undeveloped length of flow at the inlet section of microchannel increases.

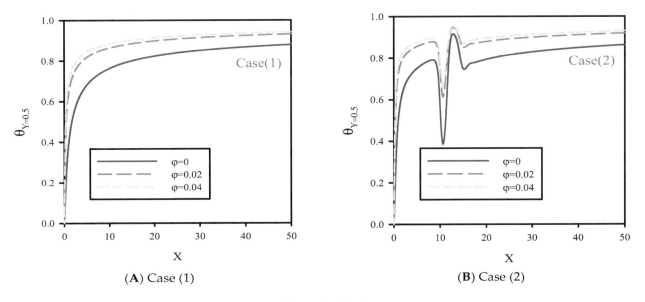

Figure 9. *Cont.*

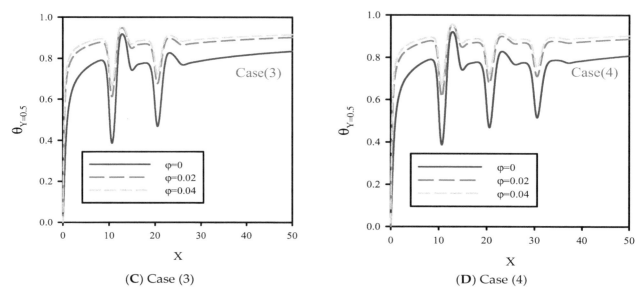

(C) Case (3)   **(D)** Case (4)

Figure 9. Temperature distribution at R e= 50, B = 0.04, and different nanoparticle volume fractions.

5.4. Axial Velocity along the Symmetric Plate

In Figure 10, dimensionless axial velocity in the flow centerline at Reynolds numbers of 10 and 50, B = 0.0, and φ=2% were investigated. Fluid motion along the microchannel was affected by velocity boundary layer caused by the solid walls. Contrary to the inlet velocity, fluid velocity was higher in case 1, which was due to the effects of the velocity boundary layer along the microchannel walls. In case 1, fluid velocity in developed areas at Re = 50 was lower than Re = 10. Fluid jet increased the volume rate of the crossing fluid downstream of the jets compared to areas upstream of the jets. Therefore, in cases 1, 2, and 3, fluid velocity increased in areas after the jets. In fact, high fluid velocity decreased the size of the velocity boundary layer. The velocity curves for fluid with Re = 50 were higher than fluid with Re = 10. In general, fluid jet eliminates the velocity boundary layer and prevents local flow development.

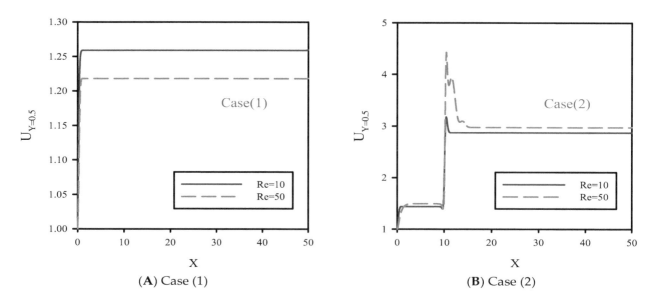

(A) Case (1)   **(B)** Case (2)

Figure 10. *Cont.*

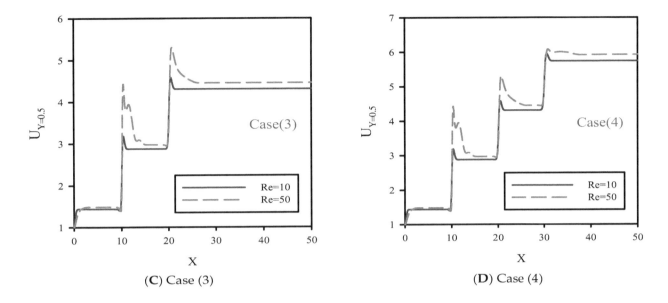

Figure 10. Axial velocity along the symmetry plane for Re = 10 and 50, B = 0, and φ = 2%.

In Figure 11, dimensionless axial velocity along the flow centerline at Reynolds numbers of 10 and 50, B = 0.08, and φ = 2% are shown. Unlike Figure 10, the slip boundary condition was applied to the solid walls which facilitated fluid motion and reduced the velocity boundary layer on the solid walls of the microchannel. Therefore, the velocity profile in the flow centerline for all cases of 1, 2, 3, and 4 were lower than Figure 10. According to Figures 10 and 11, increasing the slip velocity coefficient on the solid walls of the microchannel at Re = 10 had less effect than Re = 50. At Re = 50, the lateral mixture of flow covered more areas in the case with slip boundary conditions (BCs) compared with the no-slip BC case. According to Figure 11, this factor resulted in the microchannel contraction which increased the effect of jet injection areas and local axial velocity.

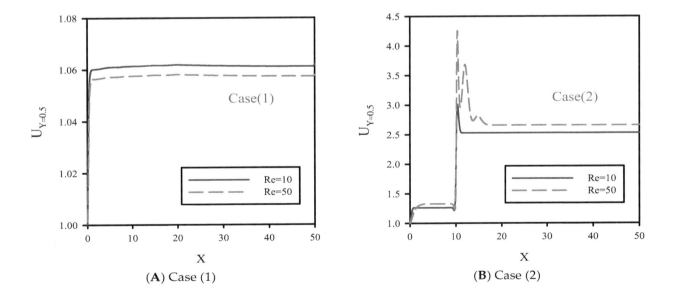

Figure 11. *Cont.*

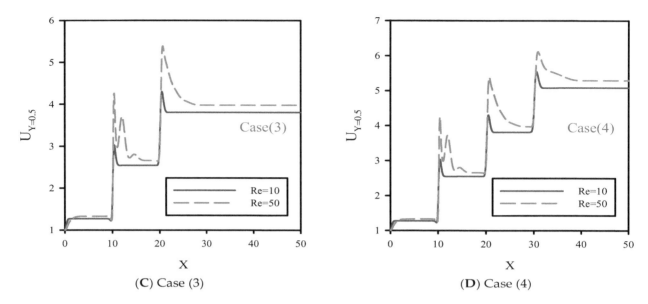

Figure 11. Axial velocity along the symmetry plane for Re = 10 and 50, B = 0.08, and φ = 2%.

5.5. Average Nusselt Number on the Heated Surface

In Figure 12, the average Nusselt number at Reynolds numbers of 10 and 50, volume fractions of 0, 2, and 4%, and B = 0.04 are shown. Using fluid jet injection with low temperature and adding solid nanoparticles to the base fluid, heat transfer was improved. Increasing the number of jets resulted in a better mixture of cold and hot flows in the microchannel, reduction of total temperature of the microchannel, and better cooling performance of the nanofluid. Adding solid nanoparticles to the base fluid with higher volume fractions resulted in better temperature distribution, and the Nusselt number increased as well. Also, increasing the Reynolds number resulted in higher fluid velocity and better temperature distribution, especially in heated areas, and also, the average Nusselt number increased. Among investigated cases, case 4 had the highest value of Nusselt number due to the maximum number of cold fluid jet injection. In case 1, due to the lack of nanofluid jet injection, the effects of thermal and velocity boundary layers became significant, and this case had the lowest heat transfer and Nusselt number.

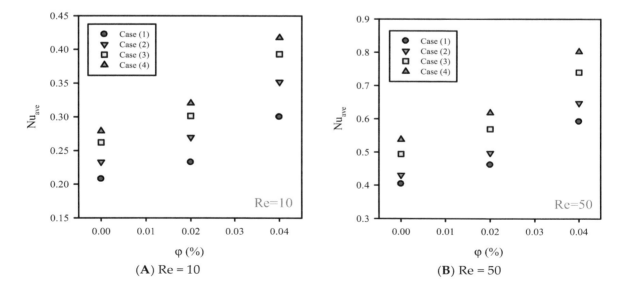

Figure 12. Average Nusselt number for Re = 10, 50; φ = 0, 2%, 4%; and B = 0.04.

5.6. Effect of Slip Coefficient on Axial Velocity

Dimensionless axial velocity in the central line of the microchannel at Re = φ = 2% and different slip coefficients was demonstrated in Figure 13. As it is seen, nanofluid jet injection at Re = 10 leads to some changes in jet areas. At this Reynolds number, increasing slip velocity coefficient results in lower fluid momentum dissipation and higher axial velocity. By increasing the slip velocity coefficient, because of the smaller velocity boundary layer, fluid moves on the microchannel walls with less friction. By increasing the jet number with a higher fluid velocity and volumetric rate of crossing fluid, the heat transfer was increased.

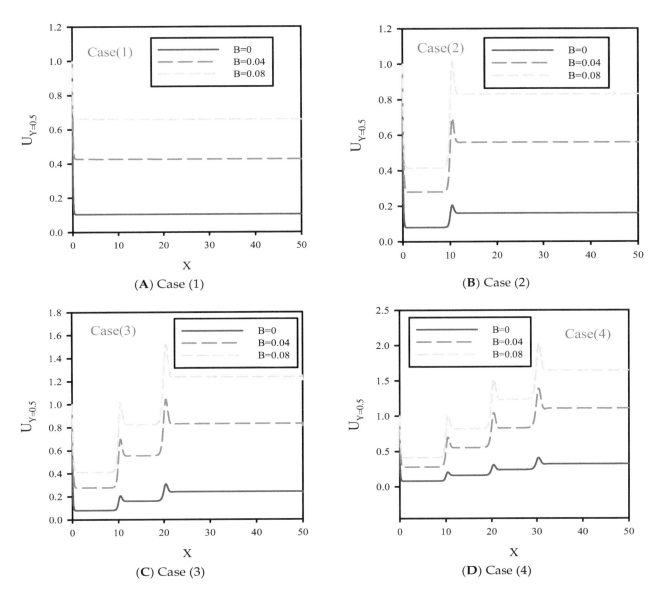

Figure 13. Dimensionless axial velocity at Re = 10 and φ = 2% with different slip coefficients on solid walls.

In Figure 14, dimensionless axial velocity in the flow centerline at Re = 50 and φ = 2% for cases 1, 2, 3, and 4 are shown. Increasing the slip coefficient on the solid surfaces parallel with the fluid direction, especially at higher Reynolds numbers, leads to the higher of fluid momentum. By increasing the slip velocity coefficient, due to the local flow contraction and significant changes in axial fluid velocity, fluid jet injection creates larger areas in the longitudinal direction. In Figure 14, because of the higher value of Reynolds number, the level of graphs are higher than those in Figure 13.

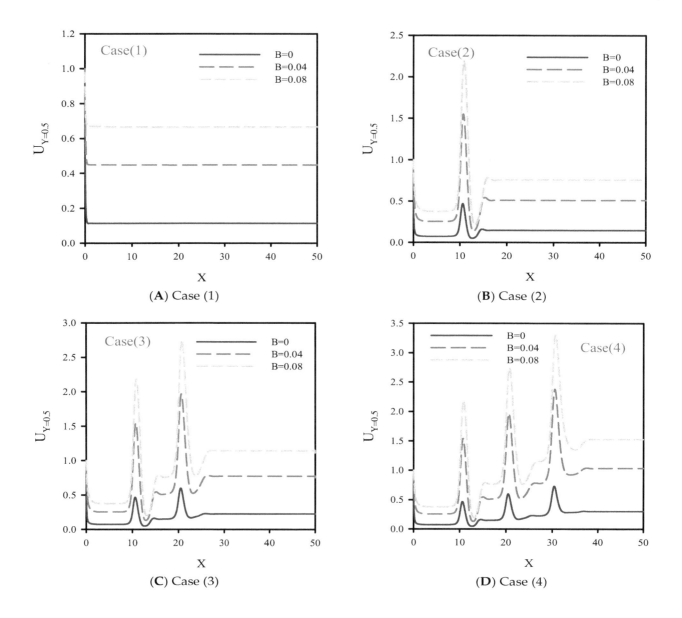

Figure 14. Dimensionless axial velocity at Re = 50 and ϕ = 2% with different slip coefficients on solid walls.

In Figure 15, dimensionless axial velocity profiles at the outlet section of the microchannel at Re = 10, ϕ = 4%, and slip coefficients of 0.0 and 0.08 are shown. Applying the slip coefficient on the solid walls resulted in a slight reduction of the axial velocity. By increasing the slip velocity coefficient, the velocity increased and the size of the velocity boundary layer on the heated surface decreased. According to the axial velocity profiles with the no-slip boundary condition, the effects of the velocity boundary layer was significant for the layers further away from the solid walls, and the velocity profile was stronger in the flow centerline. Therefore, from solid walls to flow centerline, the slope of the axial velocity variations was higher than the velocity profile with B = 0.08. Applying the slip velocity coefficient on the solid walls resulted in lower axial velocity gradients. The slope of the velocity curve from the walls to the flow centerline indicated lower gradients than the case with the no-slip boundary condition. The velocity profile in the central areas of the flow confirmed this issue and the corresponding curve had a milder slope.

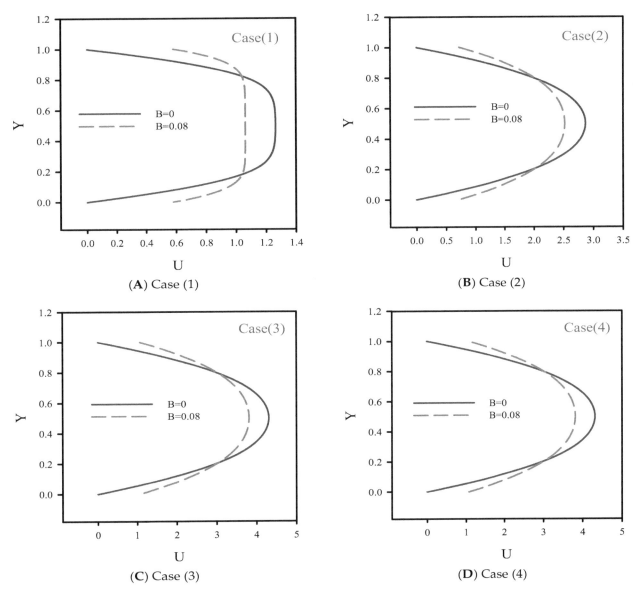

Figure 15. Effect of the slip coefficient on the outlet axial velocity for Re = 10 and ϕ = 4.

6. Conclusions

In the present numerical study, laminar flow of oil/MWCNT nanofluid inside a two-dimensional microchannel with nanofluid jet injection was investigated. This simulation was carried out in a two-dimensional domain for laminar and Newtonian flow and heat transfer of oil/MWCNT nanofluid, with volume fractions of 0–4%, at Reynolds numbers of 10 and 50. Various jet numbers were used on the insulated bottom wall of the microchannel. Results showed that nanofluid jet injection with higher velocity leads to better flow mixture through the formation of local vortexes. In Re = 50, by increasing the number of fluid jets, the thermal boundary layer thickness becomes smaller because of the better mixture of fluid. Also, the temperature line slope was milder. Heat transfer occurs due to the temperature difference between fluid and the top heated wall. Heat transfer had a direct relationship with Nusselt number and the slope of the temperature line. Fluid jets eliminated the thermal and velocity boundary layers, which was more significant in higher Reynolds numbers. Because of the temperature differences between fluid and heated surfaces, penetration of the thermal boundary layer affected all sections of the microchannel, especially areas on the flow centerline. Because of higher conductivity of the fluid layer, by increasing the nanoparticle volume fraction, the heat transfer reached the flow centerline and the fluid temperature increased in that area. By decreasing fluid velocity, due to the higher temperature penetration, the developed length of flow became shorter and the fluid developed faster. Higher fluid velocity in areas with jets eliminated the velocity boundary layer.

Unlike the case with the no-slip boundary condition, by applying the slip coefficient on the solid walls of the microchannel, the lateral mixture of flow at Re = 50 covered larger areas. Increasing the Reynolds number resulted in higher fluid velocity and better temperature distribution, especially in heated areas, and this led to higher average Nusselt number. In fluid with higher velocity, applying the slip boundary condition on solid walls significantly increases the crossing fluid momentum.

Author Contributions: All authors contributed equally.

Nomenclature

B	dimensionless slip velocity coefficient
C_p	heat capacity, J kg^{-1} K^{-1}
d	diameter, m
H, L	microchannel height and length, m
k	thermal conductivity coefficient, Wm^{-1} K^{-1}
Nu	Nusselt number
P	fluid pressure, Pa
$Pr = v_f/f$	Prandtl number
$Re = \rho_f u_c H / \mu_f$	Reynolds number
T	temperature, K
u, v	velocity components in the x-, y-directions, ms^{-1}
u_c	inlet flow velocity, ms^{-1}
$(U, V) = (u/U_0, v/U_0)$	dimensionless flow velocity in the x-, y-direction
x, y	Cartesian coordinates, m
$(X, Y = x/H, y/H)$	dimensionless coordinates

Greek symbols

α	thermal diffusivity, m^{2}s^{-1}
β *	dimensionless slip velocity coefficient
ϕ	nanoparticle volume fraction
μ	dynamic viscosity, Pa s
$\theta = (T - T_C)/(T_H - T_C)$	dimensionless temperature
ρ	density, kg m^{-3}
v	kinematics viscosity m^{2}s^{-1}

Super- and Subscripts

c	Cold
eff	Effective
f	base fluid (pure water)
h	Hot
m	Mean
nf	Nanofluid
s	solid nanoparticles

References

1.   Gorla, R.S.R.; Chamkha, A. Natural convective boundary layer flow over a vertical plate embedded in a porous medium saturated with a non-Newtonian nanofluid. *Int. J. Microscale Nanoscale Therm. Fluid Transp. Phenom.* **2011**, *3*, 1–20.
2.   Dogonchi, A.S.; Chamkha, A.J.; Seyyedi, S.M.; Ganji, D.D. Radiative nanofluid flow and heat transfer between parallel disks with penetrable and stretchable walls considering Cattaneo–Christov heat flux model. *Heat Transf. Asian Res.* **2018**, *47*, 735–753. [CrossRef]

3. Goshayeshi, H.R.; Safaei, M.R.; Goodarzi, M.; Dahari, M. Particle Size and Type Effects on Heat Transfer Enhancement of Ferro-nanofluids in a Pulsating Heat Pipe under Magnetic Field. *Powder Technol.* **2016**, *301*, 1218–1226. [CrossRef]

4. Khanafer, K.; Vafai, K. A critical synthesis of thermophysical characteristics of nanofluids. *Int. J. Heat Mass Transf.* **2012**, *54*, 4410–4428. [CrossRef]

5. Kosar, A.; Peles, Y. TCPT-2006-096. R2: Micro Scale Pin Fin Heat Sinks—Parametric Performance Evaluation Study. *IEEE Trans. Compon. Packag. Technol.* **2007**, *30*, 855–865. [CrossRef]

6. Sivasankaran, H.; Asirvatham, G.; Bose, J.; Albert, B. Experimental Analysis of Parallel Plate and Crosscut Pin Fin Heat Sinks for Electronic Cooling Applications. *Therm. Sci.* **2010**, *14*, 147–156. [CrossRef]

7. Mital, M. Analytical Analysis of Heat Transfer and Pumping Power of Laminar Nanofluid Developing Flow in Microchannels. *Appl. Therm. Eng.* **2012**, *50*, 429–436. [CrossRef]

8. Hung, T.C.; Yan, W.M.; Wang, X.D.; Chang, C.Y. Heat Transfer Enhancement in Microchannel Heat Sinks using Nanofluids. *Int. J. Heat Mass Transf.* **2012**, *55*, 2559–2570. [CrossRef]

9. Akbari, O.A.; Toghraie, D.; Karimipour, A. Impact of ribs on flow parameters and laminar heat transfer of water–aluminum oxide nanofluid with different nanoparticle volume fractions in a three-dimensional rectangular microchannel. *Adv. Mech. Eng.* **2015**, *7*, 1–11. [CrossRef]

10. Behnampour, A.; Akbari, O.A.; Safaei, M.R.; Ghavami, M.; Marzban, A.; Ahmadi Sheikh Shabani, G.R.; zarringhalam, M.; Mashayekhi, R. Analysis of heat transfer and nanofluid fluid flow in microchannels with trapezoidal, rectangular and triangular shaped ribs. *Physica E* **2017**, *91*, 15–31. [CrossRef]

11. Gravndyan, Q.; Akbari, O.A.; Toghraie, D.; Marzban, A.; Mashayekhi, R.; Karimi, R.; Pourfattah, F. The effect of aspect ratios of rib on the heat transfer and laminar water/TiO$_2$ nanofluid flow in a two-dimensional rectangular microchannel. *J. Mol. Liq.* **2017**, *236*, 254–265. [CrossRef]

12. Lee, D.Y.; Vafai, K. Comparative analysis of jet impingement and microchannel cooling for high heat flux applications. *Int. J. Heat Mass Transf.* **1999**, *42*, 1555–1568. [CrossRef]

13. Chang, S.W.; Chiang, K.F.; Chou, T.C. Heat transfer and pressure drop in hexagonal ducts with surface dimples. *Exp. Therm. Fluid Sci.* **2010**, *34*, 1172–1181. [CrossRef]

14. Elyyan, M.A.; Ball, K.S.; Diller Thomas, E.; Paul Mark, R.; Ragab Saad, A. Heat Transfer Augmentation Surfaces Using Modified Dimples/Protrusions. Ph.D. Thesis, Virginia Tech, Blacksburg, VA, USA, 2008.

15. Guo, J.; Fan, A.; Zhang, X.; Liu, W. A numerical study on heat transfer and friction factor characteristics of laminar flow in a circular tube fitted with centercleared twisted tap. *Int. J. Therm. Sci.* **2011**, *50*, 1263–1270. [CrossRef]

16. Zhang, X.; Liu, Z.; Liu, W. Numerical studies on heat transfer and flow characteristics for laminar flow in a tube with multiple regularly spaced twisted tapes. *Int. J. Therm. Sci.* **2012**, *58*, 157–167. [CrossRef]

17. Zheng, N.; Liu, W.; Liu, Z.; Liu, P.; Shan, F. A numerical study on heat transfer enhancement and the flow structure in a heat exchanger tube with discrete double inclined ribs. *Appl. Therm. Eng.* **2015**, *90*, 232–241. [CrossRef]

18. Shan, F.; Liu, Z.; Liu, W.; Tsuji, Y. Effects of the orifice to pipe diameter ratio on orifice flows. *Chem. Eng. Sci.* **2016**, *152*, 497–506. [CrossRef]

19. Zheng, N.; Liu, P.; Shan, F.; Liu, Z.; Liu, W. Effects of rib arrangements on the flow pattern and heat transfer in an internally ribbed heat exchanger tube. *Int. J. Therm. Sci.* **2016**, *101*, 93–105. [CrossRef]

20. Chin, S.B.; Foo, J.J.; Lai, Y.L.; Yong, T.K.K. Forced Convective Heat Transfer Enhancement with Perforated Pin Fins. *Heat Mass Transf.* **2013**, *49*, 1447–1458. [CrossRef]

21. Naphon, P.; Nakharintr, L. Heat Transfer of Nanofluids in the Mini-rectangular Fin Heat Sinks. *Int. Commun. Heat Mass Transf.* **2012**, *40*, 25–31. [CrossRef]

22. Jasperson, B.A.; Jeon, Y.; Turner, K.T.; Pfefferkorn, F.E.; Qu, W. Comparison of Micro-pin-fin and Microchannel Heat Sinks Considering Thermal-hydraulic Performance and Manufacturability. *IEEE Trans. Compon. Packag. Technol.* **2010**, *33*, 148–160. [CrossRef]

23. Zhuang, Y.; Ma, C.F.; Qin, M. Experimental study on local heat transfer with liquid impingement flow in two-dimensional micro-channels. *Int. J. Heat Mass Transf.* **1997**, *40*, 4055–4059. [CrossRef]

24. Gholami, M.R.; Akbari, O.A.; Marzban, A.; Toghraie, D.; Ahmadi Sheikh Shabani, G.H.R.; Zarringhalam, M. The effect of rib shape on the behavior of laminar flow of Oil/MWCNT nanofluid in a rectangular microchannel. *J. Therm. Anal. Calorim.* **2017**, 1–18. [CrossRef]

25. Raisi, A.; Aminossadati, S.M.; Ghasemi, B. An innovative nanofluid-based cooling using separated natural and forced convection in low Reynolds flows. *J. Taiwan Inst. Chem. Eng.* **2016**, 1–5. [CrossRef]

26. Aminossadati, S.M.; Raisi, A.; Ghasemi, B. Effects of magnetic field on nanofluid forced convection in a partially heated microchannel. *Int. J. Non-Linear Mech.* **2011**, *46*, 1373–1382. [CrossRef]

27. Bahmani, M.H.; Sheikhzadeh, G.; Zarringhalam, M.; Akbari, O.A.; Alrashed, A.A.A.A.; Ahmadi Sheikh Shabani, G.; Goodarzi, M. Investigation of turbulent heat transfer and nanofluid flow in a double pipe heat exchanger. *Adv. Powder Technol.* **2018**, *29*, 273–282. [CrossRef]

28. Arani, A.A.A.; Akbari, O.A.; Safaei, M.R.; Marzban, A.; Alrashed, A.A.A.A.; Ahmadi, G.R.; Nguyen, T.K. Heat transfer improvement of water/single-wall carbon nanotubes (SWCNT) nanofluid in a novel design of a truncated double layered microchannel heat sink. *Int. J. Heat Mass Transf.* **2017**, *113*, 780–795. [CrossRef]

29. Khodabandeh, E.; Rahbari, A.; Rosen, M.A.; Najafian Ashrafi, Z.; Akbari, O.A.; Anvari, A.M. Experimental and numerical investigations on heat transfer of a water-cooled lance for blowing oxidizing gas in an electrical arc furnace. *Energy Conversat. Manag.* **2017**, *148*, 43–56. [CrossRef]

30. Safaiy, M.R.; Saleh, S.R.; Goudarzi, M. Numerical studies of laminar natural convection in a square cavity with orthogonal grid mesh by finite volume method. *Int. J. Adv. Des. Manuf. Technol.* **2011**, *1*, 13–21.

31. Akbari, O.A.; Goodarzi, M.; Safaei, M.R.; Zarringhalam, M.; Ahmadi Sheikh Shabani, G.R.; Dahari, M. A modified two-phase mixture model of nanofluid flow and heat transfer in 3-D curved microtube. *Adv. Powd. Technol.* **2016**, *27*, 2175–2185. [CrossRef]

32. Safaei, M.R.; Goodarzi, M.; Akbari, O.A.; Safdari Shadloo, M.; Dahari, M. Performance Evaluation of Nanofluids in an Inclined Ribbed Microchannel for Electronic Cooling Applications. *Electron. Cool.* **2016**. [CrossRef]

33. Raisi, A.; Ghasemi, B.; Aminossadati, S.M. A Numerical Study on the Forced Convection of Laminar Nanofluid in a Microchannel with Both Slip and No-Slip Conditions. *Numer. Heat Transf. Part A* **2011**, *59*, 114–129. [CrossRef]

13

A Numerical Simulation of Silver–Water Nanofluid Flow with Impacts of Newtonian Heating and Homogeneous–Heterogeneous Reactions Past a Nonlinear Stretched Cylinder

Muhammad Suleman [1,2]**, Muhammad Ramzan** [3,4,*]**, Shafiq Ahmad** [5]**, Dianchen Lu** [1]**, Taseer Muhammad** [6] **and Jae Dong Chung** [4]

1   Department of Mathematics, Faculty of Science, Jiangsu University, Zhenjiang 212013, China; suleman@ujs.edu.cn (M.S.); dclu@ujs.edu.cn (D.L.)
2   Department of Mathematics, Comsats University, Islamabad 44000, Pakistan
3   Department of Computer Science, Bahria University, Islamabad Campus, Islamabad 44000, Pakistan
4   Department of Mechanical Engineering, Sejong University, Seoul 143-747, Korea; jdchung@sejong.ac.kr
5   Department of Mathematics, Quaid-i-Azam University, Islamabad 44000, Pakistan; ashafiq@math.qau.edu.pk
6   Department of Mathematics, Government College Women University, Sialkot 51310, Pakistan; taseer_qau@yahoo.com
*   Correspondence: mramzan@bahria.edu.pk

Abstract: The aim of the present study is to address the impacts of Newtonian heating and homogeneous–heterogeneous (h-h) reactions on the flow of Ag–H2O nanofluid over a cylinder which is stretched in a nonlinear way. The additional effects of magnetohydrodynamics (MHD) and nonlinear thermal radiation are also added features of the problem under consideration. The Shooting technique is betrothed to obtain the numerical solution of the problem which is comprised of highly nonlinear system ordinary differential equations. The sketches of different parameters versus the involved distributions are given with requisite deliberations. The obtained numerical results are matched with an earlier published work and an excellent agreement exists between both. From our obtained results, it is gathered that the temperature profile is enriched with augmented values radiation and curvature parameters. Additionally, the concentration field is a declining function of the strength of h-h reactions.

Keywords: Newtonian heating; nonlinear thermal radiation; nonlinear stretching cylinder; homogeneous/heterogeneous reactions; nanofluid

1. Introduction

In copious engineering processes, the role of poor thermal conductivity of certain base fluids is considered to be a big hurdle to shape a refined product. To overcome such snag, numerous practices such as clogging, abrasion, and pressure loss were engaged but the results were not very promising. The novel idea of nanofluid, which is an engineered amalgamation of metallic particles with size (<100 nm) and some traditional fluids like ethylene glycol, presented by Choi [1], has revolutionized the modern world. Many heat transfer applications [2] such as domestic refrigerators, microelectronics, hybrid-power engines, and fuel cells possess numerous characteristics that make them valuable because of nanofluids. In all aforementioned applications, enriched thermal conductivity is observed whenever some metallic particles are added to the ordinary base fluid [3]. The idea of a nanofluid with multiple dimensions has been conversed by many researchers and scientists in the last two decades. Amongst these, Li et al. [4] used a mixture of H_2O–CuO for the enhancement of the solidification rate. The finite element method is engaged to obtain the numerical solution of the problem. Sheikholeslami [5] pondered over the influence of radiation and magnetohydrodynamic on the Al_2O_3–H_2O mixture past a spongy semi-annulus. The numerical solution of the problem is witnessed by employing the Control Volume Finite Element Method (CVFEM). The flow of Casson nanofluid with inserted multi-walled carbon nanotubes past a swirling cylinder was deliberated by Ramzan et al. [6] using bvp4c MATLAB software. The said problem is pondered with impacts of entropy generation and melting heat transfer. The flow of micropolar nanofluid with binary chemical reaction, double stratification, and activation energy is also studied by Ramzan et al. [7]. The flow of viscoelastic nanofluid with analysis of entropy generation past an exponential stretched surface was discussed by Suleman et al. [8]. Farooq et al. [9] deliberated the flow of Newtonian fluid with the amalgamation of nanoparticles by utilizing the BVPh 2.0 technique and many therein [10–15].

The role of MHD is vital in abundant fluid flows and has many applications like medicine, aerospace, nuclear reactors, MHD generators, petroleum processes, and astrophysics. The explorations highlighting the effects of MHD may include a study by Azam et al. [16] who found the numerical solution for the time-dependent MHD Cross nanofluid flow under the influence of nonlinear thermal radiation and zero mass flux conditions. The flow of MHD nanofluid with thermal diffusion and heat generation past a permeable medium over an oscillating vertical plate is perceived by Sheikholeslami et al. [17]. Makinde and Animasaun [18] examined the MHD nanofluid flow past the upper surface of a paraboloid of revolution with impacts of quartic autocatalysis chemical reaction and nonlinear thermal radiation. Lu et al. [19] debated the rotating flow of 3D MHD Maxwell fluid with a non-Fourier heat flux and binary chemical reaction using a BVP-4c MATLAB built-in technique. The analytical technique is engaged to find the multiple solutions for MHD Jeffery-Hamel flow using the KKL nanofluid model by Rana et al. [20]. The effect of generalized Fourier and Fick laws combined with temperature-dependent thermal conductivity on 3D MHD second-grade nanofluid is considered by Ramzan et al. [21]. Yuan et al. [22] established the numerical solution of MHD nanofluid flow past a baffled U-shaped enclosure by engaging the KKL technique. The 3D flow of MHD nanofluid with varied nanoparticles including Fe_3O_4, Cu, Al_2O_3 and TiO_2 and water as the base fluid past an exponential stretched surface was discussed by Jusoh et al. [23]. Some recent investigations highlighting the importance of MHD may also be found in References [24–27].

The Newtonian heating or the conjugate convective flow is termed as a direct proportionate amid the local temperature and heat transfer rate. The role of Newtonian heating is pivotal in many processes such as heat exchanger's designing, convective flows where heat is taken from the solar radiators, and conjugate heat transfer around the fins etc. The four distinct categories of the heat transfer viz. (i) conjugate boundary condition, (ii) constant or prescribed surface heat flux, (iii) Newtonian heating, and (iv) constant or prescribed surface temperature, are given by Merkin [28]. The flow of Micropolar fluid with Newtonian heating and chemical reaction past a stretching/shrinking sheet was studied by Kamran and Wiwatanapataphee [29]. Mehmood et al. [30] examined the Oldroyd-B nanofluid flow with a transverse magnetic field and Newtonian heating. The flow of Tangent hyperbolic nanofluid under the influences of Newtonian heating, MHD and bi-convection was examined by Shafiq et al. [31]. The flow of nanofluid which is an amalgamation of the base fluid (Sodium alginate) and nanoparticles (Silver, Titanium oxide, Copper and Aluminum oxide) with effects of radiation and Newtonian heating past an isothermal vertical plate are scrutinized by Khan et al. [32]. Many researchers have undertaken the impact of Newtonian heating owing to its wide-ranging practical applications [33–36].

A close review of the literature specifies that copious research is done on the subject of nanofluids with linear/nonlinear stretching sheets in comparison to the flows past the curved stretched surfaces. This topic gets even more limited if we talk about the flows of nanofluids past cylinders stretched in a nonlinear way. So, our prime goal here is to ponder the nanofluid flow past a nonlinear stretched cylinder with nonlinear thermal radiation, h-h reactions, and Newtonian heating. The nanofluid used here is the mixture of water and silver. The numerical simulations are conducted for the proposed problem using the Runge–Kutta method by shooting technique. To corroborate the presented results, a comparison with an already published article is done and an excellent correlation between the two results is found.

2. Mathematical Modeling

Here, we assume a silver–water nanofluid incompressible flow with impacts of h-h reactions, nonlinear thermal radiation and Newtonian heating over a horizontal cylinder which is stretched in a nonlinear way. The magnetic field $B = B_0 x^{(n-1)/2}$ is applied in the radial direction. Owing to our assumption of a small Reynolds number, the induced magnetic field is overlooked (Figure 1).

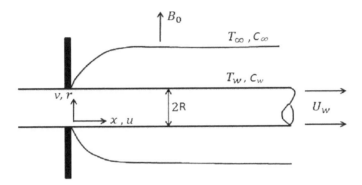

Figure 1. Diagram of the flow geometry.

The homogeneous reaction for cubic autocatalysis can be written as:

$$A_1 + 2B_1 \rightarrow 3B_1, \quad \text{rate} = k_c a b^2, \tag{1}$$

$$A_1 \rightarrow B_1, \quad \text{rate} = k_s a. \tag{2}$$

The reaction rate dies out in the outer boundary layer. The system of boundary layer equations of the subject model is given as:

$$\frac{\partial(ru)}{\partial x} + \frac{\partial(rv)}{\partial r} = 0, \tag{3}$$

$$u\frac{\partial u}{\partial x} + v\frac{\partial u}{\partial r} = \nu_{nf}\left(\frac{\partial^2 u}{\partial r^2} + \frac{1}{r}\frac{\partial u}{\partial r}\right) - \frac{\sigma_{nf} B^2(x)}{\rho_{nf}} u, \tag{4}$$

$$u\frac{\partial T}{\partial x} + v\frac{\partial T}{\partial r} = \alpha_{nf}\left(\frac{\partial^2 T}{\partial r^2} + \frac{1}{r}\frac{\partial T}{\partial r}\right) - \frac{1}{(\rho c_p)_{nf}}\frac{\partial q_r}{\partial r}, \tag{5}$$

$$u\frac{\partial a}{\partial x} + v\frac{\partial a}{\partial r} = D_A\left(\frac{\partial^2 a}{\partial r^2} + \frac{1}{r}\frac{\partial a}{\partial r}\right) - k_1 a b^2, \tag{6}$$

$$u\frac{\partial b}{\partial x} + v\frac{\partial b}{\partial r} = D_B\left(\frac{\partial^2 b}{\partial r^2} + \frac{1}{r}\frac{\partial b}{\partial r}\right) + k_1 a b^2, \tag{7}$$

Accompanied by the conditions:

$$u|_{r=R} = U_w(x), \ v|_{r=R} = 0, \ \frac{\partial T}{\partial r}\Big|_{r=R} = h_s T, \ D_A\frac{\partial a}{\partial r}\Big|_{r=R} = k_s a, \ D_B\frac{\partial b}{\partial r}\Big|_{r=R} = -k_s a,$$
$$u|_{r\rightarrow\infty} \rightarrow 0, \ T|_{r\rightarrow\infty} \rightarrow T_\infty, \ a|_{r\rightarrow\infty} \rightarrow a_0, \ b|_{r\rightarrow\infty} \rightarrow 0, \tag{8}$$

where $U_w(x) = U_0 x^n$, and $q_r = \frac{4\sigma^*}{3k^*}\frac{\partial T^4}{\partial r} = \frac{16\sigma^* T^3}{3k^*}\frac{\partial T}{\partial r}$. The numerical values of specific heat, density, and thermal conductivity of silver (Ag) and water (H$_2$O) are given in Table 1.

Table 1. Thermo-physical characteristics of the base fluid and nanoparticles [1,37].

Physical Properties	Water	Ag
C_p (J/kg K)	4179	235.0
ρ (kg/m^3)	997.1	10,500.0
K (W/mK)	0.61300	429.0

With the following characteristics:

$$\alpha_{nf} = \frac{k_{nf}}{(\rho C_p)_{nf}},\tag{9}$$

$$\frac{(\rho C_p)_{nf}}{(\rho C_p)_f} = (1 - \phi) + \phi\frac{(\rho C_p)_s}{(\rho C_p)_f},\tag{10}$$

$$\frac{\mu_{nf}}{\mu_f} = (1.005 + 0.497\phi - 0.1149\phi^2),\tag{11}$$

$$\frac{\rho_{nf}}{\rho_f} = 1 - \phi + \phi\frac{\rho_s}{\rho_f},\tag{12}$$

$$\frac{k_{nf}}{k_f} = (0.9692\phi + 0.9508),\tag{13}$$

The use of under-mentioned similarity transformations

$$\eta = \frac{r^2 - R^2}{2R}\sqrt{\frac{U_w}{\nu_f x}}, \ \psi = \sqrt{U_w \nu x} R f(\eta), \ \theta = \frac{T - T_\infty}{T_\infty}$$
$$a = a_0 h, b = a_0 g.\tag{14}$$

Satisfy the Equation (3) and convert the Equations (4)–(7) in non-dimensional form

$$(1 + 2\gamma\eta)f''' + 2\gamma f'' + (1.005 + 0.497\phi - 0.1149\phi^2)(1 - \phi + \phi\frac{\rho_s}{\rho_f})\left(\left(\frac{n+1}{2}\right)ff'' - nf'^2\right)$$
$$- (1.005 + 0.497\phi - 0.1149\phi^2)Mf' = 0,\tag{15}$$

$$(1 + 2\gamma\eta)\left(\frac{k_{nf}}{k_f} + (1 + (N_r - 1)\theta)^3\right)\theta'' + \left(\frac{1}{2K^*}\frac{k_{nf}}{k_f} + (1 + (N_r - 1)\theta)^3\right)\gamma\theta' + \frac{Pr}{4K^*}$$
$$\left(1 - \phi + \phi\frac{(\rho C_p)_s}{(\rho C_p)_f}\right)\left(\frac{n+1}{2}f\theta' - nf'\theta\right) + 2(1 + 2\gamma\eta)(N_r - 1)(1 + (N_r - 1)\theta)^2\theta'^2 = 0,\tag{16}$$

$$\frac{1}{S_c}(1 + 2\gamma\eta)h'' + 2\gamma h') + fh' - \frac{2K}{n+1}hg^2 = 0,\tag{17}$$

$$\frac{1}{S_c}(1 + 2\gamma\eta)g'' + 2\gamma g') + fg' - \frac{2K}{n+1}hg^2 = 0,\tag{18}$$

Supported by the boundary conditions

$$f(\eta) = 0, \ f'(\eta) = 1, \ \theta'(\eta) = -\lambda(1 + \theta(\eta)), \ h'(\eta) = K_s h(\eta), \ \delta g'(\eta) = -K_s h(\eta) \text{ as } \eta = 0,$$
$$\theta(\eta) \to 0, \ f'(\eta) \to 0, \ h(\eta) \to 1, \ g(\eta) \to 0, \text{ at } \eta \to \infty,\tag{19}$$

where $\lambda = h_s\left(\nu_f x/U_w\right)^{1/2}$, $M = \frac{\sigma_f B_0^2}{U_0\rho_f}$, $K^* = \frac{16\sigma^* T_\infty^3}{3k^* k_f}$, $\gamma = \left(\frac{x\nu_f}{R^2 U_w}\right)^{1/2}$, $N_r = T_w/T_\infty$, $K = \frac{a_0^2 x k_1}{U_w}$, $K_s = \frac{k_s}{D_A}\sqrt{\frac{x\nu_f}{U_w}}$ and $S_c = \frac{\nu_f}{D_A}$.

Here, it is expected that A_1 and B_1 are equivalent. From this assumption, it is inferring that D_A and D_B (diffusion coefficients) are equal i.e., $\delta = 1$, and on account of this supposition, we have

$$g(\eta) + h(\eta) = 1.\tag{20}$$

Equations (17) and (18) after the use of Equation (20) and the relevant boundary conditions take the shape

$$\frac{1}{S_c}\left((1+2\gamma\eta)h''+2\gamma h'\right)+fh'-\frac{2k_1}{n+1}h(1-h)^2=0, \tag{21}$$

$$h'(0)=k_2h(0),\, h(\infty)\to 1. \tag{22}$$

The physical quantities like Skin friction factor and Local Nusselt number in non-dimensional form are labelled as

$$C_f=\frac{2\tau_w}{\rho_f u_w^2},\; Nu_x=\frac{xq_w}{k_f(T_f-T_\infty)}, \tag{23}$$

With τ_w and q_w given by

$$\tau_w=\mu_{nf}\frac{\partial u}{\partial r}\Big|_{r=R},\, q_w=-\Big(\frac{\partial T}{\partial r}\Big)_{r=R}k_{nf}+(q_r)_{r=R}. \tag{24}$$

Equation (23), after the use of Equations (14) and (24), takes the form

$$Re_x^{1/2}C_f=\Big(\frac{1}{1.005+0.497\phi-0.1149\phi^2}\Big)f''(0),$$
$$Re_x^{-1/2}Nu_x=\frac{k_{nf}}{k_f}\lambda\Big(1+\frac{1}{\theta(0)}\Big). \tag{25}$$

3. Numerical Scheme

The numerical solution of Equations (15), (16) and (21) supported by the boundary conditions (19) and (22) is found by the Shooting scheme. In the calculation of the numerical solution of the problem, the second and third order differential equations are transformed to first order with the help of new parameters. The selection of the initial guess estimate is pivotal in the Shooting scheme as it needs to satisfy the equation and the boundary conditions asymptotically. We have selected the tolerance as 10^{-7} for this specific problem. The first order system obtained in this regard is appended below:

$$f(\eta)=y(1), \tag{26}$$

$$f'(\eta)=y(2), \tag{27}$$

$$f''(\eta)=y(3), \tag{28}$$

$$f'''(\eta)=F(3)=-\frac{1}{(1+2\gamma\eta)}\left[\begin{array}{c}(1.005+0.497\phi-0.1149\phi^2)(1-\phi+\phi\frac{\rho_s}{\rho_f})\left((\frac{n+1}{2})ff''-nf'^2\right)\\+2\gamma f''-(1.005+0.497\phi-0.1149\phi^2)Mf'\end{array}\right], \tag{29}$$

$$\theta(\eta)=y(4), \tag{30}$$

$$\theta'(\eta)=y(5), \tag{31}$$

$$\theta''(\eta)=F(5)=-\frac{1}{(1+2\gamma\eta)\left(\frac{k_{nf}}{k_f}+(1+(N_r-1)\theta)^3\right)}\left[\begin{array}{c}\left(\frac{1}{2K^*}\frac{k_{nf}}{k_f}+(1+(N_r-1)\theta)^3\right)\gamma\theta'\\+\frac{Pr}{4K^*}\left(1-\phi+\phi\frac{(\rho C_p)_s}{(\rho C_p)_f}\right)\left(\begin{array}{c}\frac{n+1}{2}f\theta'\\-nf'\theta\end{array}\right)\\+2(1+2\gamma\eta)(N_r-1)(1+(N_r-1)\theta)^2\theta'^2\end{array}\right], \tag{32}$$

$$h(\eta)=y(6), \tag{33}$$

$$h'(\eta)=y(7), \tag{34}$$

$$h''(\eta)=F(7)=-\frac{1}{(1+2\gamma\eta)}\left[2\gamma h'+S_cfh'-S_c\frac{2k_1}{n+1}h(1-h)^2\right], \tag{35}$$

and the boundary condition becomes

$$y_0(1) = 0, \ y_0(2) = 1, \ y_0(5) = -\lambda(1 + y_0(4)), \ y_0(7) = k_2 y_0(6),$$
$$y_{\text{inf}}(2) \to 0, \ y_{\text{inf}}(4) \to 0, \ y_{\text{inf}}(6) \to 0. \tag{36}$$

We have chosen $\eta_\infty = 6$, that guarantees every numerical solution's asymptotic value accurately. Here, Table 2 depicts the comparative estimates of the present model with Qasim et al. [38] in the limiting case. Both results are found in an excellent correlation.

Table 2. Nusselt number $(\text{Re}_x^{-1/2} Nu_x)$ for numerous estimates of γ and Pr with $M = 0$, $\phi = 0.0$, $\lambda = 0.0$.

γ	Pr	$\text{Re}_x^{-1/2} Nu_x$	
		Qasim et al. [38]	Present Result
0.0	0.72	1.23664	1.236651
	1.0	1.00000	1.000000
	6.7	0.33330	0.333310
	10	0.26876	0.268770
1.0	0.72	0.87018	0.870190
	1.0	0.74406	0.744070
	6.7	0.29661	0.296620
	10	0.24217	0.242180

4. Results and Discussion

In this section, we have plotted the Figures 2–11 that exhibit the influences of various parameters like volume fraction $(0.0 \le \phi \le 0.3)$, magnetic parameter $(1.0 \le M \le 4.0)$, nonlinearity exponent $(1.0 \le n \le 5.0)$, curvature parameter $(0.0 \le \gamma \le 0.4)$, conjugate parameter $(0.4 \le \lambda \le 0.7)$, radiation parameter $(0.7 \le K^* \le 1.0)$, strength of homogeneous reaction $(0.1 \le \kappa_1 \le 1.8)$, strength of heterogeneous reaction $(0.1 \le \kappa_2 \le 1.8)$ and Schmidt number $(3.0 \le S_c \le 4.5)$ on involved distributions. Figures 2–5 are drawn to portray the influence of ϕ, M, n and γ on axial velocity. The impact of ϕ is discussed in Figure 2. For growing values of ϕ, the axial velocity also augments. In Figure 3, the influence of M versus the velocity field is debated. It is perceived that the velocity field deteriorates for escalated values of M. The Lorentz force is enforced by the strong magnetic field that hinders the fluid's velocity. Figure 4 illustrates the impression of n on the axial velocity. Reduced velocity is witnessed for larger values of n. This is because higher values of n create more collision between the particles of the fluid that hinder the fluid flow and feeble velocity if perceived. In Figures 5 and 6, the behavior of the velocity and temperature fields for increasing values of γ is given. It is seen that the velocity and temperature of the fluid augment for growing estimates of γ. Larger values of γ mean a smaller radius, comparatively minimum contact region of the cylinder with the fluid and increased heat transport. That is why augmented velocity and temperature are witnessed. Figure 7 is drawn to depict the relation between the temperature of the fluid and the λ. It is detected that temperature enhances for improved values of λ. In fact, the sturdier heat transfer process is observed for larger values of λ as more heat is moved from the cylinder to the fluid. Remembering that $\lambda = 0$ means the insulated walls and $\lambda \to \infty$ constant wall temperature. Figures 8 and 9 are sketched to elaborate the influences of K^* and M on the temperature field. It is clearly perceived that temperature enhances when both K^* and M increase. The decrease in the mean absorption coefficient represents an enriched heat transfer rate and ultimately temperature is enhanced. Similarly, a stronger Lorentz force hinders the movement of the fluid, thus causing more collision between the molecules of the fluid that turns into the improved temperature. In Figures 10 and 11, the behavior of the concentration profile versus h-h reactions is depicted. The concentration diminishes for growing values of h-h reactions.

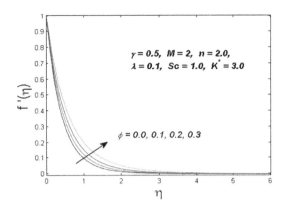

Figure 2. Diagram of axial velocity versus ϕ.

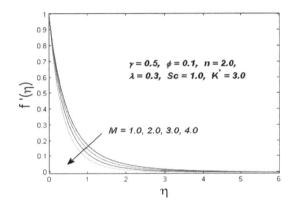

Figure 3. Diagram of axial velocity versus M.

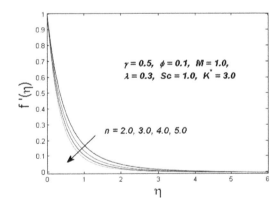

Figure 4. Diagram of axial velocity versus n.

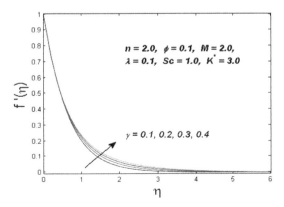

Figure 5. Diagram of axial velocity versus γ.

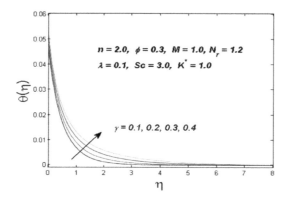

Figure 6. Diagram of the temperature field versus.

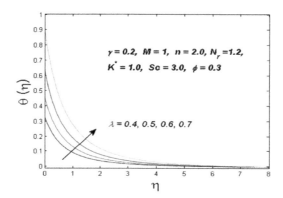

Figure 7. Diagram of the temperature field versus λ.

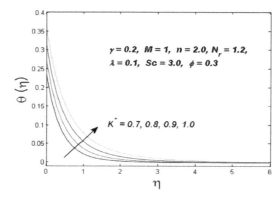

Figure 8. Diagram of the temperature field versus K^*.

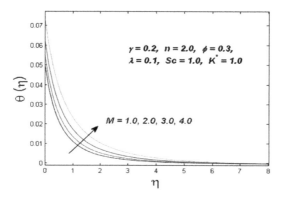

Figure 9. Diagram of the temperature field versus M.

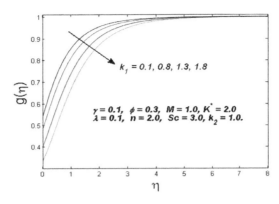

Figure 10. Diagram of concentration field versus k_1.

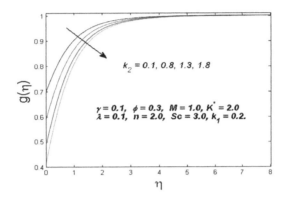

Figure 11. Diagram of concentration field versus k_2.

Table 3 depicts the numerical values of skin friction $(-\text{Re}_x^{1/2}C_f)$ and local Nusselt number $(\text{Re}_x^{-1/2}Nu_x)$ for numerous estimates of thee parameters. It is perceived that skin friction and Nusselt number upsurge for growing values of solid volume friction ϕ and M (magnetic parameter). It is also observed that for escalated values of curvature parameter γ and nonlinearity parameter n, both skin friction and local Nusselt number show opposite behavior. Moreover, Skin friction and local Nusselt number exhibit a constant trend for values of the temperature ratio parameter N_r and radiation parameter K^*.

Table 3. Numerical values of $-\text{Re}_x^{1/2}C_f$ and $\text{Re}_x^{-1/2}Nu_x$ for Ag–water with Pr $= 6.2$.

n	γ	ϕ	M	N_r	K^*	$-\text{Re}_x^{1/2}C_f$	$\text{Re}_x^{-1/2}Nu_x$
1.0	0.1	0.1	1.0	1.2	1.0	1.92700	0.32149
2.0	0.1	0.1	1.0	1.2	1.0	2.45720	0.30963
3.0	0.1	0.1	1.0	1.2	1.0	2.89050	0.30305
2.0	1.0	0.1	1.0	1.2	1.0	2.81410	0.28200
2.0	2.0	0.1	1.0	1.2	1.0	3.18730	0.26410
2.0	3.0	0.1	1.0	1.2	1.0	3.54330	0.25369
2.0	1.0	0.1	1.0	1.2	1.0	2.81410	0.28200
2.0	1.0	0.2	1.0	1.2	1.0	3.42830	0.31250
2.0	1.0	0.3	1.0	1.2	1.0	4.04270	0.34319
2.0	1.0	0.1	1.0	1.2	1.0	2.81410	0.28200
2.0	1.0	1.0	2.0	1.2	1.0	2.99960	0.28387
2.0	1.0	1.0	3.0	1.2	1.0	3.17070	0.28553
2.0	1.0	1.0	1.0	0.1	1.0	2.81410	0.28200
2.0	1.0	1.0	1.0	0.7	1.0	2.81410	0.28200
2.0	1.0	1.0	1.0	1.0	1.0	2.81410	0.28200
2.0	1.0	1.0	1.0	1.2	1.0	2.81410	0.28200
2.0	1.0	1.0	1.0	1.2	2.0	2.81410	0.28200
2.0	1.0	1.0	1.0	1.2	3.0	2.81410	0.28200

5. Final Comments

In the present exploration, we have pondered over the nanofluid flow (with base fluid water and nanoparticles as silver) past a nonlinear stretching cylinder with impacts of h-h reactions and nonlinear thermal radiation. Additional effects of Newtonian heating and magnetohydrodynamics have also been taken into account. A numerical solution of the dimensionless mathematical model is achieved via the shooting scheme. The core outcomes of the present study are as follows:

- The temperature profile is a growing function of radiation and magnetic parameters.
- For larger values of the curvature parameter, augmented velocity is observed.
- The concentration of the fluid decreases for growing values of homogeneous–heterogeneous reactions.
- For escalated values of the magnetic parameter, velocity and temperature distributions show the opposite trend.
- The skin friction and local Nusselt number show opposite behavior for curvature and nonlinearity parameters.

Author Contributions: Data curation, J.D.C.; Formal analysis, S.A.; Funding acquisition, M.S.; Investigation, D.L.; Methodology, T.M.; Resources, J.D.C.; Software, M.S.; Supervision, M.R.; Validation, S.A.; Visualization, T.M.; Writing—original draft, M.R.; Writing—review & editing, D.L.

Acknowledgments: This research was funded by National Natural Science Foundation China (No. 11571140 and 11671077), Faculty of Science, Jiangsu University, Zhenjiang, China.

Nomenclature

U	along x-axis fluid velocity [m/s]
V	along r-axis fluid velocity [m/s]
U_w	Stretching velocity [m/s]
U_e	Free stream velocity [m/s]
M	Magnetic parameter
ρ_f, ρ_s	Density of fluid and solid particle respectively [kg/m^3]
ϕ	nanofluid volume fraction
τ_w	surface shear stress [N/m^2]
T_∞	Ambient temperature [K]
λ	conjugate parameter
σ_{nf}	electric conductivity of fluid and nanofluid respectively [S/m]
γ	curvature parameter
k_f	thermal conductivities of fluid [J/mKs^{n-2}]
k_s	thermal conductivities of nanomaterial [J/mKs^{n-2}]
δ	ratio of mass diffusion coefficients
q_r	radiative heat flux [kg/m^2]
h_f	convective heat transfer coefficient
C_f	Skin friction coefficient
Nu_x	Nusselt number
T	Temperature [K]
T_f	convective fluid temperature [K]
h_s	heat transfer coefficient
μ_{nf}	Nanofluid dynamic viscosity [kg/ms^{n-2}]
Re_x	local Reynolds number
k^*	mean absorption coefficient
σ^*	Stefan-Boltzmann constant

D_A, D_B	diffusion coefficients [m^2/s]
q_w	surface heat flux [W/m^2]
S_c	Schmidt number
k_c, k_s	Rate constants
K^*	Radiation parameter
μ_f	Fluid dynamic viscosity [kg/ms^{n-2}]
α_{nf}	Nanofluid thermal diffusivity [m^2/s^{n-2}]
k_1	Strength of homogeneous reaction
ν_{nf}	nanofluid kinematic viscosity [m^2/s]
k_2	strength of heterogeneous reaction
R	radius of cylinder [m]
B	Magnetic field strength [A/m]
N_r	temperature ratio parameter
A_1, B_1	concentrations of chemical species a, b
n	Nonlinearity exponent

References

1. Choi, S.U.S. Enhancing conductivity of fluids with nanoparticles, ASME Fluid Eng. *Division* **1995**, *231*, 99–105.
2. Minkowycz, W.J.; Sparrow, E.M.; Abraham, J.P. (Eds.) *Nanoparticle Heat Transfer and Fluid Flow*; CRC Press: Boca Raton, FL, USA, 2012; Volume 4.
3. Kakac, S.; Pramuanjaroenkij, A. Review of convective heat transfer enhancement with nanofluids. *Int. J. Heat Mass Transf.* **2009**, *52*, 3187–3196. [CrossRef]
4. Li, Z.; Sheikholeslami, M.; Shafee, A.; Ramzan, M.; Kandasamy, R.; Al-Mdallal, Q.M. Influence of adding nanoparticles on solidification in a heat storage system considering radiation effect. *J. Mol. Liq.* **2019**, *273*, 589–605. [CrossRef]
5. Sheikholeslami, M.; Shafee, A.; Ramzan, M.; Li, Z. Investigation of Lorentz forces and radiation impacts on nanofluid treatment in a porous semi annulus via Darcy law. *J. Mol. Liq.* **2018**, *272*, 8–14. [CrossRef]
6. Lu, D.; Ramzan, M.; Ahmad, S.; Chung, J.D.; Farooq, U. Upshot of binary chemical reaction and activation energy on carbon nanotubes with Cattaneo-Christov heat flux and buoyancy effects. *Phys. Fluids* **2017**, *29*, 123103. [CrossRef]
7. Ramzan, M.; Ullah, N.; Chung, J.D.; Lu, D.; Farooq, U. Buoyancy effects on the radiative magneto Micropolar nanofluid flow with double stratification, activation energy and binary chemical reaction. *Sci. Rep.* **2017**, *7*, 12901. [CrossRef] [PubMed]
8. Suleman, M.; Ramzan, M.; Zulfiqar, M.; Bilal, M.; Shafee, A. Entropy analysis of 3D non-Newtonian MHD nanofluid flow with nonlinear thermal radiation past over exponential stretched surface. *Entropy* **2018**, *20*, 930. [CrossRef]
9. Farooq, U.; Lu, D.C.; Ahmed, S.; Ramzan, M.; Chung, J.D.; Chandio, F.A. Computational analysis for mixed convective flows of viscous fluids with nanoparticles. *J. Therm. Sci. Eng. Appl.* **2019**, *11*, 021013. [CrossRef]
10. Ramzan, M.; Sheikholeslami, M.; Saeed, M.; Chung, J.D. On the convective heat and zero nanoparticle mass flux conditions in the flow of 3D MHD Couple Stress nanofluid over an exponentially stretched surface. *Sci. Rep.* **2019**, *9*, 562. [CrossRef] [PubMed]
11. Ramzan, M.; Bilal, M.; Chung, J.D.; Mann, A.B. On MHD radiative Jeffery nanofluid flow with convective heat and mass boundary conditions. *Neural Comput. Appl.* **2019**, *30*, 2739–2748. [CrossRef]
12. Ramzan, M.; Sheikholeslami, M.; Chung, J.D.; Shafee, A. Melting heat transfer and entropy optimization owing to carbon nanotubes suspended Casson nanoliquid flow past a swirling cylinder—A numerical treatment. *AIP Adv.* **2018**, *8*, 115130. [CrossRef]
13. Lu, D.; Ramzan, M.; Ullah, N.; Chung, J.D.; Farooq, U. A numerical treatment of radiative nanofluid 3D flow containing gyrotactic microorganism with anisotropic slip, binary chemical reaction and activation energy. *Sci. Rep.* **2017**, *7*, 17008. [CrossRef] [PubMed]

14.   Ramzan, M.; Bilal, M.; Kanwal, S.; Chung, J.D. Effects of variable thermal conductivity and non-linear thermal radiation past an Eyring Powell nanofluid flow with chemical reaction. *Commun. Theor. Phys.* **2017**, *67*, 723. [CrossRef]

15.   Muhammad, T.; Lu, D.C.; Mahanthesh, B.; Eid, M.R.; Ramzan, M.; Dar, A. Significance of Darcy-Forchheimer porous medium in nanofluid through carbon nanotubes. *Commun. Theor. Phys.* **2018**, *70*, 361. [CrossRef]

16.   Azam, M.; Shakoor, A.; Rasool, H.F.; Khan, M. Numerical simulation for solar energy aspects on unsteady convective flow of MHD Cross nanofluid: A revised approach. *Int. J. Heat Mass Transf.* **2019**, *131*, 495–505. [CrossRef]

17.   Sheikholeslami, M.; Kataria, H.R.; Mittal, A.S. Effect of thermal diffusion and heat-generation on MHD nanofluid flow past an oscillating vertical plate through porous medium. *J. Mol. Liq.* **2018**, *257*, 12–25. [CrossRef]

18.   Makinde, O.D.; Animasaun, I.L. Bioconvection in MHD nanofluid flow with nonlinear thermal radiation and quartic autocatalysis chemical reaction past an upper surface of a paraboloid of revolution. *Int. J. Therm. Sci.* **2016**, *109*, 159–171. [CrossRef]

19.   Lu, D.; Ramzan, M.; Bilal, M.; Chung, J.D.; Farooq, U. A numerical investigation of 3D MHD rotating flow with binary chemical reaction, activation energy and non-Fourier heat flux. *Commun. Theor. Phys.* **2018**, *70*, 089. [CrossRef]

20.   Rana, P.; Shukla, N.; Gupta, Y.; Pop, I. Analytical prediction of multiple solutions for MHD Jeffery–Hamel flow and heat transfer utilizing KKL nanofluid model. *Phys. Lett. A* **2019**, *383*, 176–185. [CrossRef]

21.   Ramzan, M.; Bilal, M.; Chung, J.D.; Lu, D.; Farooq, U. Impact of generalized Fourier's and Fick's laws on MHD 3D second grade nanofluid flow with variable thermal conductivity and convective heat and mass conditions. *Phys. Fluids* **2017**, *29*, 093102. [CrossRef]

22.   Yuan, M.; Mohebbi, R.; Rashidi, M.M.; Yang, Z.; Sheremet, M.A. Numerical study of MHD nanofluid natural convection in a baffled U-shaped enclosure. *Int. J. Heat Mass Transf.* **2019**, *130*, 123–134.

23.   Rahimah, J.; Nazar, R.; Pop, I. Magnetohydrodynamic boundary layer flow and heat transfer of nanofluids past a bidirectional exponential permeable stretching/shrinking sheet with viscous dissipation effect. *J. Heat Transf.* **2019**, *141*, 012406.

24.   Benos, L.; Sarris, I.E. Analytical study of the magnetohydrodynamic natural convection of a nanofluid filled horizontal shallow cavity with internal heat generation. *Int. J. Heat Mass Transf.* **2019**, *130*, 862–873. [CrossRef]

25.   Sajjadi, H.; Delouei, A.A.; Izadi, M.; Mohebbi, R. Investigation of MHD natural convection in a porous media by double MRT lattice Boltzmann method utilizing MWCNT–Fe_3O_4/water hybrid nanofluid. *Int. J. Heat Mass Transf.* **2019**, *132*, 1087–1104. [CrossRef]

26.   Zhao, G.; Wang, Z.; Jian, Y. Heat transfer of the MHD nanofluid in porous microtubes under the electrokinetic effects. *Int. J. Heat Mass Transf.* **2019**, *130*, 821–830. [CrossRef]

27.   Ibrahim, M.G.; Hasona, W.M.; ElShekhipy, A.A. Concentration-dependent viscosity and thermal radiation effects on MHD peristaltic motion of Synovial Nanofluid: Applications to rheumatoid arthritis treatment. *Comput. Methods Programs Biomed.* **2019**, *170*, 39–52. [CrossRef] [PubMed]

28.   Merkin, J.H. Natural-convection boundary-layer flow on a vertical surface with Newtonian heating. *Int. J. Heat Fluid Flow* **1994**, *15*, 392–398. [CrossRef]

29.   Kamran, M.; Wiwatanapataphee, B. Chemical reaction and Newtonian heating effects on steady convection flow of a micropolar fluid with second order slip at the boundary. *Eur. J. Mech. B/Fluids* **2018**, *71*, 138–150. [CrossRef]

30.   Mehmood, R.; Rana, S.; Nadeem, S. Transverse thermopheroptic MHD Oldroyd-B fluid with Newtonian heating. *Results Phys.* **2018**, *8*, 686–693. [CrossRef]

31.   Shafiq, A.; Hammouch, Z.; Sindhu, T.N. Bioconvective MHD flow of tangent hyperbolic nanofluid with Newtonian heating. *Int. J. Mech. Sci.* **2017**, *133*, 759–766. [CrossRef]

32.   Khan, A.; Khan, D.; Khan, I.; Ali, F.; Karim, F.U.; Imran, M. MHD flow of Sodium Alginate-based Casson type nanofluid passing through a porous medium with Newtonian heating. *Sci. Rep.* **2018**, *8*, 8645. [CrossRef] [PubMed]

33.   El-Hakiem, M.A.; Ramzan, M.; Chung, J.D. A numerical study of magnetohydrodynamic stagnation point flow of nanofluid with Newtonian heating. *J. Comput. Theor. Nanosci.* **2016**, *13*, 8419–8426. [CrossRef]

34. Ramzan, M.; Yousaf, F. Boundary layer flow of three-dimensional viscoelastic nanofluid past a bi-directional stretching sheet with Newtonian heating. *AIP Adv.* **2015**, *5*, 057132. [CrossRef]

35. Shehzad, S.A.; Hussain, T.; Hayat, T.; Ramzan, M.; Alsaedi, A. Boundary layer flow of third grade nanofluid with Newtonian heating and viscous dissipation. *J. Cent. South Univ.* **2015**, *22*, 360–367. [CrossRef]

36. Ramzan, M. Influence of Newtonian heating on three dimensional MHD flow of couple stress nanofluid with viscous dissipation and Joule heating. *PLoS ONE* **2015**, *10*, e0124699. [CrossRef] [PubMed]

37. Upreti, H.; Pandey, A.K.; Kumar, M. MHD flow of Ag-water nanofluid over a flat porous plate with viscous-Ohmic dissipation, suction/injection and heat generation/absorption. *Alex. Eng. J.* **2018**, *57*, 1839–1847. [CrossRef]

38. Qasim, M.; Khan, Z.H.; Khan, W.A.; Shah, I.A. MHD boundary layer slip flow and heat transfer of ferrofluid along a stretching cylinder with prescribed heat flux. *PLoS ONE* **2014**, *9*, e83930. [CrossRef] [PubMed]

Permissions

The contributors of this book come from diverse backgrounds, making this book a truly international effort. This book will bring forth new frontiers with its revolutionizing research information and detailed analysis of the nascent developments around the world.

We would like to thank all the contributing authors for lending their expertise to make the book truly unique. They have played a crucial role in the development of this book. Without their invaluable contributions this book wouldn't have been possible. They have made vital efforts to compile up to date information on the varied aspects of this subject to make this book a valuable addition to the collection of many professionals and students.

This book was conceptualized with the vision of imparting up-to-date information and advanced data in this field. To ensure the same, a matchless editorial board was set up. Every individual on the board went through rigorous rounds of assessment to prove their worth. After which they invested a large part of their time researching and compiling the most relevant data for our readers.

The editorial board has been involved in producing this book since its inception. They have spent rigorous hours researching and exploring the diverse topics which have resulted in the successful publishing of this book. They have passed on their knowledge of decades through this book. To expedite this challenging task, the publisher supported the team at every step. A small team of assistant editors was also appointed to further simplify the editing procedure and attain best results for the readers.

Apart from the editorial board, the designing team has also invested a significant amount of their time in understanding the subject and creating the most relevant covers. They scrutinized every image to scout for the most suitable representation of the subject and create an appropriate cover for the book.

The publishing team has been an ardent support to the editorial, designing and production team. Their endless efforts to recruit the best for this project, has resulted in the accomplishment of this book. They are a veteran in the field of academics and their pool of knowledge is as vast as their experience in printing. Their expertise and guidance has proved useful at every step. Their uncompromising quality standards have made this book an exceptional effort. Their encouragement from time to time has been an inspiration for everyone.

The publisher and the editorial board hope that this book will prove to be a valuable piece of knowledge for researchers, students, practitioners and scholars across the globe.

List of Contributors

Ahmad Fakhari
Institute for Polymers and Composites, Department of Polymer Engineering, Campus of Azurém, University of Minho, 4800-058 Guimarães, Portugal

Se Min Park, Jong Wook Kim, Byung Il Park and Sai Kee Oh
LG Electronics, Changwon 2nd Factory, Seonsan-dong, Changwon-city, Keongsang Namdo 51554, Korea

Seo-Yoon Ryu, Cheolung Cheong and Young-Chull Ahn
School of Mechanical Engineering, Pusan National University, Busan 46241, Korea

Muhammad Altaf Khan and Waqas Noman
Department of Mathematics, City University of Science and Information Technology, Peshawar 25000, Pakistan

Taza Gul
Department of Mathematics, City University of Science and Information Technology, Peshawar 25000, Pakistan
Department of Mathematics, Govt. Superior Science College Peshawar, Khyber Pakhtunkhwa, Pakistan

Ilyas Khan
Faculty of Mathematics and Statistics, Ton Duc Thang University, Ho Chi Minh City 72915, Vietnam

Tawfeeq Abdullah Alkanhal
Department of Mechatronics and System Engineering, College of Engineering, Majmaah University, Majmaah 11952, Saudi Arabia

Iskander Tlili
Energy and Thermal Systems Laboratory, National Engineering School of Monastir, Street Ibn El Jazzar, 5019 Monastir, Tunisia

Anwar Saeed, Saeed Islam and Zahir Shah
Department of Mathematics, Abdul Wali Khan University, Mardan, Khyber Pakhtunkhwa 23200, Pakistan

Abdullah Dawar
Department of Mathematics, Qurtuba University of Science and Information Technology, Peshawar 25000, Pakistan

Poom Kumam
KMUTTFixed Point Research Laboratory, Room SCL 802 Fixed Point Laboratory, Science Laboratory Building, Department of Mathematics, Faculty of Science, King Mongkut's University of Technology Thonburi (KMUTT), Bangkok 10140, Thailand
KMUTT-Fixed Point Theory and Applications Research Group, Theoretical and Computational Scienc Center (TaCS), Science Laboratory Building, Faculty of Science, King Mongkut's University of Technology Thonburi (KMUTT), Bangkok 10140, Thailand
Department of Medical Research, China Medical University Hospital, China Medical University Taichung 40402, Taiwan

Waris Khan
Department of Mathematics, Kohat University of Science and Technology, Kohat 26000, Pakistan

Mutaz Mohammad
Department of Mathematics & Statistics, College of Natural and Health Sciences, Zayed University, 144543 Abu Dhabi, UAE

Muhammad Bilal
Department of Mathematics, University of Lahore, Chenab Campus, Gujrat 50700, Pakistan

Fares Howari
College of Natural and Health Sciences, Zayed University, 144543 Abu Dhabi, UAE

Ali Saleh Alshomrani and Malik Zaka Ullah
Department of Mathematics, Faculty of Science, King Abdulaziz University, Jeddah 21589, Saudi Arabia

J. Prakash
Department of Mathematics, Avvaiyar Government College for Women, Karaikal 609602, Puducherry-U.T., India

Dharmendra Tripathi
Department of Mathematics, National Institute of Technology, Uttarakhand 246174, India

Abhishek Kumar Tiwari
Department of Applied Mechanics, MNNIT Allahabad, Prayagraj, Uttar Pradesh 211004, India

Sadiq M. Sait
Center for Communications and IT Research, Research Institute, King Fahd University of Petroleum & Minerals, Dhahran 31261, Saudi Arabia

Rahmat Ellahi
Center for Modeling & Computer Simulation, Research Institute, King Fahd University of Petroleum & Minerals, Dhahran 31261, Saudi Arabia
Department of Mathematics & Statistics, Faculty of Basic and Applied Sciences (FBAS), International Islamic University (IIUI), Islamabad 44000, Pakistan

Muhammad Jawad
Department of Mathematics, Abdul Wali Khan University, Mardan, Khyber Pakhtunkhwa 23200, Pakistan

Jihen Majdoubi
Department of Computer Science, College of Science and Humanities at Alghat Majmaah University, Al-Majmaah 11952, Saudi Arabia

I. Tlili
Department of Mechanical and Industrial Engineering, College of Engineering, Majmaah University, Al-Majmaah 11952, Saudi Arabia

Ahmed Zeeshan
Department of Mathematics & Statistics, Faculty of Basic and Applied Sciences (FBAS), International Islamic University (IIUI), Islamabad 44000, Pakistan

Farooq Hussain
Department of Mathematics & Statistics, Faculty of Basic and Applied Sciences (FBAS), International Islamic University (IIUI), Islamabad 44000, Pakistan
Department of Mathematics, Faculty of Arts and Basic Sciences (FABS), Balochistan University of Information Technology, Engineering, and Management Sciences (BUITEMS), Quetta 87300, Pakistan

Yongou Zhang and Aokui Xiong
Key Laboratory of High Performance Ship Technology (Wuhan University of Technology), Ministry of Education, Wuhan 430074, China
School of Transportation, Wuhan University of Technology, Wuhan 430074, China

Nawaf N. Hamadneh
Department of Basic Sciences, College of Science and Theoretical Studies, Saudi Electronic University, Riyadh 11673, Saudi Arabia

Waqar A. Khan
Department of Mechanical Engineering, College of Engineering, Prince Mohammad Bin Fahd University, Al Khobar 31952, Saudi Arabia

Ali S. Alsagri
Mechanical Engineering Department, Qassim University, Buraydah 51431, Saudi Arabia

Esmaeil Jalali
Department of Mechanical Engineering, Najafabad Branch, Islamic Azad University, Najafabad, Iran

Omid Ali Akbari
Young Researchers and Elite Club, Khomeinishahr Branch, Islamic Azad University, Khomeinishahr, Iran

M. M. Sarafraz
Centre for Energy Technology, School of Mechanical Engineering, The University of Adelaide, South Australia, Australia

Tehseen Abbas
Department of Mathematics, University of Education Lahore, Faisalabad Campus, Faisalabad 38000, Pakistan

Mohammad Reza Safaei
Division of Computational Physics, Institute for Computational Science, Ton Duc Thang University, Ho Chi Minh City, Vietnam
Faculty of Electrical and Electronics Engineering, Ton Duc Thang University, Ho Chi Minh City, Vietnam

Dianchen Lu
Department of Mathematics, Faculty of Science, Jiangsu University, Zhenjiang 212013, China

Muhammad Suleman
Department of Mathematics, Faculty of Science, Jiangsu University, Zhenjiang 212013, China
Department of Mathematics, Comsats University, Islamabad 44000, Pakistan
Department of Mathematics, COMSATS University, Islamabad 45550, Pakistan

Muhammad Ramzan
Department of Computer Science, Bahria University, Islamabad Campus, Islamabad 44000, Pakistan
Department of Mechanical Engineering, Sejong University, Seoul 143-747, Korea

Jae Dong Chung
Department of Mechanical Engineering, Sejong University, Seoul 143-747, Korea

Shafiq Ahmad
Department of Mathematics, Quaid-i-Azam University, Islamabad 44000, Pakistan

Taseer Muhammad
Department of Mathematics, Government College Women University, Sialkot 51310, Pakistan

Index

Printed in the USA
CPSIA information can be obtained
at www.ICGtesting.com
JSHW051401091023
49903JS00006B/226